D-NAG Special Publication

Perspectives in REGIONAL GEOLOGICAL SYNTHESIS

Planning for *The Geology of North America*

Edited by
A. R. Palmer
Centennial Science Program Coordinator
Geological Society of America
P.O. Box 9140
Boulder, Colorado 80301

Published by
Geological Society of America
3300 Penrose Place, P.O. Box 9140, Boulder, Colorado 80301

Publication of this volume has been made possible by
a grant from the Geological Society of America Foundation, Inc.

Published by The Geological Society of America, Inc.
3300 Penrose Place, P.O. Box 9140, Boulder, Colorado 80301
Printed in U.S.A.

Library of Congress Cataloging in Publication Data
Main entry under title:

Perspectives in regional geological synthesis.

(D-NAG special publication ; 1)
Includes bibliographies.
1. Geology—North America. I. Palmer, Allison R.
II. Series.
QE71.P45 1982 557 82-9331
ISBN 0-8137-5201-9 AACR2

Contents

Continental Margins and Adjacent Oceanic Regions

Introduction

A. R. Palmer
Geological Society of America
P.O. Box 9140
Boulder, Colorado 80301

The Decade of North American Geology involves one of the most massive attempts at geological synthesis ever attempted. Its justification is the Centennial decade of the Geological Society of America. Its goal is to bring together knowledge about all aspects of the geology of North America and its surrounding oceanic regions—really the North American Plate and its environs—in a variety of documents all compiled within a period of a few years. These documents will include tectonic, geologic, magnetic anomaly, gravity anomaly, and lineament maps of the entire region at a common scale of 1:5,000,000 and on a common computer-compatible base; geological/geophysical transects from continental to oceanic crust along 26 corridors around the margins of the continent: at least sixteen regional tectono-stratigraphic correlation charts; and at least twenty-nine 400-page volumes of regional and topical geological synthesis in a series titled *The Geology of North America.* This book represents a part of the process of creating the synthesis volumes for *The Geology of North America.*

Project leaders for the 20 regional geology volumes of the synthesis series were asked to prepare short manuscripts presenting plans for synthesis of geological knowledge about their areas or topics. The goal for each chapter—a text of 5 published pages, with references kept to a minimum—was achieved with varying degrees of success. Some of the manuscripts have taken the form of annotated outlines; others are regional overviews with topics arranged in the general order that they will be treated in the synthesis volume; and most, in one way or another, address the questions and problems that have to be considered during the synthesis process.

This book should serve two very important purposes in the development of the syntheses for *The Geology of North America.* Contributors to each volume will know how syntheses for other volumes are being approached, and the remainder of the geological community is provided with a thoughtful and authoritative series of statements about the regional geology of North America as well as a chance to interact with the planning process for the synthesis.

Three of the chapters—the Appalachian/Ouachita, Gulf Basin, and Conterminous U.S. Cordillera—have benefited from planning workshops held in January and February 1982, at which time the collaborators at the chapter level for each of these syntheses had a chance to polish the appropriate chapter texts and synthesis plans presented here. Workshops for other synthesis volumes will be held later in 1982 and in 1983, and the related chapters in this book will be the focus for discussion and polishing of details for those syntheses.

This process of iteration at increasing levels of detail began with a gathering of all the project leaders for the synthesis volumes in April 1981 to discuss overall objectives of the synthesis. It is continuing at the volume level with the workshops cited above, and will continue further as those responsible at the chapter level identify and interact with their co-workers. Additional workshops for most of the synthesis volumes are planned as the syntheses near completion to be sure that the diverse mass of information in each synthesis is well integrated.

The products of this effort should be the best possible statements that can be made in the decade of the 1980s about the state of knowledge of North American geology. This book is a step toward that goal and should be read for what it is—a statement of plans and a stimulus for thinking about the large-scale problems that must be considered in an undertaking of this magnitude.

The chapters have been grouped into regional geologic clusters. The first six chapters deal with the Precambrian basement and cover rocks of the craton; the next seven chapters deal with the principal terrestrial orogenic regions; and the final seven chapters deal with the continental margins and adjacent oceanic regions. This is not necessarily the order of publication of the volumes, but is the probable order in which they will be identified by letters, with subscripts for related volumes, in the series. The first volumes should appear in 1985, and if all schedules are maintained, the last volume will be published before the Centennial meeting of the Geological Society of America in 1988.

Acknowledgment

Special thanks and appreciation go to John Wheeler, Geological Survey of Canada, Vancouver, B.C., for his behind-the-scenes contributions to the chapters on Canadian geology. In addition to encouraging the Canadian authors to persevere on their text assignments, John critically and constructively reviewed the drafts of all of those chapter texts, to the benefit of us all.

The Precambrian: Canada and Greenland

P. F. Hoffman
K. D. Card
A. Davidson
Precambrian Geology Division
Geological Survey of Canada
588 Booth Street
Ottawa, Ontario K1A OE4, Canada

Introduction

Phanerozoic tectonics have not affected the basic integrity of the North American continent. From time to time, sialic crustal elements have been rifted off, accreted onto, and slipped along the continent's margins, but the vast continental interior has remained intact. Furthermore, the dimensions of the continent are little different now than they were in early Paleozoic time.

No such assertions can be made for the Precambrian of North America. The structure of the shield (Fig. 1) and buried basement of the interior platform (Fig. 2) is part of a continent-wide "tectonic collage" (Helwig, 1974) consisting of (1) five or more old cratonic elements having crystalline basement rocks of Archean age (3.8 to 2.5 b.y.) and remnants of Early Proterozoic (2.5 to 1.6 b.y.) cratonic cover: (2) anastomosing Early Proterozoic orogenic belts, mostly culminating between 1.9 and 1.8 b.y., that enclose and marginally affect the Archean cratons; (3) intracratonic igneous and sedimentary rocks, mainly of Middle Proterozoic age (1.6 to 0.9 b.y.), that document consolidation of the various Archean cratons and Early Proterozoic orogens to form a vast composite craton here called "Proto-Laurentia;" and (4) the Grenville Orogen, culminating at about 1.1 b.y., which developed along the presumed southeast margin of Proto-Laurentia and may contain elements exotic to it. In addition, the bounding Phanerozoic orogens contain Middle and Upper Proterozoic rocks deposited marginal to Proto-Laurentia, as well as Precambrian rocks outboard of Phanerozoic sutures that may not belong to Precambrian North America at all.

An apparent implication of this Precambrian collage is that the amalgamation and consolidation of about one-half of the North American continent (including Greenland) occurred in Early Proterozoic time, mainly between 1.9 and 1.8 b.y. (Van Schmus and Bowring, 1980). This is not a uniquely Precambrian phenomenon, however, and has parallels with the even grander assembly of Eurasia north of the Alpine belts between late Paleozoic and early Mesozoic time.

Scope and Organization

This volume will focus on the rifted Precambrian shield of Canada and Greenland. However, in order to view the shield in as broad a context as possible, we will discuss aspects of the extensions of the shield in the United States, the buried basement of the interior platform and basins, and inliers of Precambrian rocks within and peripheral to the bounding Phanerozoic orogens.

We tentatively plan to organize the volume into seven chapters, reflecting the tectonic elements and history of the shield. An introductory chapter will deal with such general topics as physiography, history of investigation, tectonic classification and terminology, and the Precambrian time scale.

The second chapter will introduce the main tectonic elements of the collage and outline their principal geologic and geophysical characteristics. The next four chapters will contain the main description and synthesis of the Archean cratons, Early Proterozoic orogens, Proto-Laurentia, and Grenville Orogen in turn. We wish to make as holistic a synthesis as possible, endeavoring to thoroughly integrate data from all aspects of geology, geophysics, metallogeny, and so forth. To achieve full integration, we wish to avoid topical divisions in the text based on fields of specialization.

The final chapter will summarize the tectonic history of the shield and compare it with areas of younger tectonism in order to assess the possible significance of secular variation in tectonic styles and processes. It is important that this discussion *not* be limited to North America, so as

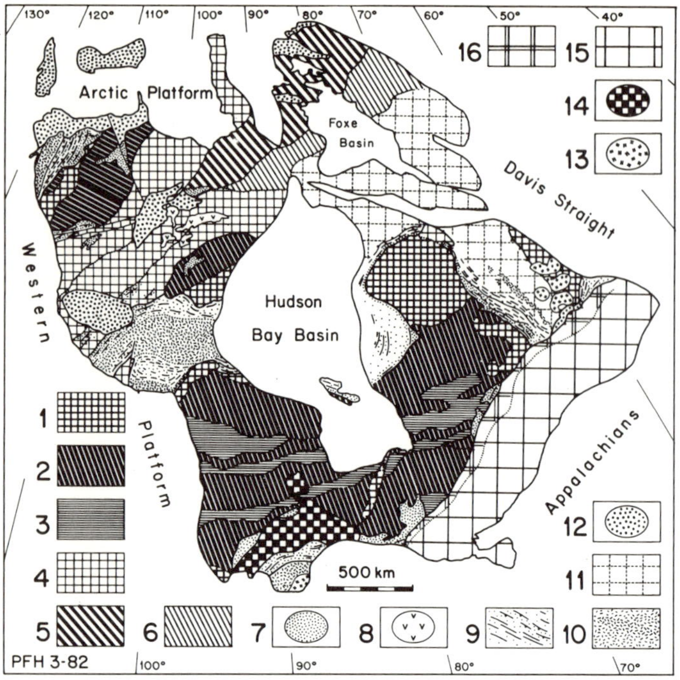

Figure 1. Some major tectonic elements of the Canadian Shield: (1) Archean high-grade gneiss terranes, including Middle and/or Early Archean components; (2) Late Archean low-grade "granite-greenstone" belts; (3) Archean high-grade "granite-metasedimentary gneiss" belts; (4) probable Archean high-grade rocks reactivated in the Early Proterozoic; (5) possible Middle Archean craton reactivated in the Early Proterozoic; (6) foldbelt involving high-grade Early Proterozoic sedimentary rocks and Archean basement; (7) Early Proterozoic epicratonic rocks, mostly sedimentary; (8) Early Proterozoic epicratonic rocks, mostly volcanic; (9) Early Proterozoic orogenic belts, mainly metasedimentary, metavolcanic, and anatectic granitoids; (10) major calc-alkaline magmatic belts within Early Proterozoic orogens; (11) high-grade rocks, in part Archean, in the hinterland of Circum-Ungava Orogen (relation to unit 4 uncertain but unroofing age may be younger); (12) mainly Middle Proterozoic epicratonic sedimentary and volcanic rocks; (13) "Elsonian" (1.45 b.y.) anorthosite-mangerite and granite intrusions; (14) mafic volcanic and intrusive rocks, and minor sediments, related to the 1.2 to 1.1-b.y. Keweenawan rift (CNAR); (15) Grenville Orogen, exclusive of (16) Grenville Front Tectonic Zone.

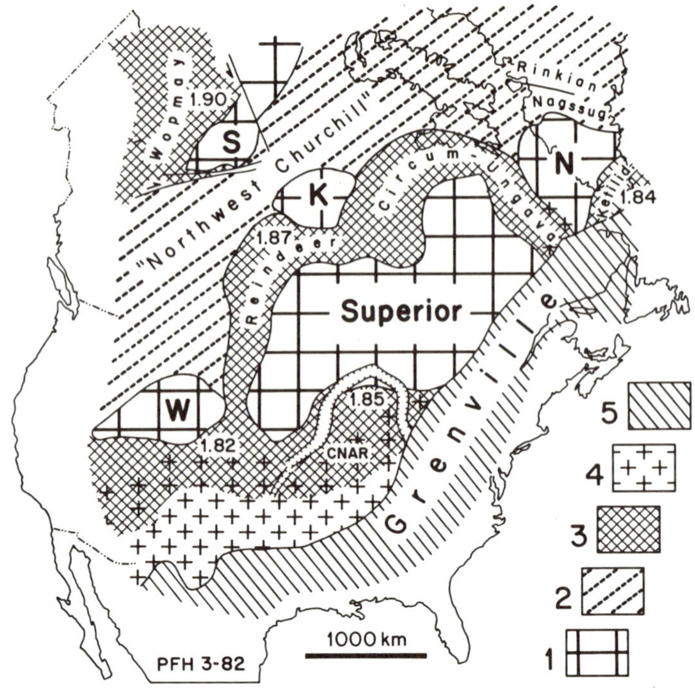

Figure 2. Some major tectonic elements of the North American Precambrian tectonic collage: (1) Archean cratons (S, Slave; K, Kaminak; W, Wyoming; N, Nutak; (2) composite of pre-1.9-b.y. Proterozoic orogens and reactivated Archean cratonic elements; (3) 1.9 to 1.6 b.y. Proterozoic orogens with ages of maximum convergence estimated from U-Pb zircon ages; (4) 1.6 to 1.3 b.y. high-level granite and anorthosite-mangerite intrusions and related volcanics; and (5) Grenville Orogen, strongly affected by 1.1-b.y. orogeny. Greenland has been restored to pre-drift position by fitting adjacent continental slopes. CNAR, Central North American Rift.

to avoid compounding one biased sample (North American Precambrian) with another (North American Phanerozoic).

Archean Cratons

Four or five Archean cratons are recognized within the shield. Each craton will be described separately, their geologic histories synthesized, and then a comparative analysis made of them all. The Superior Craton is by far the largest and therefore the most informative area of relatively undisturbed Archean crust. It consists partly of alternating Late Archean (3.0 to 2.5 b.y.) belts of low-grade volcanic and high-grade metasedimentary rocks, intruded by high- and intermediate-level granitoids, respectively, and partly of very high-grade metamorphic complexes, some of which involve Middle Archean (3.4 to 3.0 b.y.) or even older crust. The contemporaneity of short-lived magmatic episodes over distances of 800 km along strike in the volcano-plutonic belts of Superior Craton, as documented by ultra-high-precision zircon U-Pb dating (Krogh and others, 1982), seems more consistent with a subduction-dominated than a hotspot-dominated tectonic regime for the Late Archean.

The Kaminak and Slave cratons resemble Superior Craton in lithostructural style and, perhaps, in timing, but volcanic rocks of intermediate composition are most abundant in the Archean of Slave Craton, which also contains more low-grade sedimentary rocks. The Nutak, or "North Atlantic," Craton consists mainly of high-grade rocks, but events spanning the entire Archean (3.8 to 2.5 b.y.) have been documented. A fifth possible craton in north-central Churchill Province (Fig. 1, unit 5) contains relatively low-grade volcanic and sedimentary rocks possibly of Middle Archean age.

Early Proterozoic Orogens

Early Proterozoic igneous and sedimentary rocks on the Archean cratons will be handled together with the orogenic belts of the same age. The organizational strategy is to focus on various craton/orogen boundaries, starting with related intracratonic basins, where possible, and working outward across the margin of the craton into the orogen itself. The orogens are divisible into numerous tectonic zones based on lithologic-stratigraphic and structural integrity, and contain major structural discontinuities across which genetically unrelated tectonic zones may be juxtaposed. Gravimetric and aeromagnetic maps will help locate and establish the polarity of collision zones and crustal thrusts (Gibb and others, 1978). As K-Ar cooling ages lack precision and whole-rock Rb-Sr ages are reset in most places by later hydrothermal events, heavy reliance will be placed on U-Pb zircon dating, the impact of which is just beginning to be felt. This dating will undoubtedly revolutionize tectonic concepts of many of these orogens, and we are acutely aware of the need to retain flexibility in our planning.

Wopmay Orogen, in the northwest, has a well-defined tectonic zonation, an accurately dated tectonic evolution, and associated geophysical anomalies that permit the interpretation to be extended to the Cordilleran front. The Penokean Orogen, south of Superior Craton, and the Ketilidian Orogen of south Greenland are also relatively well mapped and dated. The "Reindeer Orogen" (provisional name) and the southern foreland of Circum-Ungava Orogen are both well mapped but as yet poorly dated. The Wopmay, Reindeer, and Penokean orogens have been interpreted, at least by some, as arc-continent or microcontinent-continent collision zones, and comparable conclusions have been reached for the Circum-Ungava Orogen on the basis of Bouguer and magnetic anomalies. Although connected beneath Hudson Bay Basin, Reindeer Orogen is best approached from the northwest, whereas Circum-Ungava Orogen should be approached from Superior Craton.

The "Northwest Churchill Province" is still very poorly understood, mainly because of totally inadequate geochronology. Much of it consists of high-grade rocks, locally known to be Archean, in places overlain by epicratonic Early Proterozoic sedimentary rocks, which themselves may be strongly deformed and metamorphosed. Although strongly reactivated by events in Wopmay and the Reindeer–Circum-Ungava orogens, the main orogenies in Northwest Churchill Province appear to be older. Geophysical data suggest great structural complexity, and it will be a challenge to make sense of this province in time for the volume deadline.

In our view, the structure of these orogens appears consistent with the operation of actualistic plate tectonics and, increasingly, the craton margins are being interpreted as ancient continental margins, although the nature and extent of intervening crust is disputed.

Proto-Laurentian Craton

The intracratonic igneous and sedimentary rocks of Proto-Laurentia, mostly Middle Proterozoic but ranging from late Early (1.8 to 1.6 b.y.) to Late Proterozoic in age, fall into four categories: (1) the mainly sedimentary Athabasca, Thelon, Borden, and Amundsen basins in the northwest, and possible correlatives in the Canadian Cordillera; (2) various "anorogenic" igneous suites occurring predominantly in the southeast, such as the "Elsonian" (1.45 b.y.) anorthosite-mangerite intrusions of Labrador, the 1.2-b.y. Gardar Intrusions of southern Greenland, and possible correlatives in the southern midcontinent region; (3) a plethora of dyke swarms over most of the craton, excluding those older than about 1.8 b.y. which will have been discussed in the previous chapter; and (4) volcanic and sedimentary deposits of Middle Proterozoic age peripheral to Grenville Orogen, principally the Keweenawan, or "Central North American Rift," and Naskaupi Foldbelt, or "Central Mineral Belt," of Labrador. The significance of these rocks is difficult to ascertain because we see so little of the margin of Proto-Laurentia.

Grenville Orogen

This is the only Middle (or Late) Proterozoic orogen in the shield and perhaps the only exposed margin of Proto-Laurentia. Continuing from the previous chapter (the two may be combined), this one will begin with the Grenville Front Tectonic Zone, a strip less than 100 km wide along the northwest margin of the orogen, which is the only part known for certain to involve the crust of Proto-Laurentia. The other tectonic zones are oblique to the Grenville Front, being more northerly in trend, and will be described from southwest to northeast. More or less deformed anorthosite massifs are common in the eastern two-thirds of the orogen, but how many are cogenetic with the undeformed Elsonian anorthosites of Labrador is moot.

Metamorphic and structural evidence suggests that large parts of the orogen experienced severe crustal imbrication and thickening, followed by deep erosion, during what appears to have been a major collision event at about 1.1 b.y. Appropriately, the eastern Grenville Front Tectonic Zone coincides with the deepest negative Bouguer anomaly (−120 mgal) in the shield. The major unknowns are how much of the exposed orogen was always part of Laurentia and how much may have been accreted at 1.1 b.y. or since 1.8 b.y.

Late Proterozoic

Late Proterozoic (0.9 to 0.57 b.y.) rocks are almost absent from the shield. Cursory reference will be made to the important sequences of this age exposed in the Appalachian and Cordilleran orogens as they will be fully treated in other volumes. A distinction will be made between those rocks deposited on or adjacent to Laurentia (Windermere Supergroup for example) and those accreted in the Phanerozoic.

Comparative Tectonics

The question of secular variation in tectonic style consequent on the Earth's declining heat production has long bedevilled geologists of the Precambrian. One empirical approach to the problem is to compare the products of Precambrian orogeny with those of more recent times, especially the Cenozoic. As part of our tectonic synthesis, we wish to explore this approach by comparing the Precambrian orogens of the shield with areas of Cenozoic plate convergence, in particular the Proterozoic with southern Eurasia and the Archean with southeast Asia. By comparing such large areas, we hope to see past the local tectonic idiosyncracies inherent in plate tectonics.

Summary

The rifted Precambrian shield of Canada and Greenland was formed by the convergence and consolidation of at least four Archean cratons, each with a complex history, within an anastomosing system of Early Proterozoic orogens, probably the sites of former oceans. The resulting composite craton, Proto-Laurentia, was apparently assembled in less than 100 m.y., between 1.9 and 1.8 b.y. In the Middle Proterozoic, the southeast margin of the craton was extensively intruded by batholiths and related volcanics, culminating with the Grenville orogeny at about 1.1 b.y., which has been interpreted as the Andean side of a continental collision (Dewey and Burke, 1973). Late Proterozoic rocks, almost absent from the shield, include deposits of the Laurentian margin in the Appalachian and Cordilleran orogens, and exotic rocks unrelated to Laurentia, such as the Pan-African Avalon Zone of the Appalachians.

References Cited

Dewey, J. F., and Burke, K.C.A., 1973, Tibetan, Variscan, and Precambrian basement reactivation: Products of continental collision: Journal of Geology, v. 81, p. 683–692.

Gibb, R. A., Thomas, M. D., Kearey, P., and Tanner, J. G., 1978, Gravity anomalies at structural province boundaries of the Canadian Shield: Products of Proterozoic plate interaction: GeoSkrifter, v. 10, p. 21–57.

Helwig, J., 1974, Eugeosynclinal basement and a collage concept of orogenic belts, *in* Dott, R. H., Jr., and Shaver, R. H., eds., Modern and ancient geosynclinal sedimentation; Society of Economic Paleontologists and Mineralogists Special Publication 19, p. 359–376.

Krogh, T. E., Davis, D. W., Nunes, P. D., and Korfu, F., 1982, Archean evolution from precise U-Pb isotopic dating: Geological Association of Canada, Program with Abstracts, v. 7 (in press).

Van Schmus, W. R., and Bowring, S. A., 1980, Chronology of igneous events in the Wopmay Orogen, Northwest Territories, Canada: Geological Society of America Abstracts with Programs, v. 12, p. 540.

Precambrian of the Conterminous United States

John C. Reed, Jr.
U.S. Geological Survey
Denver, Colorado 80225

Paul K. Sims
U.S. Geological Survey
Denver, Colorado 80225

Robert S. Houston
Department of Geology and Geophysics
University of Wyoming
Laramie, Wyoming 82070

Leon T. Silver
Division of Geological and Planetary Science
California Institute of Technology
Pasadena, California 91125

Douglas W. Rankin
U.S. Geological Survey
Reston, Virginia 22092

Mitchell W. Reynolds
U.S. Geological Survey
Reston, Virginia 22092

In these anxious days, when economics and politics disturb even the minds of scientists, perhaps you will share with me a certain feeling of relief in turning to a consideration of the old pre-Cambrian, where the unsolved problems are perplexing enough, but at least not political

C. K. Leith
Geological Society of America Presidential Address, 1933

Introduction

Precambrian rocks are at or near the surface in only about 10 percent of the conterminous United States, but it can reasonably be inferred that they comprise the continental crust beneath about 90 percent. They are missing or unrecognized in the exotic terranes along the Pacific margin of North America, but they probably form significant parts of the crust in exotic terranes or continental fragments accreted to the eastern part of the continent during Paleozoic time. Thus, the total area of Precambrian rocks to be considered in this volume is comparable to that of the exposed Precambrian of the Canadian Shield. It is important to remember that in spite of the enormous lateral extent of the craton the volume of the continental crust is almost insignificant. The width of the North American craton is more than half the radius of the planet, but the thickness of the continental crust is less than one hundredth of the planetary radius (fig. 1). The volume of the crust is less than 2 percent of the volume of the mantle beneath the U.S. part of North America.

Precambrian rocks contain the only available record of the assembly and evolution of the fragile continental raft that we know as North America during more than four fifths of geologic time. Of the areas of exposed Precambrian rocks in the conterminous United States, about half have been covered by modern reconnaissance geologic mapping (scale 1:250,000 or larger); less than a quarter have been covered by detailed modern mapping (scale 1:62,500 or larger). About 60 percent of the concealed Precambrian has been at least sketchily explored by drilling at spacings ranging from a few kilometers to a few tens of kilometers. Speculation on the nature of the remainder must depend on indirect geophysical and geochemical data, on studies of inclusions brought to the surface by later igneous intrusions, and on projections from outcrops or better-explored subcrop areas.

In spite of these limitations, our knowledge of the Precambrian rocks of the conterminous United States has increased by orders of magnitude in the century since the founding of the Geological Society of America, and the past two decades have witnessed an explosion of knowledge as vastly increased numbers of workers have applied new techniques to the study of Precambrian rocks.

Probably the most important advance in the study of Precambrian rocks has been the development of the techniques of isotope geochemistry and their application to

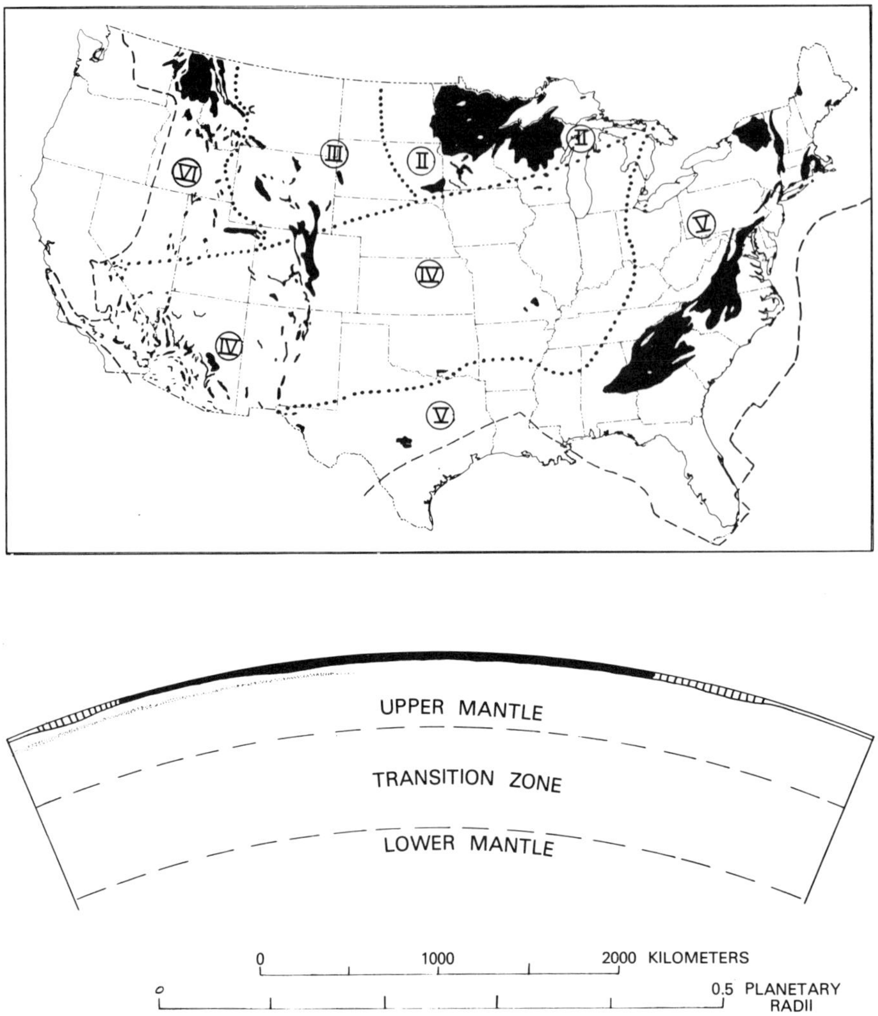

Figure 1. Index map showing outcrop areas of Precambrian rocks in the conterminous United States after King (1976). Black areas are Precambrian outcrops; dashed lines indicate limits of Precambrian continental crust as inferred from geophysical evidence on the east and south and from strontium isotopic compositions in the west. Dotted lines indicate approximate areas discussed in the various chapters of the volume. Roman numerals indicate chapter numbers in the outline. Cross-section shows thickness of Precambrian continental crust (black) at same scale as map. Vertical lined areas indicate continental crust in areas of exotic terranes; stippled zone in western part of section is the low-velocity zone marking the boundary between lithosphere and asthenosphere.

geochronology. At the turn of the century, Lord Kelvin's view that the age of the earth must be less than 10 m.y. seemed irrefutable. By the close of the first decade of the century it was realized that radioactivity provided both a heat source that invalidated Kelvin's cooling-rate calculations and a tool that could make possible direct measurements of radiometric ages of minerals. Development of isotope dilution techniques and improvements in mass spectrometry during and after World War II has made possible precise and meaningful dating of a wide variety of minerals and rocks and, by inference, of a whole spectrum of geologic events involving reequilibration of isotopic systems. This has made it possible to base the correlation of Precambrian rock units on radiometric age rather than on similarities in lithology or degree of metamorphism or deformation. It has become clear that the time involved in the bewildering collage of events recorded in Precambrian rocks is forty times longer than was realized at the turn of the century! In addition, studies of the ratios of radiogenic to nonradiogenic isotopes of lead, strontium, and other elements have yielded new insights into the origins of magmas, sources of sediments, and evolution of the crust and mantle.

A second set of powerful new tools has been provided by developments in experimental petrology and physical chemistry. Theoretical analyses and laboratory studies of phase equilibria and new techniques for determining the exact compositions and structural states of phases in rock systems have made it possible to make quantitative estimates of temperature, pressure, and activities of volatile components that prevailed in various parts of the crust during metamorphic and tectonic events. Applications of these techniques to the study of Precambrian rocks have provided new insights into the relations between metamorphic terranes, the processes involved in generation and emplacement of magmas, and the growth and evolution of the crust during Precambrian time.

New and improved geophysical techniques and the use of computers to process geophysical data have also made vast contributions to the understanding of Precambrian terranes. In most areas of "exposed" Precambrian rocks shown on Figure 1, real outcrops generally comprise less than 5 percent and commonly far less than one tenth of 1 percent of the surface area. Even reliable float is a rarity in many places. Aeromagnetic and aeroradioactivity surveys have become indispensible adjuncts to modern geologic mapping in many poorly exposed Precambrian terranes because they provide the only reliable methods, short of close-spaced drilling, of projecting contacts between outcrops or areas of reliable float. Even more importantly, aeromagnetics, gravity, and seismic techniques can be used to trace Precambrian units beneath the Phanerozoic cover in the midcontinent, permitting the data from deep wells and widely scattered outcrops to be integrated into a rational regional picture. Multichannel seismic profiling is being applied to the resolution of some of the structural complexities of the deformed belts flanking the craton, so that understanding of the distribution of Precambrian basement rocks and their involvement in Phanerozoic deformations can be vastly improved. Finally, seismic and gravity techniques provide the great bulk of the information from which models of the present structure of the continental crust are derived, and it is these models that represent the starting place for speculations on the origin and evolution of the crust through geologic time.

Of equal significance with the development of new tools and techniques has been the development of more advanced conceptual models. Such models serve as frameworks in which to organize and interpret old data, as guides for new research, and as challenges to be tested by new data and new approaches. Most fundamental of these conceptual models are those that deal with the origin and early history of the earth as deduced from lunar and planetary exploration in the past two decades. Of special interest is the recognition that the very early history of the earth may best be read on the moon and other planets, and that geologic and tectonic processes on the planets have important implications for interpretation of processes on earth. For example, the failure to find rocks older than 3.8 b.y. on earth, and the end of the initial period of widespread impact cratering on the moon and terrestrial planets can hardly be coincidental.

The discovery of microfossils in Precambrian rocks in many parts of the world and the ability to date these rock sequences by isotopic methods has led to the realization of the antiquity of life. While fossils are too rare and too primitive to allow detailed zonation of Precambrian rocks, studies of their paleobiology have led much closer to an understanding of the origin of life and have allowed development of models for the evolution of the atmosphere. Such models go far toward explaining why certain classes of sediments such as major banded iron-formations and gold-uraninite conglomerates are unique to Precambrian rocks of a rather limited span of ages.

The development of the concepts of plate tectonics and recognition of the complex interrelations between sedimentation, tectonics, and igneous and metamorphic processes associated with plate interactions have revolutionized the interpretation of Phanerozoic geology. Plate tectonics has undoubtedly operated throughout much of the Proterozoic and provides a powerful set of conceptual models for the interpretation of Proterozoic rocks. The questions of how plate tectonics was initiated, how processes operated during the Archean, and what role plate-tectonic processes played in the evolution of the early crust are among the fundamental problems of Precambrian geology.

Finally, it is important to realize that in spite of refinements in techniques and concepts and increased efforts in Precambrian research, some of the questions that we pose in this volume may never be satisfactorily answered, and many must await the attention of our successors who will write the GSA bicentennial volume. Probably the questions that will ultimately emerge as the most fundamental will prove to be among those that we are not yet astute enough to ask!

A meaningful synthesis of the current state of knowledge of the U.S. Precambrian (or rather, the state of knowledge as it will exist half a decade from now) requires an array of expertise far beyond that of any individual. We therefore plan a volume consisting of several major sections, each the responsibility of an editor who will solicite input from a wide spectrum of individual experts. Such input may range from major authored chapters or short sections to brief but carefully acknowledged personal communications. The general areas encompassed by each of the major sections are indicated on Figure 1, but obviously there will have to be considerable overlap between discussions of adjoining provinces. Special attention will be given to integration and synthesis of U.S. Precambrian terranes with their extensions and equivalents to the north and south—for example, the extensions of the Belt Supergroup into Canada and the Grenville terrane into both Mexico and Canada. Every effort will be made to adopt a uniform Precambrian time scale to facilitate these discussions.

The volume will be accompanied by a geologic tectonic map of Precambrian terranes both exposed and in subcrop beneath Phanerozoic rocks. The map will be on the same 1:5 million base as the new geologic and tectonic maps of North American and in areas where Precambrian rocks are buried or are poorly mapped will depend heavily on newly compiled gravity and aeromagnetic maps of the United States.

Tentative Outline of the Volume

I. Summary Introduction (Reed, Sims, Silver, Houston, Rankin, Reynolds)
  - A. History of Precambrian studies
  - B. Special problems in dealing with Precambrian rocks
  - C. Precambrian time scale
  - D. Organization of the volume

II. Lake Superior Region (P. K. Sims, U.S. Geological Survey)
  - A. Introduction and overview of geology and geophysics
  - B. History of investigations
  - C. Archean rocks
    1. Gneiss terrane (3,500 m.y., at least in part)—sedimentation, volcanism, plutonism, tectonism. Reactivation during later events
    2. Greenstone-granite terrane (2,600–2,750 m.y.)—sedimentation, volcanism, plutonism
    3. Basement rocks of Wisconsin—relation to well-dated gneisses of southern Minnesota and northern Michigan. Are basement rocks in Wisconsin a discrete block?
    4. Evolution and interrelationship of the two Archean terranes. Nature of the boundary between them—what does COCORP data show? What tectonic environments do they represent?
    5. Subsurface extensions of Archean rocks to the south
    6. Mineral resources—description and tectonic environments
  - D. Early Proterozoic rocks
    1. Synthesis of regional sedimentation and volcanism
       - a. Pre-Penokean rocks (pre 1,760–1,850 m.y.)—Animikie and Mille Lacs Groups, Marquette Range Supergroup, and equivalent rocks
       - b. Post-Penokean rocks (1,760–1,850 m.y. and younger)—rhyolite, epizonal granites, Sioux and Baraboo formations
    2. Geochemistry of volcanic and granitoid rocks
    3. Tectonic and depositional environments. What was the environment of emplacement of the dominantly subaerial post-Penokean rhyolite and cogenetic granites? Of the rapakivi-type Wolf River batholith? Origin of thick post-Penokean quartzite sequences and their relation to the volcanic and granitic rocks of the central midcontinent
    4. Reworking of Archean basement by Penokean and later events. What is the nature, extent, and cause of post-Penokean disturbances?
    5. Mineral resources
  - E. Middle and Late Proterozoic rocks
    1. 1,500 m.y. intrusive rocks—distribution, character
    2. Midcontinent rift system
       - a. Volcanic, sedimentary, and gabbroic rocks
       - b. Subsurface extensions
       - c. Tectonic evolution—relation to the Grenville orogeny
       - d. Mineral resources
  - F. Review of granitic magmatism through time in the Lake Superior region. Sources of magma; primitive versus reworked crustal materials; lateral versus vertical accretion of early crust

III. Wyoming Province (R. S. Houston, University of Wyoming)
  - A. Introduction—definition, location, extent, history of investigations
  - B. Archean rocks
    1. Gneissic terrane
       - a. Description
       - b. Origin and tectonic setting. Is the southern gneissic terrane a distinct unit or a geochronologic artifact?
    2. Greenstone belts
       - a. Description—Owl Creek Mountains, southern Wind River Mountains, Granite Mountains, Casper and Seminoe Mountains, Laramie and Medicine Bow Mountains, Sierra Madre
       - b. Tectonic setting—comparison with greenstone belts of the Canadian Shield. Do the Wyoming greenstone belts represent deeper levels of exposure? Why are they generally more deformed and at higher metamorphic grade?
    3. Metasedimentary and metavolcanic successions of miogeoclinal affinity
       - a. Description—Southwest Montana, Beartooth Mountains, southern Idaho, northern Utah, Sierra Madre, Medicine Bow Mountains, Hartville Uplift, Black Hills
       - b. Origin and tectonic significance. Why do these successions tend to be at the margins of the province? The successions seem to have greenstone belt and early Proterozoic characteristics but Archean ages. Are they transitional between Archean and Proterozoic?
    4. Archean igenous rock
       - a. Mafic rocks—Stillwater Complex and related rocks
       - b. Granitic rocks—description, distribution, geochemistry
       - c. Origin and significance. How are the granitic rocks related to the layered sequences?
    5. Mafic dikes cutting Archean rocks
       - a. Character and distribution
       - b. Age—geochronology is very suspect. Is the indicated wide span of ages real? How many generations of dikes can be distinguished with confi-

dence?
    c. Tectonic significance. Do the mafic dikes represent tensional fractures related to plate tectonics? Can Precambrian stress directions be inferred from them?

C. Proterozoic metasedimentary successions on Archean basement
    1. Description—Medicine Bow Mountains, Sierra Madre, Black Hills
    2. Origin and tectonic significance
D. Cheyenne belt
    1. Definition and description
    2. Tectonic setting and interpretation of southern boundary. Is the Cheyenne belt a Proterozoic suture, rift, wrench fault, or some combination?
E. Proterozoic rocks south of the Cheyenne belt with no known Archean basement
F. Proterozic igneous rocks of the Wyoming province
G. Subsurface extensions of the Wyoming province
H. Regional structure, including involvement in Phanerozoic deformations
I. Summary of geochronology
J. Relationship of Wyoming province and the Canadian Shield
K. Review of concepts of the geologic evolution of the Wyoming province
    1. Evidence for accretion and for plate-tectonics processes in the Archean gneiss terrane and greenstone belts
    2. Significance of geochronologic break in central Wyoming and evidence for remobilization of Archean basement to the south
    3. Evidence for multiple orogenies in late Archean and early Proterozoic time. Relations between orogenies and with events in the Lake Superior region

IV. Transcontinental Proterozoic Province (L. T. Silver, California Institute of Technology)
  A. Major component terranes
    1. 1,800–1,900 m.y. Penokean volcanic-plutonic complexes adjacent to and overlapping Lake Superior Archean rocks—possible craton margin volcanic arc system
    2. 1,700–1,800 m.y. stratified volcanic-sedimentary sequences intruded by associated batholithic suites. Extends from California across southern and central Cordillera to midcontinent subsurface and reappears in Lake Superior regions of Wisconsin and Ontario
    3. 1,620–1,700 m.y. eugeosynclinal sedimentary packages with intercalcated calc-alkaline volcanic rocks. Intruded by synorogenic and postorogenic calc-alkaline batholiths. Extends from southern Arizona across New Mexico and in the subsurface at least to Iowa and northern Missouri
    4. 1,400–1,480 m.y. anorogenic batholithic system, locally (as in St. Francois Mountains, Missouri) with preserved high-level volcanic expression. Extends from California to Labrador
    5. Isolated supracrustal sedimentary basins generally ranging from 1,100 to 1,400 m.y.
    6. Additional anorogenic igneous suites
        a. 1,320–1,380 m.y. felsic granites of southern midcontinent
        b. 1,100–1,200 m.y. Keweenawan diabase and flood basalts of the midcontinent rift system and local basins in Arizona, California, and west Texas
        c. 1,000 m.y. Pikes Peak batholith in Colorado
        d. 550–575 m.y. felsic and alkaline intrusives
    7. Mixed metasedimentary and igneous packages of orogenic character ranging from 1,000 to 1,200 m.y. in the Llano Uplift and Van Horn Mountains, Texas
    8. Granulite-anorthosite terrane of southern California ranging from 1,200 to 1,700 m.y. Appears to be an allochthonous block
  B. Major recognized orogenic belts or tectono-stratigraphic packages—Penokean (1,800–1,900 m.y., Lake Superior); Yavapai (1,700–1,800 m.y. California and Sonora to Iowa); San Gabriel (1,400–1,700 m.y., Arizona and Sonora to Ontario); Mazatzal (1,620–1,700 m.y., southern California); Texan (1,000–1,200 m.y., west central Texas)
  C. Perimeter relations to adjacent provinces
    1. Are contacts with Archean terranes to the north tectonic (faulted, accretionary) or accumulative on an older basement?
    2. What is the relation to younger orogenic overlay of the Grenville province? How far into that province did lower and middle Proterozoic packages extend? Does the orogenic package of Grenville age extend along the entire southern margin of the Lower and Middle Proterozoic belts? Were percursor rifts with ocean-floor formation followed by renewed suturing of younger Precambrian rocks?
    3. When (and how many times) was the western margin of the old orogenic belts truncated? How do the Middle and Upper Protorezoic sedimentary basins of the Cordillera relate to this history of truncation?
  D. Igneous activity
    1. Geometry and timing major igneous elements. Can major magmatic arcs be defined and their trends and asymmetries be sufficiently analyzed to provide insight into magmatic or tectonic processes?
    2. Primary chemical and isotopic characteristics of the major magmatic elements—comparison with modern magmatic suites
    3. Basis for distinction between orogenic and anorogenic suites. Can these suites be placed in an orderly-scheme of crustal development and modification?
    4. Are there old ophiolite components? If not, why not?
  E. Sedimentation and relations to tectonic setting and source terranes
    1. Distribution of relict sedimentary packages and their relation to volcanic sequences. Major basins and source regions
    2. Sedimentary characteristics of best-preserved sections, including sources, transport mechanisms, transport directions, and tectonic settings
    3. Evidence for biologic activity and the nature of the oceans and atmosphere
  F. Deformation and metamorphism
    1. Major tectonic elements—expression in space and time, regional metamorphic characteristics, gradients, and trends
    2. Relation of tectonic elements to preexisting continental and oceanic crust. Evidence for plate-tectonic processes. Can island arcs, continental margin arcs, transform boundaries, collision boun-

daries, incipient or extended rifting, and old oceanic lithosphere be identified and evaluated?

3. History of cratonization. What role has vertical tectonics played in the tectonic and sedimentation history?

G. Ore deposits, metallogenesis, and geochemical anomalies. What is the distribution of volcanic arc assemblages favorable for Proterozoic massive sulfide deposits? What other significant mineralization can be related to the principal Precambrian orogenic cycles? To what extent can broad Precambrian orogenic and geochemical patterns be correlated with Phanerozoic metallogenic provinces such as the Arizona–New Mexico copper province, the Colorado Plateau uranium province, and the midcontinent lead-zinc province?

H. Synthesis

V. Proterozoic Rocks East and Southeast of the Grenville Front (D. W. Rankin, U.S. Geological Survey)

A. Overview

Dated Precambrian rocks in eastern United States are all of late Proterozoic age (mostly younger than 1,300 m.y.). They include both crustal and supracrustal rocks and constitute three terranes:

1. Crustal rocks metamorphosed in the Grenville orogeny and possibly some supracrustal cover that together were part of Proterozoic North America and have not been through an orogeny since the beginning of Phanerozoic time
2. Crustal rocks metamorphosed in the Grenville orogeny and supracrustal rocks that together were part of Proterozoic North America and have been deformed in one or more Paleozoic orogenic events
3. Crustal and supracrustal rocks that probably were not part of Proterozoic North America.

Terrane 1 is exposed in the Adirondack massif and and in a number of regions of Texas such as the Llano uplift. It is also present in the subsurface between New York and Texas but may be interrupted by the Ouachita reentrant. Terrane 2 is exposed in the western part of the Appalachian belt and may be present in the subsurface underpart of the Ouachita orogenic belt. Terrane 3 is exposed in the eastern part of the Appalachian orogenic belt and is presumably present in the subsurface south of the Ouachita orogenic belt.

B. Adirondacks massif

1. Stratified rocks—character, age, correlation
2. Plutonic rocks—age, origin, signficance
3. Geophysics—gravity, magnetics, heat flow, COCORP line
4. Mineral resources
5. Significant problems
   a. What is the signficance of anorthosite and associated rocks as keys to structure of the crust?
   b. What are the relations between plutonic and stratified rocks?
   c. Conditions of metamorphism. What goes on in granulite facies metamorphism? What is the significance of the present distribution of metamorphic zones?
   d. Nature of the Grenville orogeny. What were the mechanics of formation of the very large folds in the Adirondacks? Can the Grenville orogeny be interpreted in a plate-tectonics context? Is the Himalayan model viable? If so, where is the suture? Nature of the Grenville front?
   e. What is the relation of the Adirondacks to the Grenville province of the Canadian Shield?
   f. What is the origin of the Adirondack Dome and the modern seismicity?

C. Southeastern extension of the Grenville province

1. Subsurface between the Adirondack massif and Black Warrior basin
   a. Presentation of data: location of holes; lithology of basement rocks; geochemical and isotopic data; gravity, magnetic, heat-flow, and seismic studies; basement tectonics as reflected by younger rocks
   b. Interpretation of the Proterozoic geology
   c. Nature of the Grenville front
   d. Relationship of the Grenville orogeny to the coeval rifting event of the mid-continent
   e. Late Proterozoic–early Paleozoic post-Grenville rifting event
2. Western extensions—Texas and beyond
   a. Introduction, geologic setting, and significance of area
   b. Exposed Precambrian (Proterozoic) of Texas
      (1) Llano uplift
         (a) Lithologies: metamorphosed stratified rocks, gabbro-ultramafic complexes, and younger granitic plutons
         (b) Structures
         (c) Metamorphism (low pressure) and isotopic ages
      (2) Van Horn uplift
         (a) Lithologies: metamorphosed stratified rocks and younger granodiorites
         (b) and (c) As above
         (d) Metamorphism (up to amphibolite facies) and isotopic ages
3. Subsurface Precambrian geology and its relationship to the exposed area: data, interpretation
4. Regional problems
   a. Timing of events: deposition, deformations, and metamorphism
   b. Proterozoic tectonic environments of the Llano area: back-arc versus fore-arc basin
   c. Relationship of the Buchan metamorphic and possibly ophiolitic rocks of the Llano area to the Barrovian metamorphic and typical shelf facies sedimentary rocks of the Van Horn area
   d. Relationship of the Texas Proterozoic to the eastern North American Grenville orogenic belt
   e. Possible extensions of the Grenville orogenic belt

D. The Appalachian orogen

1. Overview
   a. Regional setting
   b. Sketch of Phanerozoic history
   c. Differentiating Proterozoic and Phanerozoic rocks in the crystalline Appalachians
2. Precambrian rocks of pre-Taconic North America
   a. Crustal rocks affected by Grenville orogeny
      (1) Main massifs of the Blue-Green-Long axis
         (a) Character, Precambrian structure and metamorphism, geochemistry, and isotopic ages

(b) Origin
(c) Paleozoic deformation and tectonic setting
(2) Gneiss domes (= continental fragments?)—as (a)–(c) above
b. Post-Grenville supercrustal rocks
(1) Sedimentary and volcanic rocks flanking the main massifs—distribution and character, age and correlation, environment of deposition (including evidence for Proterozoic glaciation), Paleozoic deformation
(2) Sedimentary and volcanic rocks on gneiss domes—as (1) above
(3) Intrusive and volcanic rocks—distribution and character, age, geochemistry
c. Development of the pre-Taconic margin of North America—history of late Proterozoic rifting. What can be learned from sedimentation, igneous activity, and structural relations? What is the nature of the Proterozoic-Cambrian boundary?
3. Precambrian rocks of exotic terranes
a. Central belt
(1) New England
(a) Basement rocks—Chain Lakes massif, Bronson Hill anticlinorium, Merrimack synclinorium, Nashoba block
(b) Younger stratified rocks
(2) Problematic terranes of the southeast—Inner Piedmont and Charlotte belts
b. Avalonian platform
(1) New England: crustal and supracrustal rocks
(2) Southern Appalachians: supracrustal rocks
(a) Carolina volcanic slate belt
(b) Eastern terranes—Raleigh belt, Eastern slate belt, Kiokee belt, Uchee belt
c. Florida platform—subsurface Precambrian rocks
d. Regional problems
(1) Where is the eastern basement south of New England?
(2) Where is the Iapetus suture?
(3) What is the nature of the Proterozoic-Cambrian boundary in the exotic terranes?

VI. Middle and Late Proterozoic Sequences of the Northern Rocky Mountains and Basin and Range (M. W. Reynolds, U.S. Geological Survey)
A. Overview
B. Belt Supergroup
1. Distribution and structure
2. Stratigraphy and stratigraphic variations
3. Age and correlation
4. Tectonic significance
C. Possible southern equivalents of the Belt (Uinta Mountain Group, Big Cottonwood Formation, Unkar Group, Apache Group, Pahrump Groups, etc.)—each as 1–4 above
D. Windermere Group and possible equivalents (including Mineral Fork and Mutual Formations, Chuar Group, etc.)—each as 1–4 above
E. Economic significance
F. Significant regional problems
1. Tectonic interpretation and relation to the Proterozoic continental margin
2. Regional correlation of isolated units
3. Extent and significance of tillite and its potential as a marker horizon
4. Significance and reliability of the paleomagnetic record
5. Base of the Cambrian in the southern Basin and Range

VII. Topical Studies
A. Application of gravity and magnetic techniques to Precambrian studies
B. New developments in basement tectonics
C. Isotopic "remote sensing" and crustal evolution
D. Seismic profiling and the nature of the Precambrian crust
E. Paleomagnetic studies as applied to Precambrian plate motions and to the dating and correlation of Precambrian rocks

VIII. Discussion and Synthesis
A. Early evolution of the craton and the cratonic margins. Can a coherent geologic history of the major individual components of the craton be established? Can the order and manner of their assembly be determined?
B. Thickness and character of the continental crust beneath the craton. Can the nature of the major dynamic processes in the crust and deeper lithosphere be inferred? How do these control igneous activity, sedimentation, and tectonics?
C. Comparison of history and evoluton of the North American craton with other cratonic areas. Can the development of the North American craton be placed in an overall view of continental evolution and the global processes that operated throughout the history of the earth? If so, was accretion or growth continuous or episodic? Are provinces of similar character and age to those in North America contained in other continents? If so, are they dismembered extensions or independent contemporaries?
D. Frontiers of research in the Precambrian of the conterminous United States

Reference Cited

King, P. B., 1976, Precambrian geology of the United States: An explanatory text to accompany the geologic map of the United States: U.S. Geological Survey Professional Paper 902, 85 p.

Continental platforms and basins of Canada

D. F. Stott
J. D. Aitken
Institute of Sedimentary and Petroleum Geology
Geological Survey of Canada
3303 33rd Street NW
Calgary, Alberta
Canada T2L 2A7

Introduction

This chapter is an extended annotated outline of the planned contents for the volume *Continental Platforms and Basins of Canada* that will form a part of the sixth edition of the *Geology and Economic Minerals of Canada* as well as part of the series on *The Geology of North America* that is being prepared in connection with the Decade of North American Geology. Sections preceded by arabic numbers will be chapters in the synthesis volume. Information on the Hudson Platform was provided by A. W. Norris, and information on the St. Lawrence platform was provided by B. V. Sanford. Their assistance is gratefully acknowledged.

Volume Contents

1. Introduction

This chapter will provide a brief outline of the geologic development of the following tectonostratigraphic elements in Canada: Interior Platform—Western Canada, Mackenzie, and Williston Basins; Hudson Platform—Hudson Bay and Moose River Basins; and St. Lawrence Platform—Michigan, Quebec, and Anticosti Basins, and Ottawa Embayment. The Canadian Shield is considered only as basement to the mainly Phanerozoic sedimentary accumulations; for its anatomy and historical development, the reader will be referred to the volume on the Canadian Shield. Similarly, the Arctic Platform is ceded to the volume on the Innuitian Orogen and Arctic Platform. Although detailed consideration of the Cordilleran and Appalachian Orogens is left to the appropriate volumes, the depositional histories of the Western Canada Basin and the St. Lawrence Lowlands are inextricably linked with the histories of the mountain chains that form the outer boundaries of these two elements; accordingly, the salient features of the two mountain chains are outlined briefly. Within the Interior Platform, the Western Canada and Mackenzie Basins are essentially continuous and have much in common, but differences are such that discussion is facilitated by treating them as separate elements.

The geology of fossil fuel deposits, nonmetallic mineral deposits, and groundwater within the Interior Platform and the frontal areas of its bounding mountain systems will be outlined in some detail. Deposits of metallic minerals will be considered at length in the volume on the Cordillera and will receive only brief mention here in their stratigraphic context.

Historical Summary

The Canadian part of the North American craton has undergone differentiation in both space and time (Figs. 1, 2) since consolidation of the Canadian Shield. Both the Atlantic and Pacific margins of the continent were defined by continental separation at about the beginning of Cambrian time, following a precursory period of rifting with associated sedimentation and local volcanism in very late Precambrian time. The newly defined margins transected structural provinces of the Canadian Shield.

The strata deposited on the young, subsiding, passive western margin bear many similarities to those of the eastern margin. However, the passive-margin behaviour of the Pacific region continued through early Paleozoic time at least to the beginning of Devonian time, whereas that of the Atlantic margin was terminated by the Taconic Orogeny in early Middle Ordovician time. In the northwest, tectonism (Ellesmerian Orogeny?) caused a marked change in sedimentation in the Mackenzie Basin. Destruction of the con-

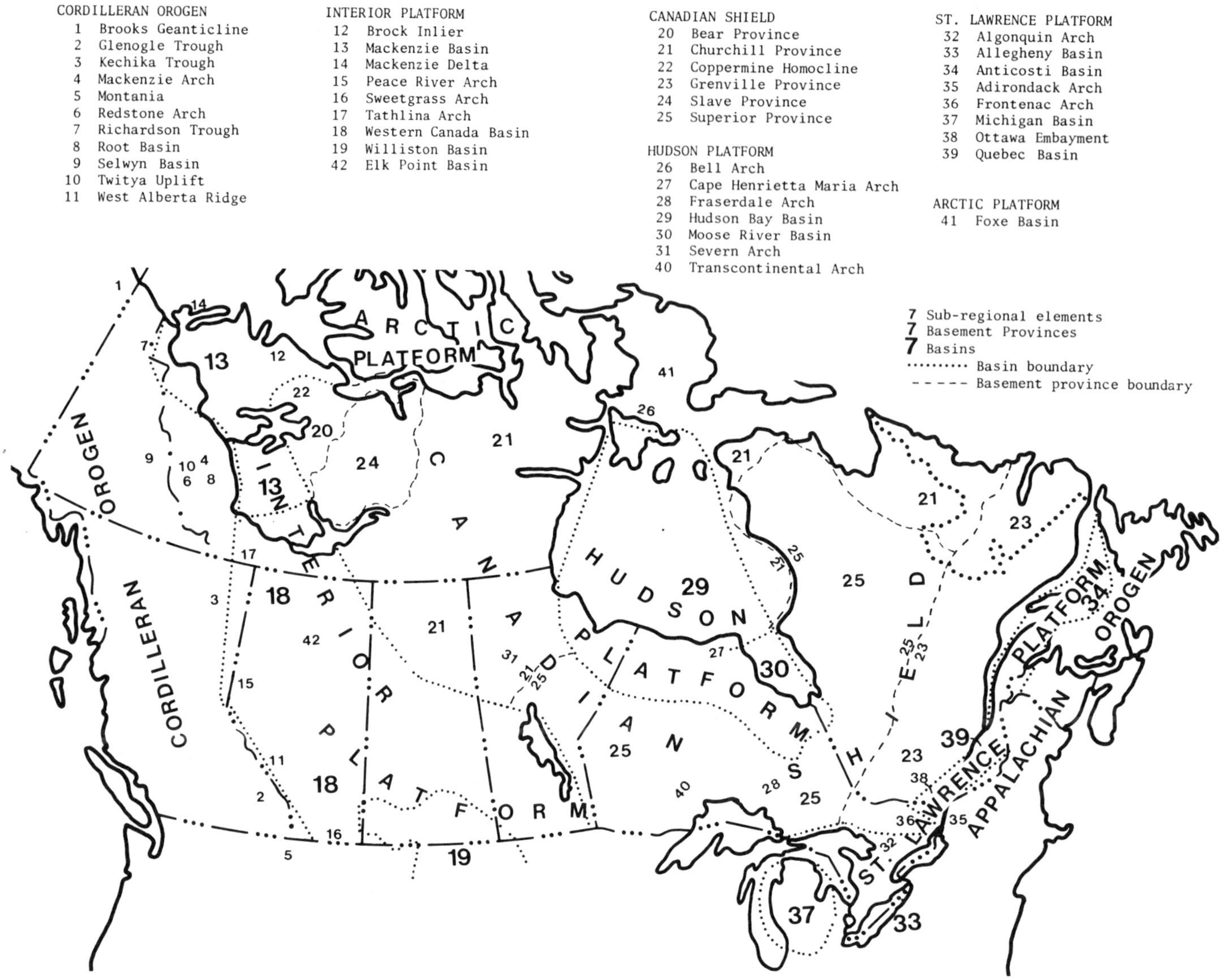

Figure 1. Tectonic elements of the Canadian interior platforms and basins.

tinental terrace wedge and creation of a foredeep receiving the detritus from the young mountain chain commenced in the proto-Appalachians in early Middle Ordovician time, continued in a pulsatory manner through the Taconic and Acadian Orogenies into Late Carboniferous time, and resulted in a stacked succession of northwestward prograding molasse deposits. In the Cordillera, similar events, delayed until Middle Jurassic time, continued into Eocene time through the Columbian and Laramide Orogenies.

Subsidence of the intracratonic basins was essentially independent from that of the continental margins. The commencement of subsidence in both the Michigan and Williston Basins in early Middle Ordovician time suggests a common mechanism, but one that is not understood. Although subsidence began late within the intracratonic basins, their main phases of near-continuous subsidence ceased early, by mid-Carboniferous time for the Michigan Basin and by Late Carboniferous time for the Williston Basin. Subsidence was renewed within the Williston Basin during Jurassic time, but basin development was terminated during latest Jurassic time. Emergence at that time possibly was related to the Columbian Orogeny occurring farther west. The briefest history of all was that of the Hudson Platform, extending only from Late Ordovician through Late Devonian time. Brief episodes of continental clastic sedimentation during Middle Jurassic and Early Cretaceous times in the Moose River Basin appear more closely linked to Mesozoic uplift of the Canadian Shield than to the earlier subsidence.

Deposition of the continental terrace wedge in the west ended with the Columbian Orogeny, followed by the development of eastward migrating foredeeps and the thick

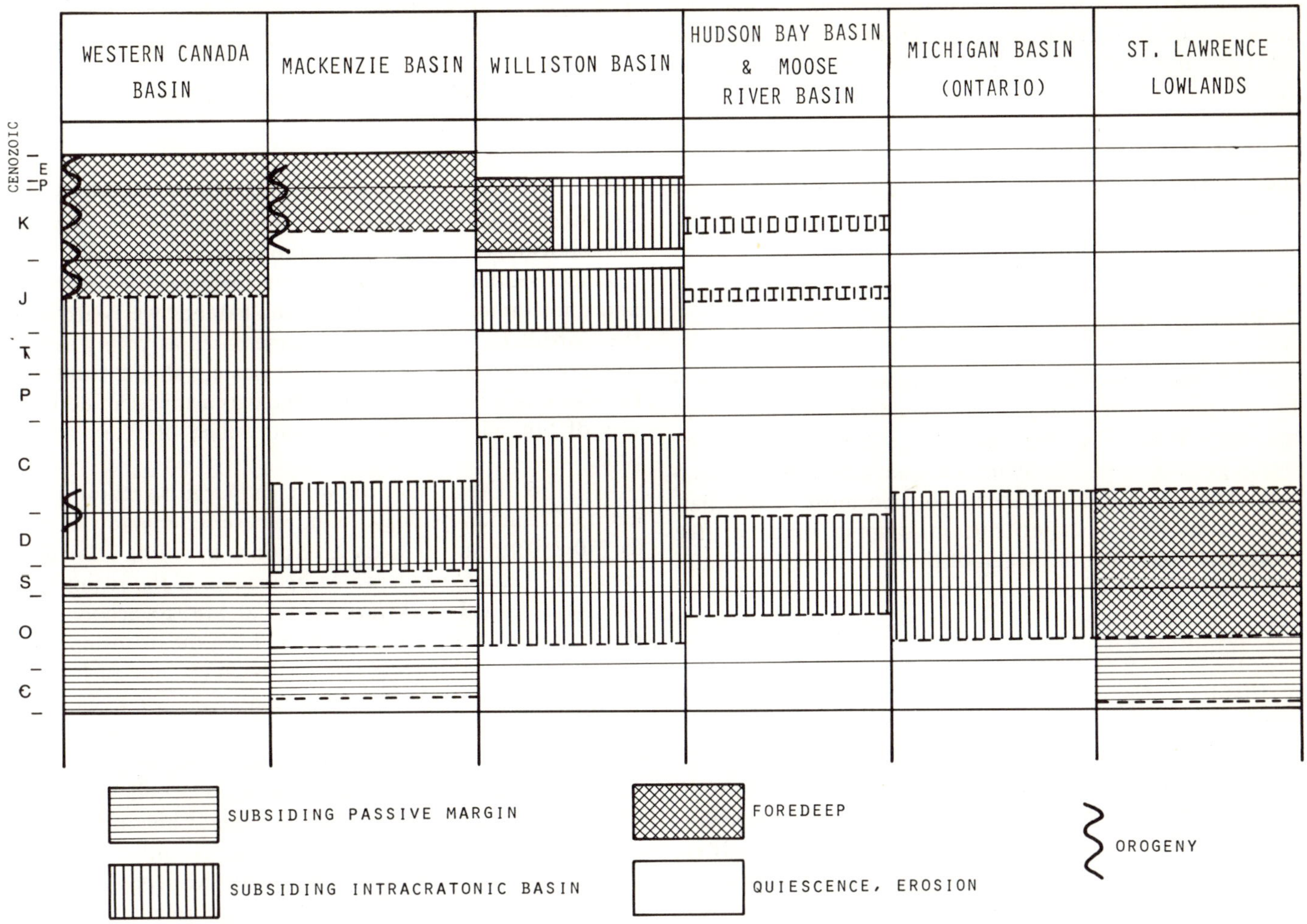

Figure 2. Summary diagram of the tectonic development of the various sectors of the interior platforms and basins of Canada.

accumulation of clastic sediments in the Western Canada Basin. Two major phases, the first during Late Jurassic and earliest Cretaceous times, and the second largely within Albian time, had pulses defined by thick wedges of sandstone and coal-bearing sediments. The Jurassic–Early Cretaceous foredeep was narrowly restricted, and associated sediments did not overlap far eastward on the underlying Paleozoic succession. Only during the later Albian phase did the boreal embayment flood the continental interior, gradually extending over much of the cratonic cover, southeastward across the ancient Williston Basin, and eventually joining with the southern embayment advancing from the Gulf of Mexico.

The Western Canada Basin was relatively quiescent during the early (but not earliest) part of Late Cretaceous time, when most of the region was submergent and blanketed by black shales. With the beginning of the Laramide Orogeny towards the end of Cretaceous time, thick wedges of nonmarine clastic sediments were spread far eastward across the basin until well into Tertiary time. With cessation of orogeny and emergence of the craton, sedimentation ended.

No definitive record is present in the Mackenzie Basin and northern Yukon of orogenic movements similar to the early phase of the Columbian Orogeny farther south, although the sedimentary regime appears to change in Late Jurassic time. Not until late Early Cretaceous time is there a record of a developing foredeep in front of the rising Mackenzie Mountains and Brooks Range to the north. Deformation and consequent molassic deposition continued into Eocene time.

The continental interior underwent extensive erosion after early Tertiary time, and the clastic detritus was transported northward, building the Tertiary and Holocene deltas of the Mackenzie River. The Mississippi Delta also received sediments from the Canadian Interior. Earlier erosion of the craton had occurred in Cretaceous time with transport of the detritus across the Hudson Bay region and the Labrador Shelf and through the St. Lawrence region to the Atlantic offshore.

The widespread continental glaciations of Pleistocene age, which affected the topography and left thick sediments, will be summarized in the volume on Quaternary geology of Canada.

2. Physiography

The old, low-lying Precambrian Shield is surrounded by lowlands, plains, and plateaus of generally flat-lying sedimentary rocks. The physiography has been controlled by drainage developed largely in Cenozoic time and by glaciation. The Laurentide ice sheets covered most of the platforms and had a major role in shaping much of the Holocene land surface. Much of this chapter will be summarized from the Quaternary volume.

3. History of Exploration

Although some geologic investigations had been made earlier in Eastern Canada, systematic geologic exploration of the sedimentary basins of Canada began with the establishment of the Geological Survey of Canada in 1842. Much of the early investigation of the St. Lawrence Lowlands and Gaspé was undertaken by Logan, Murray, Richardson, and others. The first oil discovery in that region was made in Ontario in 1861. Drilling in the St. Lawrence Platform has continued to the present, and wells are now being drilled offshore in Lakes Ontario and Erie. The geology of the Hudson Bay region was examined during the 1870s by R. Bell, followed by many others. The first well for oil and gas was drilled in 1966–67, and subsequent drilling provides information from the offshore central part of Hudson Bay Basin. In western Canada, a large part of the plains was visited in the middle of the nineteenth century by Hector of the Palliser Expedition, but geologic exploration lagged until the building of the railway in the 1880s. In the late nineteenth century, exploration by such pioneers as Selwyn, Dawson, and McConnell provided some of the first geologic assessments of the country. Systematic investigations in all regions by both federal and provincial agencies followed in the first half of the twentieth century, but progress generally was slow. Major oil and gas discoveries in Western Canada in the late 1940s and 1950s resulted in a rapid escalation of surface and subsurface exploration and geophysical activity. The unique availability of much of the accumulated data allows very detailed analyses of large parts of the regions.

4. Evolution of the Tectonic Framework

I. Interior Platform and Eastern Cordillera

A. The Canadian Shield: buried structural provinces, truncated at the western margin of the continent. The western shield was largely cratonized by the close of the Hudsonian Orogeny at about 1,700 to 1,800 Ma. Younger sedimentary and volcanic cover is little deformed and unmetamorphosed and in part extends westward into the Cordillera. Dyke swarms, some dated as young as 625 Ma, are the youngest events in the western shield. Contrary to what might be assumed, the shield as a source of detritus was of limited significance during much of Phanerozoic time.

B. Creation of the western margin: a major problem in Canadian geology. Windermere sedimentation, 570 to 770(850?) Ma, is clearly a response to rifting related to creation of the Cordilleran Miogeocline. If the rifting episode is extended backward to 1,500 Ma, to account for Purcell sedimentation (1,500–1,350 Ma), the rifting episode seems unreasonably long and spans an episode of folding and metamorphism (East Kootenay Orogeny). Purcell-aged rocks in the northern Cordillera are not positively identified; the most likely candidates are the Wernecke Assemblage and the older succession at Tuchodi Lakes. Purcell equivalents onlapping the older Canadian Shield form a northwest-facing succession apparently truncated to the southwest by the Cordilleran Miogeocline. The Mackenzie Mountains Supergroup, intermediate in age between the Purcell and Windermere Supergroups, has the geometry but not the lithic character of a continental terrace wedge and may also be truncated by the Cordilleran Miogeocline. Numerous problems of chrono-correlation bedevil these questions; the assumption of synchroneity of glaciations (the earlier following closely upon early Windermere rifting) brings some order. Whether the western margin, which truncates the southwestward extension of the Churchill structural province, was defined by rifting in Purcell or Windermere times, the paucity of volcanic rocks is an anomaly.

C. Commencement of drift—570 Ma? (Based on subsidence curve for Cambrian.) Problem: extraordinarily long rift-drift lag.

D. Early Paleozoic history: fundamentally, a record of passive margin subsidence and formation of rift basins near the outer edge of the continent at various times (Selwyn Basin, Glenogle Trough, Kechika Trough, Richardson Trough), and the appearance, from time to time, of basins and uplifts on the craton (Elk Point Basin, Root Basin, Montania, West Alberta Ridge, Peace River Arch, Mackenzie Arch, Tathlina Arch).

E. Throughout post-Early Cambrian Paleozoic

time, the low-lying Canadian Shield, increasingly blanketed by the deposits of successive marine transgressions, was a subordinate and decreasing source of detritus. The main sources of clastic material were local uplifts and, near the edge of the continent, the high blocks flanking rift basins.

F. In Late Devonian time, a new, major source of detritus appeared in the northern Yukon and Alaska giving rise to renewed clastic deposition in the Mackenzie Basin. It persisted into Mesozoic time.

G. Especially in late Paleozoic time, oceanic and island-arc successions to the west are unrelated to the platformal and miogeoclinal, autochthonous and parautochthonous successions of the Interior Platform and Eastern Cordillera. The former are allochthonous, having been "docked" against North America in a long series of events, currently only poorly understood.

H. The rocks of Triassic, and Lower and Middle Jurassic ages, although mainly clastics, are similar to the underlying Paleozoic strata in their general lithology, westward thickening, and eastern source of detritus.

I. The appearance of coarse, westerly derived clastics at the beginning of Late Jurassic time records westward movement of North America as the Atlantic Ocean began to open; the western margin became convergent, and the earliest preserved depositional record of Phanerozoic mountains in Western Canada was laid down. Farther west in the Cordillera, evidence of convergence appears earlier, in Middle Jurassic time.

J. With convergent tectonism occurring in the west, loading of the crust by tectonic stacking, as well as loading by the rapidly accumulating clastic wedge derived from the young mountains, created a new tectonic feature, the Cordilleran Foredeep. In the Western Canada Basin, two major phases of the Columbian Orogeny occurred, and several pulses were marked by coarse clastic tongues, the first in Late Jurassic–Early Cretaceous (Berriasian-Valanginian) time and the second in late Early Cretaceous (Aptian-Albian) time. During early Late Cretacous time, when deformation was apparently less intense along the western margin, a boreal embayment advanced southward to join with the northward-advancing embayment from the Gulf of Mexico. The later Laramide Orogeny, beginning about Maastrichtian time, resulted in the deformation of the western margin of the basin while the eastward prograding wedge of clastic sediments filled the basin. Convergent tectonism continued until Eocene time. The approximate synchrony of the terminal movements and the youngest marine sediments is not coincidence but cause and effect.

K. In the northern Cordillera, no folds older than mid-Cretaceous age are identified. However, no pre-Aptian foredeep deposits are known in the Mackenzie Basin. There, molasse deposition proceeded episodically from Aptian-Albian to Eocene time.

II. Hudson Platform

A. The boundary between the Superior and Churchill structural provinces of the Canadian Shield passes beneath the southern part of the Hudson Platform, coinciding with the Cape Henrietta Maria Arch that separates the Hudson Bay Basin, largely on a Churchill Province basement, from the Moose River Basin, on a Superior Province basement.

B. Initiation of the Hudson platformal sequence began with widespread submergence of the Canadian Shield during Late Ordovician time. Uplift of the Fraserdale Arch produced coarse clastics. Later transgression connected the Hudson, St. Lawrence, Interior, and Arctic Platforms. Limestone accumulated along the margin, thin evaporites accumulated in the Moose River Basin, and widespread salt accumulated in the central Hudson Bay Basin.

C. Silurian epeirogeny affected the Fraserdale Arch and caused block faulting on the Bell Arch. Reef development took place during Late Silurian time. The Severn and Fraserdale Arches were emergent during Late Silurian and into Early Devonian time, providing detritus to the Moose River and Hudson Bay Basins.

D. Early Devonian deposits are largely terrigenous and are succeeded by Middle Devonian carbonates and evaporites. Upper Devonian clastics are derived in part locally, and in part from the Acadian Orogen. The Transcontinental Arch caused pronounced provincialism of the benthonic fauna, and Early and Middle Devonian faunas formed part of the endemic eastern Realm. Breaching of the arch was initiated in early Middle Devonian time.

E. Epeirogeny during late Paleozoic and early Mesozoic times resulted in uplift and stripping of sedimentary cover bordering the basinal areas.

F. Deposition of thin continental Jurassic and Lower Cretaceous sediments took place in the southern part of the Moose River Basin.

G. Renewed uplift of the Fraserdale Arch in Early Cretaceous time is associated with faulting and emplacement of lamprophyric dykes and sills in the southern part of the Moose River Basin.

III. St. Lawrence Platform

A. The Canadian Shield: deformed Lower Proterozoic sedimentary and intrusive rocks beneath the Michigan Basin are truncated eastward at the Grenville Front by Middle Proterozoic older metasedimentary and intrusive rocks that underlie the entire length of the platform. The Grenville Province was cratonized at about 1,000 Ma. The youngest recorded events include Upper Proterozoic dyke swarms and volcanic flows, and Upper Proterozoic clastic redbeds that are relatively undeformed and unmetamorphosed. Positive tectonic elements at the beginning of Paleozoic time include the Algonquin Arch, the Frontenac-Adirondack Arches, and the southern margin of the Canadian Shield.

B. Initial continental rifting probably began in latest Proterozoic time with spreading in progress by Early Cambrian time. The St. Lawrence Platform represents the northwestern foreland of the Appalachian Orogen. The geographical boundary between the platform and the miogeocline is variable through Paleozoic time and largely obscured beneath allochthonous slices emplaced during Taconian and Acadian orogenic events.

C. Cambrian to lower Middle Ordovician sediments were deposited on the continental shelf of a passive margin, with orthoquartzites derived from bordering arches and the southern shield. Lower Middle Ordovician to Devonian sediments were deposited in troughs (foredeeps) marginal to tectonic highlands on the south. A major depocentre, the intracratonic Michigan Basin, developed in Early to Middle Ordovician time and prevailed to the end of Late Carboniferous time.

D. Extensional block faulting in widely separated parts of the platform and exposed shield is related to more intensive phases of the Taconian and Acadian Orogenies in late Middle Ordovician, Late Silurian/Early Devonian, and Late Devonian/Mississippian times.

E. The marginal troughs (foredeeps) were foreshortened from time to time by stacking of allochthonous slices along the Appalachian Front to eventually form the Allegheny Basin (west division), Quebec Basin and adjoining Ottawa Embayment (central division), and Anticosti Basin (eastern division).

F. Faulting, and emplacement of dykes and alkalic intrusions (i.e., Monteregian) in widely separated parts of the platform and southern craton are related to separation of the African and European plates from the North American plate in Early Jurassic and Early Cretaceous times, respectively.

5. Geophysical Characteristics of the Interior Platform and Basins

This chapter will be presented as much as possible in the context of basement provinces. The summary will consider gravity, electrical conductivity, and aeromagnetic patterns, deep seismic profiles, seismicity and heat flow, and crustal models.

6. Description of Geology by Geological Subprovince

I. Interior Platform: Western Canada and Mackenzie Basins

A. Basement rocks: summary statement only (see Tectonic Framework [4] and Geophysical Characteristics [5]). Most of the Western Canada Basin lies on thoroughly cratonized, Precambrian basement of the Churchill Province. Sedimentary strata of the exposed Precambrian sequences of the Cordillera (see below) are fully involved in the structures of the Eastern Cordillera. For the Mackenzie Basin, little is known of "true" or crystalline basement because of the difficulty of projecting westward the Bear and Slave Provinces of the shield. Pseudo-basement is composed of sedimentary and volcanic rocks of the Coppermine Homocline succession (1,400–800 Ma±).

B. Older Proterozoic sedimentary sequence: summary from the Cordilleran volume. Content: classical Purcell; "Helikian" strata of the Tuchodi Lakes area; in the north, the "Wernecke Assemblage" of the Cordillera and pre-Rae Group strata of the Coppermine Homocline. The older sequence reaches the mountain front at Waterton and Tuchodi Lakes in the Rockies and Margaret Lake in the Mackenzie Mountains; it may underlie the Phanerozoic rocks southwest of Great Bear Lake. Mineral potential of the older sequence is in copper, and in the Coppermine Homocline, uranium as well.

C. Intermediate Proterozoic sequence: not present in the Western Canada Basin. The Mackenzie Mountains Supergroup (1,200–800 Ma±) is involved in the frontal Mackenzie Mountains structures and underlies the Phanerozoic rocks in much of the Mackenzie Basin. It is bevelled northeastward beneath the sub-Cambrian unconformity, but surfaces in the Brock Inlier and Victoria Island. The principal mineral potential is in zinc.

D. Younger sequence, or Windermere Supergroup: summary from Cordilleran volume. Theme: sedimentation in response to rifting (pre-drift?). Content: in the southern Cordillera, the type Windermere, Miette Group, and "Grit Unit"; in the north, the Rapitan Group and younger Precambrian units, the "Grit Unit," and, tentatively, the 'copper cycle' of the Mackenzie Mountains. The younger sequence nowhere reaches the mountain front; its existence must be recognized to establish the magnitude of the sub-Cambrian hiatus over the platform. Problem: Why the 200 m.y.(?) delay between rift and drift? Mineral potential: in the north, iron, copper, lead, zinc, silver; south of 60°, limited potential.

E. Early Paleozoic

Themes:

i. Sedimentation on a young, passive margin (continental separation initiated about the beginning of Cambrian time).

ii. The Great Lower Paleozoic Transgression.

iii. Three facies belts surrounding the exposed craton; establishment of vast carbonate platforms.

iv. Paleolatitudes and their effect on deposition.

v. Strongly expressed cyclicity due to interaction of subsidence and variable rates of sea-level rise.

vi. Birth of rift-bounded troughs near the

edge of the continent; Early Cambrian in the Mackenzie Mountains, Early Ordovician in southern British Columbia.

Phases:

i. Sub-Cambrian, "worldwide" unconformity.

ii. Depocentres for thick Lower Cambrian sandstones; carbonate platforms rather local. Landward shift of littoral environments. In the Mackenzie Basin deposits are strongly transgressive; early supply of shield-derived quartz sand fails before end of Early Cambrian time, and a diachronous carbonate sheet extends well onto the craton. The depositional platform is extremely broad.

iii. Middle Cambrian. Renewed transgression above an unconformity. Sharply diminished supply of coarse detritus. Probably non-shield sources of detritus. Establishment of abrupt platform to basin transitions; Middle Cambrian is "starved" in the basins. At the end of Middle Cambrian time, evaporites and redbeds in the north and south reflect low paleolatitudes. In the Cordillera to the southwest, the abrupt transition between platformal carbonates and basinal shale and limestone, established in Early Cambrian time, persists. Upper Middle/lower Upper Cambrian evaporites accumulate in a short-lived basin behind the rejuvenated Mackenzie Arch. The Middle Cambrian record is poor, except in basinal areas, because of deep, sub-Upper Cambrian erosion. The Middle Cambrian has economic potential for natural gas.

iv. Late Cambrian. Continued waning of detrital supply in the west; carbonates dominate. Deadwood clastics are trapped against the craton, inside carbonate shoals. In the Selwyn Basin, deposition was probably uninterrupted although a major unconformity at the base of the Upper Cambrian succession bevels the Mackenzie Arch deeply. Truncated folds are known only deep in the Mackenzie Fold Belt. The extremely broad carbonate platform extends northeastward almost into contact with Precambrian rocks of the shield.

v. Early Ordovician. Continuation of Upper Cambrian style on the platform. Subsidence of the Glenogle and Kechika Troughs; sedimentation of black shales with, probably, a non-shield source (e.g., Mount Forster high). In the north, black shales in the Selwyn Basin (well inside the fold belt) must be supplied longitudinally rather than across the platform. The basinal, shaly sediments are hosts to lead, zinc, silver, and barite mineralization.

vi. Middle Ordovician. Very limited extent on the platform due to sub-Upper Ordovician unconformity? Continuous sedimentation in the troughs.

vii. Late Ordovician and Early Silurian. A single formation in many areas; isofacial over the platform because this transgression, possibly the greatest of the Phanerozoic, overlapped the shield almost completely. Contemporaneous shales in the outer-platform troughs had a non-shield source. The first phase of pronounced subsidence of the Williston Basin begins in Late Ordovician time.

viii. Middle Silurian (Wenlock). Known only from the shale troughs; stripped from the platform by "sub-Devonian" erosion, except possibly in the Hudson Bay area.

F. Late Silurian through Late Devonian

Themes:

i. "Sub-Devonian" (sub-Upper Silurian?) unconformity.

ii. Thermal subsidence of the continental margin no longer the dominant factor.

iii. Patterns of sedimentation dominated by eustatic variation in sea level and intracratonic uplifts and basins.

iv. In the south, pronounced uplift of the Peace River Arch and the West Alberta Ridge. In the north, first uplift of the Keele Arch and Twitya Uplift and renewed uplift of the Redstone Arch.

v. With the shield mostly blanketed by lower Paleozoic carbonates, a weak clastic supply, mostly from local uplifts, and mud from a northern source, are demonstrable on the platform and inferred in the basins. Chemical sediments dominate on the platforms.

vi. Widespread evaporites lie along belts of low paleolatitude.

vii. Mineral deposits include salt, potash, lead, and zinc. The Middle and Upper Devonian are hosts to important pools of oil and gas.

Phases:

i. Early Devonian. The Elk Point Basin, east of the West Alberta Ridge, was first flooded in late Early Devonian time, resulting in the first redbeds-to-carbonates-to-evaporites cycles. Valley-filling redbeds on the ridge may be of this age. At this time, such positive features as the Peace River Arch, Tathlina Arch, and West Alberta Ridge were most strongly expressed. The Delorme Formation and equivalents in the Mackenzie Basin are fish-bearing clastic rocks developed locally (as deltas?). Elsewhere, littoral carbonate rocks, dark shale and limestone are developed in the Selwyn Basin and Richardson Trough. Post-Delorme, Early Devonian formations on the platform are peritidal carbonate rocks, passing to evaporites toward the craton and onlapping northeastward.

ii. Middle Devonian. The Elk Point Basin continues to be evaporitic, with salt in Alberta and salt and potash in Saskatchewan and Manitoba. The basin developed a fringing carbonate platform with outlying reefs. Toward the end of Middle Devonian time, the Peace River and other uplifts contract because of continued marine onlap. Middle Devonian anhydrite and gypsum and open-marine carbonates are preserved west of the West Alberta Ridge. Mineral deposits include salt, potash, and zinc-lead.

The Middle Devonian is host to important pools of oil and gas. In the Mackenzie Basin, open marine environments succeed restricted evaporitic ones on the platform, and some terrigenous mud appears, apparently from the north. The first Devonian reefs appear on the outer parts of the platform, and later, extensive reefal carbonate platforms appear more widely. The Early Devonian uplifts are by now submerged. The Middle Devonian provides the only current production of Paleozoic oil and gas in the Mackenzie Basin.

iii. Late Devonian. Upper Devonian sediments were deposited upon a regional disconformity. Major carbonate platforms and reefs retreat southward to central Alberta in a series of steps in front of increasing mud supply from an old and a new source of detritus in the north. Evaporite deposition, greatly reduced in area, continues in Saskatchewan. Following post-Frasnian regression and the great extinctions at the Frasnian-Famennian boundary, carbonates of the Wabamun Formation formed a regional blanket in the south, backed by evaporites to the east. The Wabamun platform lacks an abrupt edge against the basin to the northwest, in which shales of the Besa River Formation accumulated and reefs are lacking. The Upper Devonian is Western Canada's dominant source of conventional oil and gas. In the Mackenzie Basin, there is a subtle disconformity at the base of the Upper Devonian succession. In addition to strengthening of an established northern source of mainly fine-grained detritus from the shield and/or the Innuitian Fold Belt, a new source of fine to coarse detritus—the Brooks Range Anticlinorium—appears, giving rise to the Devono-Carboniferous "black clastic" unit in the Selwyn Basin; these are responses to the Ellesmerian Orogeny in the north.

G. Late Paleozoic

Themes:

i. With the Exshaw transgression, a return to behavior like that of a passive margin.

ii. In Early Carboniferous time, the supra-regional setting resembles that of Late Devonian time, but the northern area of clastic deposition has expanded southward; reefs and abrupt platform edges have not been recognized. Hydrocarbons produced from Carboniferous rocks occur in structural traps developed during Mesozoic time, rather than in stratigraphic traps as in Devonian rocks.

iii. In the absence of a reef-building community, barriers now consist of echinoderm-bryozoan, and to a lesser extent, ooid shoals, which make good reservoirs.

iv. By late Famennian, the Peace River Arch had become inactive, and was buried by sediments. By the Tournaisian, it had become an area of accelerated subsidence, the Peace River Embayment. This subsiding area received mainly terrigenous muds, with carbonates around its margins, until the Viséan, when carbonate deposition expanded across it. In Late Viséan, deposition of clastic rocks (sandstone and shale) recommenced, with minor amounts of carbonates, continuing into the Namurian.

v. Continued subsidence of the Williston Basin, including development of carbonate and evaporitic facies.

vi. The Upper Carboniferous and Permian successions are characterized by numerous disconformities and gaps in the record. The Middle Namurian to low Moscovian (Lower and Middle Pennsylvanian) record is one of both erosion and deposition of thick sandstones and subordinate carbonates. Younger Carboniferous rocks are missing from the platform, because of truncation at the sub-Permian unconformity. The condensed, Permian succession, characterized by chert and impure, phosphatic carbonates, displays remarkable persistence of thin lithostratigraphic units, from the U.S.A. to the southern Mackenzie Mountains. These units undergo a gradual, westward facies change to basinal mudrocks. Sedimentary phosphate is of economic interest in the Permian.

vii. There is a marked contrast between the autochthonous and parautochthonous sedimentary rocks of the Eastern Cordillera and Interior Platform and allochthonous, volcanics-bearing, oceanic and island-arc, accreted terranes of the Western Cordillera.

viii. Much of the Mackenzie Basin lacks Paleozoic rocks younger than Late Devonian. Coaly, sandstone-rich, deltaic facies of the Carboniferous at the northern end of the Western Canada Basin and eastward onlap in Richardson and White Mountains suggest that this absence is due to nondeposition as well as subsequent erosion. In the northwestern part of the basin (Peel Basin), Carboniferous clastic rocks with coal, having a northwest depositional strike, are preserved beneath Mesozoic strata. In contrast, Permian rocks of the same region have a northeast depositional strike (the Sverdrup Basin trend) and onlap southeastward above a basal unconformity. Both Carboniferous and Permian successions are strongly bevelled eastward by Triassic, Jurassic, and Cretaceous strata, thus their extent in the subsurface, if any, is limited and mainly beyond the present reach of drilling.

Phases:

i. Early Carboniferous. In the south, the platform remained stable with widespread deposition of black shale (includes uppermost Devonian) followed by extensive sheets of bioclastic carbonates that intertongue westerly and northwesterly with cherty limestone and black shale, and easterly and southeasterly with evaporites. At the north end of the Western Canada Basin, a westward facies change to partly turbiditic, clastic and carbonate rocks.

ii. Late Carboniferous. In the south, an abrupt change to mainly sandy deposits took place in late

Early Carboniferous and persisted into middle Late Carboniferous. Whether this influx of sand was from the shield or from the new source in the north is problematical. Deep erosion followed.

iii. Permian. Several relatively thin, condensed marine clastic and cherty to phosphatic carbonate sequences are separated by disconformities. A remarkably continuous but thin chert and siltstone unit overlaps older beds along the length of the Interior Platform.

H. Triassic and Early and Middle Jurassic

Themes:

i. Continuation of deposition on a stable shelf.

ii. Development of a restricted basin with associated redbeds and evaporites.

iii. Return to marine deposition under stable shelf conditions.

iv. East-striking Triassic rocks of the Prudhoe Bay region of north coastal Alaska pinch rapidly southward and probably do not enter the Mackenzie Basin. Jurassic rocks of the northern Yukon have a similar geometry but greater southward extent.

Phases:

i. Early and Middle Triassic. Stable shelf with embayments and platforms. Basins and subsidence patterns are related to those of the Carboniferous and Permian. Sediments have a cratonic source. Deposits include shoreline sandstone and black, clastic, basinal sediments with associated phosphate deposits. Important pools of oil and gas are associated with epineritic sandstone.

ii. Middle to Late Triassic. Restricted basin, including redbeds, dolomitized skeletal limestone and evaporites. Important oil occurrences are in dolomitized limestone.

iii. Late Triassic. Marine deposition on a stable shelf; deposition of marine carbonate and fine-grained clastics. Important gas occurrences.

iv. Early to Middle Jurassic. Stable shelf, fine black clastics and phosphatic chert were deposited west of thin shoreline sandstones. The Williston Basin subsides with deposition of redbeds, evaporites, marine limestone, and shale.

I. Late Jurassic and Cretaceous

Themes:

i. The Columbian Orogeny starts in the west and a foredeep develops in the east. Molasse deposits extend along the length of the Western Canada Basin. Thick alluvial-deltaic and piedmont deposits accumulate in the south, and marine sediments accumulate in the Peace River region.

ii. Continued subsidence of the Williston Basin and deposition of marine shales.

iii. Renewed influence of structures associated with the ancient Peace River Arch.

iv. Deposition of cratonic detritus in a Jurassic epicontinental sea, disrupted by local epeirogenic uplifts, is recorded in northern Yukon but only locally in the Mackenzie Basin.

v. A major tectonic reorganization in the Mackenzie Basin commences in late Early Cretaceous time with a rising western highland and a north-south-trending foredeep.

vi. The Laramide Orogeny affected the entire length of the Interior Platform. Development of a second molasse basin took place in front of the rising Rocky Mountains and mountains along the western flank of the Mackenzie Basin.

Phases:

i. Late Jurassic. Marine deposition begins in a newly created foredeep; there is renewed subsidence in the Williston Basin.

ii. Berriasian-Valanginian. The major episode of the Columbian Orogeny, reflected by molasse deposition in the foredeep and thick commercial coal seams in the southern Cordillera. Development of a thick marine succession north of the Peace River Arch and coal-bearing beds to the south.

iii. Hauterivian-Barremian rocks are largely missing owing to nondeposition and/or pre-Aptian erosion.

iv. Barremian-Aptian. Boreal marine transgression with associated alluvial-deltaic sedimentation. Low barriers through the interior result in several basins. Oil sand and heavy oil deposits are characteristic.

v. Albian-Early Cenomanian. A new phase of the Columbian Orogeny; foredeep is migrating eastward; molasse deposition and commercial coal deposits are characteristic. Continued pattern of transgressive-regressive cycles and renewed influence of the Peace River Arch. A flyschoid depositional phase along the northwest flank of the Mackenzie Basin.

vi. Late Cenomanian-Campanian. Widespread marine flooding of the continental interior and the joining of boreal and gulf seas. Numerous transgressive-regressive cycles within a dominantly shale sequence and development of epineritic sands (Turonian, Santonian). Molasse deposition in the Mackenzie Basin accompanied by marine inundations followed by northward alluvial deltaic progradation. Major oil pools are associated with sandstone in west-central Alberta.

vii. Maastrichtian. Initial phases of the Laramide orogeny. Widespread molassic, eastward-prograding, alluvial-deltaic deposits with associated coal beds and dinosaur bones. Transgression of the Bearpaw Sea.

J. Cenozoic

Themes:

i. Continuation of the Laramide Orogeny; development of the easternmost Cordilleran structures.

ii. Orogenic and epeirogenic uplift; erosion

and resultant deposition of coarse gravels.

iii. The major phase of Mackenzie Delta development as the western margin of the Interior Platform continues to be deformed.

Phases:

i. Paleocene-Eocene. Late Cretaceous and Paleocene formations are involved in the easternmost structures as the Laramide Orogeny reaches its culmination. Post-orogenic normal faults. Thick accumulations of continental beds with major coal deposits are final deposits of the Cordilleran foredeep. The Cannonball Sea is the final marine invasion. Development of the Sweetgrass Arch and accompanying intrusions. Thick accumulation of clastics in the Mackenzie Delta.

ii. Oligocene and younger. Sediments within structural basins of the Rocky Mountains. Gravels and conglomerates of Oligocene, Miocene and Pliocene age.

II. Hudson Platform

A. Basement rocks

i. The Moose River Basin and the southern fringe of the Hudson Bay Basin are underlain by mainly Archean, crystalline rocks of the Superior Province.

ii. Most of the Hudson Bay Basin is underlain by rocks of the Churchill Province. These are crystalline Archean rocks reworked during the Hudsonian Orogeny and a miogeoclinal succession of platform carbonates to "eugeosynclinal" greywackes and volcanics.

B. Paleozoic

Themes:

i. Sedimentation within a cratonic basin.

ii. An extensive Paleozoic transgression with chemical sediments dominating. Limited clastic deposition is associated with redbeds at times of uplift.

Phases:

i. Sub-Upper Ordovician unconformity.

ii. Upper Ordovician strata record an initial transgression onto the central Canadian Shield, commonly with basal orthoquartzitic sandstone. Development of intracratonic carbonates with minor anhydrite and salt and, finally, biohermal reefs.

iii. Silurian. Absence of Lower Silurian rocks. The Middle and Upper Silurian succession includes stromatolitic limestones, bioclastic carbonate bank deposits, and fringing reefs.

iv. Devonian. Apparently unconformable on Upper Silurian redbeds. Lower Devonian continental clastic rocks are succeeded by carbonates and evaporites.

C. Mesozoic

Theme:

i. Erosion of the cratonic cover and development of local basins in Jurassic and Cretaceous times. Some renewed uplift and intrusion of dykes and sills.

Phases:

i. Jurassic. Continental clay and quartzose sand overlap different Paleozoic formations in a small area within the Moose River Basin.

ii. Lower Albian nonmarine sediments in the south-central part of the Moose River Basin unconformably overlap Silurian and Devonian rocks and lap on to Precambrian rocks of the Fraserdale Arch.

iii. Lamprophyric dykes and sills emplaced in Lower and Middle Devonian strata near the southern margin of the Moose River Basin are dated as 128 Ma (K-Ar) and record renewed uplift of the Fraserdale Arch.

III. St. Lawrence Platform

A. Basement Rocks. Middle Proterozoic or older crystalline rocks, stabilized in Grenvillian (1,000 Ma) time border and underlie the platform, except for the Michigan Basin, which is floored by Lower Proterozoic sedimentary and intrusive rocks stabilized by Hudsonian orogeny (1,700–1,800 Ma).

B. Clastic redbeds unconformably overlie Lower Proterozoic basement along the northern rim of the Michigan Basin. Similar redbeds, with volcanic flows, overlie Middle Proterozoic and older basement rocks at the northeastern extremity of the Anticosti Basin. K-Ar dates of similar rocks in northwest Newfoundland (605± Ma) are indicative of a latest Proterozoic age. Opening of Iapetus?

C. Cambrian to Early Middle Ordovician

Themes:

i. Mirror-image with the west. Sedimentation on a stable passive margin, with transport of coarse clastics from positive basement elements southeastward onto and across a mainly carbonate shelf to the slope and rise.

Phases:

i. Sub-Paleozoic unconformity.

ii. Lower/Middle Cambrian red and grey orthoquartzitic sandstone, shale, and carbonate are confined to the southeastern margin of the Anticosti Basin.

iii. Upper Cambrian/Lower Ordovician rocks are present along the entire length of the platform and consist of basal transgressive sandstone succeeded by carbonate. Regional disconformity at the top of the Lower Ordovician.

iv. Lower Middle Ordovician rocks are confined to the central and eastern divisions. These are carbonates in Quebec Basin. Sandstone and shale in the Ottawa Embayment were derived from the nearby emergent Frontenac and Adirondack Arches and the southern shield. Carbonates in the Anticosti Basin intertongue eastward with clastics derived from an allochthon advancing towards the platform. Allochthons, preserved as klippen, were emplaced into the Anticosti Basin at the end of early Middle Ordovician time during the Taconic Orogeny. Rocks of this age are important economically for oil, gas, lead, zinc, and construction materials.

D. Late Middle Ordovician to Late Silurian

Themes:

i. Long period of transport and sedimentation of clastics from the proto-Appalachian highlands northwestward into foredeeps to intertongue with platformal carbonates and intermittently overlap the latter during the most intense pulses of the Taconic and Acadian Orogenies.

ii. Widespread marine transgressions caused episodic interconnection with the Hudson, Interior, and Arctic Platforms.

Phases:

i. Along the entire length of the platform, upper Middle and Upper Ordovician carbonates intertongue with shale on the outer margins of the platform; these are superseded by shales and sandstones derived from tectonic lands to the southeast. There is a widespread disconformity(?) on segments of the shelf and southern craton, with associated faults, at the end of Middle Ordovician.

ii. In the western division, Lower to lower Middle Silurian orthoquartzitic sandstones and shale bordering the Allegheny Basin intertongue with carbonates along the eastern rim of the Michigan Basin. Carbonates are dominant in the Anticosti Basin, with minor shales.

iii. Upper Middle Silurian carbonates are present in both the western and eastern divisions; thick crinoid bank and barrier reef facies border the Michigan Basin. Numerous patch and pinnacle reefs are present within the confines of the barrier.

iv. Onset of evaporite deposition during Late Silurian time, together with shale and carbonate deposition at the borders of the Michigan and Allegheny Basins. Clastic redbeds with minor carbonates in the Anticosti Basin. Widespread disconformity at the top of the Silurian in Ontario and evidence of extensive faulting and salt dissolution. Significant for oil, gas, lead, zinc, salt, and construction materials.

E. Late Silurian through Middle Devonian

Themes:

i. "Sub-Devonian" unconformity.

ii. Platform-type sedimentation superseded by basinal clastics related to the prograding Catskill Delta.

Phases:

i. Lower and lower Middle Devonian carbonate, including minor biohermal development, and gypsum/anhydrite are present only in the western division of the platform, except for remnants preserved in a diatreme breccia in the Quebec Basin.

ii. The Catskill Delta prograded into the Acadian foredeep, producing the upper Middle and Upper Devonian shale of the Allegheny Basin and extending across Ontario into the Michigan Basin where the clastics intertongue with carbonates. Oil, gas, lead, zinc, and gypsum are the main mineral deposits.

F. Late Paleozoic and Mesozoic

Theme:

i. Erosion of the cratonic cover.

Phases:

i. Post-Devonian uplift; faulting and subaerial erosion are widespread along the entire length of the platform.

ii. Dyke emplacement in the central and eastern divisions, and alkalic intrusions in the Quebec Basin (84–123 Ma).

7. Evolutionary Models and Tectonic Comparisons

I. Interior Platform: Western Canada and Mackenzie Basins

A. Purcell, or older Proterozoic Sequence, and Intermediate Sequence. Platformal aspect compared with: platformal Riphaean of the Soviet Union; Sinian of North China type; Middle and Upper Proterozoic of Adelaide Geosyncline (Australia). Platform-to-basin transition (Purcell only) compared with: Pethei Group of Athapuscow Aulacogen; Lower Proterozoic of Coronation Geosyncline; Lower Proterozoic of Labrador Trough; Cambrian of southern Rocky Mountains.

B. Windermere, or younger Proterozoic Sequence. Coarse fill of rift-bounded basin compared with: Tertiary of California Borderland; Jurassic Whitehorse Trough; Tertiary of Magdalena River Valley, Colombia.

C. Upper Precambrian through Lower Ordovician. Classical passive-margin successions compared with: Upper Proterozoic through Lower Ordovician of southern Appalachians; post-Triassic of submerged Atlantic margin of North America.

D. Middle Ordovician to Middle Silurian. Carbonate-dominated platform passing outward to slope deposits of a passive margin compared with: Arctic Platform of Canada.

E. Lower and Middle Devonian of the Elk Point Basin. Supracratonic, areally restricted basins with (commonly) central evaporites and (commonly) fringing carbonate complexes, compared with: Zechstein Basin; Permian Basin of Texas.

F. Middle and Upper Devonian. Reef development compared with: western Australia, Belgium, North Africa (all Devonian), Edwards Formation (Cretaceous) of Texas, modern Bonaparte Gulf.

G. Famennian and Carboniferous. Essentially carbonate cover of extensive platforms compared with:

Upper Paleozoic of south China; Proterozoic platform successions listed under (A).

H. Permian of the Rocky Mountains. Extremely condensed, supracratonic, carbonate-mature clastic successions: no examples(?).

I. Triassic. Extensive platforms with mainly mudrock deposits compared with: (?)Lower Paleozoic of Baltic Shield. Mixed clastic–evaporite platformal successions compared with: evaporites and redbeds at the Middle/Upper Cambrian boundary in the Canadian Cordillera and for a sedimentological analogue, the modern Colorado River delta.

J. Lower and Middle Jurassic. Black shale basin compared with the Chattanooga Shale.

K. Upper Jurassic and Cretaceous. Molasse deposits and clastic wedge compared with: Devonian of Appalachians, Tertiary of South American Llanos, Molasse basin of Alps, Cretaceous of Western United States.

L. Eocene. Molasse deposits compared with the Cordillera Oriental and Llanos of Colombia and Venezuela.

II. Hudson Platform

A. Ordovician. Carbonates of intracratonic basins compared with the Michigan Basin in Ontario and the Williston Basin in Manitoba.

B. Silurian. Bioclastic and reef carbonates compared with the Michigan Basin and, with less similarity, to the Williston Basin of Manitoba.

C. Devonian. Compared with the Michigan Basin of Ontario.

III. St. Lawrence Platform

A. Cambrian to early Middle Ordovician. Platform carbonates and clastics compared with the Paleozoic of the Western Interior Platform and the Cambrian to Late Silurian of Arctic Platform.

B. Late Middle Ordovician to Devonian. Platform to foredeep transition compared with the Upper Jurassic to Paleogene of the Interior Plains.

C. Michigan Basin Paleozoic. Intracratonic basin compared with the Paleozoic of Illinois, Hudson Bay, Williston, and Elk Point Basins.

8. Economic Geology

I. Interior Platform: Western Canada Basin and Mackenzie Basin

A. Metallic deposits—lead-zinc: Devonian platform carbonates; iron: Rapitan (Upper Proterozoic); Cretaceous of Clear Hills; gold: placer.

B. Nonmetallic deposits—barite and witherite: widespread in Lower Devonian; magnesite: in Purcell and Middle Cambrian; gypsum: mainly Devonian, Jurassic in Williston Basin; limestone and dolomite: Paleozoic; clay: Upper Cretaceous; bentonite: Upper Cretaceous; phosphate: Mississippian to Jurassic; sodium and magnesium: modern lake deposits; salt: mainly Devonian; and potash: Devonian.

C. Fossil fuels—coal: Lower Carboniferous to Eocene; petroleum and natural gas: mainly Devonian, Carboniferous, Triassic, and Cretaceous. Hydrocarbon associates—sulfur: mainly Devonian and Carboniferous; and helium and nitrogen: Upper Cambrian (Saskatchewan).

II. Hudson Platform

A. Nonmetallic deposits—limestone and dolomite: Ordovician, Silurian, and Devonian; anhydrite/gypsum: Devonian Moose River Formation, minor deposits in Ordovician and Silurian; halite: Ordovician, Silurian, and Devonian; clay: Lower Cretaceous Mattagami Formation; and sands: Jurassic and Cretaceous, Moose River Basin.

B. Fossil fuels—coal: Lower Cretaceous lignite deposit, Onakawana, Moose River Basin; petroleum and natural gas: potential in Paleozoic carbonates but none discovered to date.

III. St. Lawrence Platform

A. Metallic deposits—lead and zinc: Lower Ordovician and Middle Silurian carbonates, Anticosti and Michigan Basins.

B. Nonmetallic deposits—salt: Upper Silurian, Michigan Basin; gypsum: Upper Silurian, Allegheny Basin; limestone and dolomite: Ordovician, Silurian, Devonian—Anticosti Basin, Quebec Basin/Ottawa Embayment, and Michigan and Allegheny Basins; clay: Ordovician, Silurian, and Devonian—Quebec Basin/Ottawa Embayment, and Michigan and Allegheny Basins; and silica sand: Cambrian/Ordovician, Allegheny Basin.

C. Fuels—petroleum and natural gas: Cambrian, Ordovician, Silurian, Devonian—Michigan and Allegheny Basins; some potential in Quebec and Anticosti Basins.

The Midcontinent Province: United States

L. L. Sloss
Department of Geological Sciences
Northwestern University
Evanston, Illinois 60201

Introduction

Although Charles J. Mankin, Peter R. Vail, and L. L. Sloss constitute the group charged with planning, guiding, and expediting preparation of a volume titled *The Midcontinent Province: United States* for the series, *The Geology of North America*, this paper is a highly personal document put together without benefit of input from Mankin and Vail; they cannot be held responsible for the many sins of commission and omission that are recorded in the pages that follow.

Readers, especially those who are asked to consider individual regions within the Midcontinent province, should exercise an appropriate degree of compassion where outright errors of fact are found; such mistakes are inevitable when one person attempts to cover the Phanerozoic record of a major part of a continent, but all recognizable errors require early identification and correction. It should be noted further that the stratigraphic classification and correlation implicit in this preliminary treatment are not integrated with COSUNA (Correlation of Stratigraphic Units of North America) charts; these will be available shortly(?) and will be the framework for final texts.

Scope of Proposed Volume

The subject matter to be covered in the volume on *The Midcontinent Province: United States* is here defined as the pre-Pleistocene Phanerozoic geology of that part of the North American craton lying within the conterminous United States (Fig. 1). "Geology" is used broadly to include stratigraphy, structure, regional geophysics, tectonic synthesis, and those aspects of economic geology (especially water, coal, oil, and gas) and geomorphology not subsumed by parallel volumes. "Craton" identifies the nuclear region of North America underlain by unmodified and relatively undeformed continental crust. Cratonic margins, in this construct, are time variable, expanding when new granitoid crust is accreted, contracting when rifting and accompanying mafic intrusion grossly alter the thickness, petrology, and geophysical parameters of preexisting continental crust at divergent continental margins. Such a revision of the classical definition of craton—continental terrain inboard of miogeosynclines (Kay, 1951)—is introduced in recognition of the degree of similarity, if not identity, existing in the sedimentary facies and syndepositional tectonics shared by continental-shelf prisms (miogeoclines) and adjoining continental-interior regions. The redefinition, in practical terms, obviates the necessity of arbitrarily separating consideration of the stratigraphy and depositional tectonics on either side of a line (e.g., the Wasatch line of Kay, 1951) drawn through a continuum on the basis of rate-of-change of thickness, alone. (Equivalent rates of thickness change are observed at the margins of many basins of unequivocal cratonic position, and it is common to find thicknesses that increase nearly linearly from cratonic interior to shelf prism). Thus, depositional events responsible for accumulation of Appalachian ridge-and-valley strata and of the pre-Jurassic rocks of the Sevier thrust belt are appropriately discussed in the context of the continental interior while the younger deformation and metamorphism of these areas are treated in the volumes on the Appalachians and the Cordillera.

Chronologic Subdivisions

The envisaged synthesis of the continental interior is cast in historical terms, proceeding from latest Precambrian to Quaternary by chronologic increments defined by interregional unconformities. The defining unconformities are those almost everywhere present in the cratonic interior where the appropriate chronostratigraphy is preserved and which are capable of extrapolation to surfaces of apparent conformity at the former positions of continental slopes and in the interior of certain basins. The major divisions are the sequences of Sloss (1963) with further subdivision derived from the coastal onlap chart of Vail et al (1977). At this writing it is unclear, and not of first-order importance, whether the units should be called "sequences" and "subsequences" or "supersequences" and "sequences"; for the purposes of this preliminary paper, sequences and subsequences are applied. Regardless of terminology, the major units and many of the designated subdivisions are

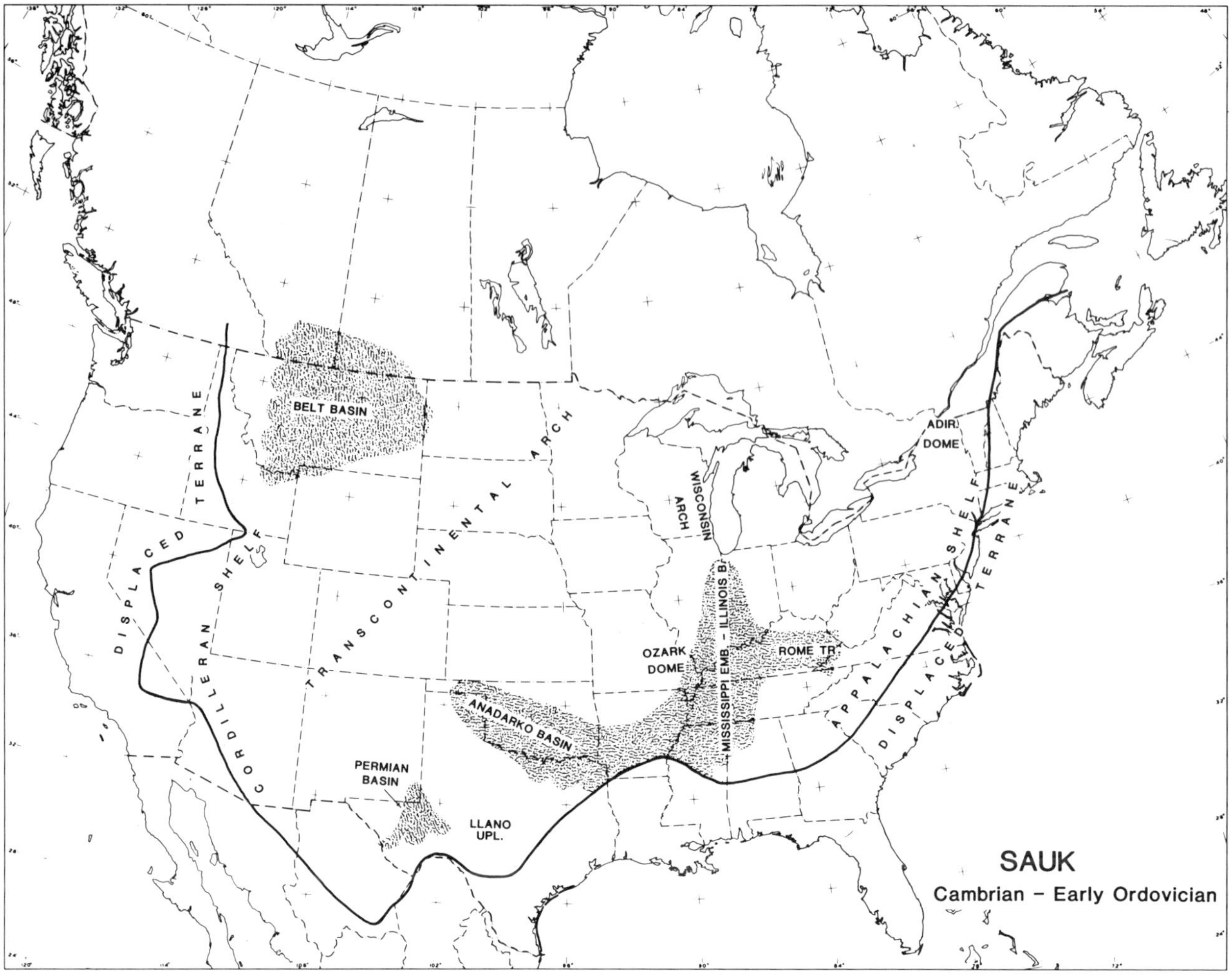

Figure 1. Major tectonic elements of the craton: Sauk sequence.

applicable throughout the craton and permit display and discussion of the stratigraphic-tectonic record from region to region within a common chronologic framework that is intrinsic to cratonic history and without involvement of often debilitating debate on correlation.

Table 1 presents the sequences and roman-numbered sub-sequences in relation to accepted chronostratigraphic units. Isotopic ages at the end of corresponding chronological units (ages where such are recognized) are given, as are best estimates of the points in time representing the minimal lacunas or surfaces of actual conformity between sub-sequences where such surfaces are preserved. The isotopic time scale used is a provisional product, from many sources, and is subject to modification as stratotypes are defined, the correlation of magnetostratigraphic, radiometric, biostratigraphic, and chronostratigraphic units adjusted, and controversies among radiometric laboratories arbitrated.

Precambrian Prologue

The North American craton is an ancient block, well consolidated in something approaching its present geometry by Middle Proterozoic (~ 1,800 Ma) time, with significant Precambrian crustal additions and modifications completed by Grenville events (~ 1,000 Ma). Over perhaps 90 percent of the craton in the U.S., Sauk, and younger Phanerozoic sequences rest on igneous and metasedimentary basement rocks that indicate deep burial before exhumation and exposure at the sub-Sauk surface. Onlap of cratonic cover rocks was initiated on the Cordilleran margin, and in aulacogen-like invaginations of that margin, by sediments and minor volcanics of the Belt Sequence (~ 1,350 Ma), including massive quartz arenites derived from the cratonic interior; Belt rocks were relatively unaffected by pre-Phanerozoic deformational or thermal episodes. These rocks are considered to represent deposition on a rifted passive margin (Stewart, 1972). Late Proterozoic rifting of the cratonic interior was responsible for accumulation and preservation of the volcanics and sediments of the Keweenaw Sequence (~ 1,100 Ma) in the Lake Superior Basin and its extension marked by the Midcontinent Gravity Anomaly; similar rocks, structures, and perturbations of the gravity field are present on the western flank of the Michigan Basin and are provisionally assigned to the Keweenaw Sequence. Finally, in latest Proterozoic

Table 1. Chronostratigraphy of stratigraphic sequences

System	Series	Stage	Ma B.P.		Sequence
Quaternary		Pleistocene	0		
Tertiary	Neogene	Pliocene	2		Tejas III
		Miocene	7		
	Paleogene	Oligocene	26	29	
		Eocene	38	49	Tejas II
		Paleocene	54	60	Tejas I
Cretaceous	Upper	Maastrichtian	64		
		Campanian	70		
		Santonian	82		
		Coniacian	86		Zuni IV
		Turonian	87		
		Cenomanian	89	91	
	Lower	Albian	94		
		Aptian	104		
		Barremian	111		Zuni III
		Hauterivian	117		
		Valanginian	123		
		Berriasian	129	129	
Jurassic	Upper	Portlandian	135		
		Kimmeridgian	138		
		Oxfordian	143		Zuni II
	Middle	Callovian	149	150	
		Bathonian	156		
		Bajocian	165		
	Lower	Aalenian	171		Zuni I
		Toarcian	174		
		Pliensbachian	178		
		Sinemurian	183	185	
		Hettangian	189		
Triassic	Upper	Rhaetian			
		Norian	205		
		Carnian			Absaroka III
	Middle	Ladinian	218		
		Anisian	224		
	Lower	Scythian	230	232	
Permian	Ochoa	Tartarian	245		
	Guadalupe	Kazanian	250		
	Leonard	Kungurian	258		Absaroka II
		Artinskian	273		
	Wolfcamp	Sakmarian	286	286	
Pennsylvanian (Carboniferous)	Virgil	Stephanian	300		
	Missouri	Westphalian D	307		
	Des Moines	Westphalian C			Absaroka I
		Westphalian B	317		
	Atoka	Westphalian A			
	Morrow	Namurian C			
		Namurian B			
Mississippian (Carboniferous)	Chester	Namurian A		335	
	Valmayer	Viséan	338		Kaskaskia II
	Kinderhook	Tournaisian	358		
Devonian	Upper	Famennian	370	372	
		Frasnian	376		
	Middle	Givetian	380		
		Eifelian	384		Kaskaskia I
		Emsian	388		
		Siegenian	392		
	Helderberg	Gedinnian	401	401	
Silurian	Upper	Pridolian	405		
		Ludlovian	409		Tippecanoe II
	Middle	Wenlockian	413		
		Llandoverian	422		
	Lower			431	
Ordovician	Upper	Ashgillian	431		
		Caradocian	434		
	Middle	Llandeilian	447		Tippecanoe I
		Llanvirnian	449	458	
	Lower	Arenigian	458		
		Tremadocian	478		
Cambrian	Upper	Trempealeau	498		Sauk III
		Franconia		515	
		Dresbach			Sauk II
	Middle		523	548	
	Lower		548		Sauk I
Precambrian	Upper Proterozoic		575	600	

time (~850–650 Ma) renewed passive-margin rifting and the accumulation of thick craton-derived sediments, including diamictites of possible glacial origin, took place at both the Cordilleran and Appalachian margins (Stewart, 1976). Each of the Proterozoic successions (Belt, Keweenaw, and Windermere) is enclosed between major unconformities representing kilometers of erosion.

Sauk Sequence

The base of the Sauk Sequence rests on any of the stratigraphic sequences or basement rocks noted above; age of the basal strata ranges from "Ediacaran" and oldest Cambrian at the cratonic margins to Early Ordovician at places on the edge of the Canadian Shield. The top of the sequence is Early Ordovician or any pre-Llanvirnian chronostratigraphic unit preserved below the sub-Tippecanoe and younger unconformities.

Sub-Sauk Surface

The unmetamorphosed Proterozoic sedimentary and volcanic rocks below the Sauk Sequence are confined to the cratonic margins or occupy limited rift terrains of the interior; as a consequence, most Sauk and younger Phanerozoic sequences rest on igneous and metasedimentary basement rocks that indicate deep placement or burial before exhumation and exposure at the sub-Sauk surface over much of the craton. The ubiquitous sub-Sauk surface represents an immensely long erosional episode (the Lipalian interval of Walcott) forming a stratigraphic lacuna occupying a minimum of 100 Ma in the cratonic interior (the time between the initiation of onlap by Sauk I and the maximum of submergence in Sauk III). The surface is properly considered a peneplain of low relief over major areas but, in spite of the dimensions of the pre-Sauk lacuna, thick paleosols are not commonly preserved. Numerous departures from a state of uniform low relief are known; examples are found in the quartzite ridges of central and southern Wisconsin and the "porphyry" knobs that interrupt the Sauk stratigraphy of the Ozark Dome.

Tectonic Mode

Depositional tectonics of the Sauk Sequence are more closely related to patterns developed in Late Proterozoic time than to the younger Phanerozoic. Sauk isopach and facies studies reveal little evidence of typical craton-interior basins; instead, the craton (Fig. 1) appears to have behaved as a very stable, gently subsiding shelf penetrated from the margins by broad embayments such as the Belt Basin of Montana and by aulacogen-like features such as the Anadarko Basin, the Mississippi Embayment–Illinois Basin, and the Rome Trough(?). Certain of the latter elements are usually considered to be dominated by broadly flexural subsidence, but this view may be the product of low definition caused by insufficient subsurface detail. Evidence accumulates to suggest vertical motion along fault systems, accompanied by igneous activity. Elsewhere, there are outcrop manifestations of local brittle deformation during Sauk deposition, ranging from minor dislocations in western Montana and northern Illinois to uplifts of greater scale in the Upper Peninsula of Michigan. The sheer volume of clastic detritus preserved in the Sauk Sequence, the persistence of such detritus throughout the sequence, and the absence of Sauk sediments in basins of the Canadian Shield (e.g., James Bay) suggest continued uplift of the Shield, presumably an accompaniment of faulting.

Sauk I

A major part of the volume of clastics noted above is present in latest Proterozoic and Early Cambrian quartz arenites of the Sauk I subsequence on the Cordilleran and Appalachian cratonic margins where thicknesses up to 3 km are preserved. I do not subscribe to the conventional-wisdom view that these arenites are the products of highly selective quarrying of older sandstones (without involvement of feldspathic rocks, micaceous gneisses and schists, greenstones, and slates) and the transport of quartz grains alone to sites of Sauk deposition. Rather, it seems more reasonable to relate the mineralogic and textural maturity of Sauk sands to the slow pace of Sauk onlap during which maturation was accomplished by multiple reworking on subtidal bars, barrier islands, beaches, and dune fields lacking stabilizing vegetative cover. Regardless of their genesis, thick quartz arenites are everywhere present at the base of Sauk I and dominate the entire subsequence on what was the Cordilleran continental shelf; in the Appalachian area Sauk I basal sandstones give way to intercalations of carbonate and shale. Among the aulacogen-like troughs that penetrate the Sauk craton, only the Rome Trough was active during Sauk I deposition.

Sauk II and III

Pre-Dresbachian Sauk II deposition extended inboard from the Cordilleran shelf as far as central Wyoming, but equivalent sediments encroached only moderate distances beyond the zero edge of Sauk I in the Appalachian area and are lacking in the Anadarko Basin and adjoining areas of the southern Midcontinent. From Dresbachian time on, Sauk II and III represent the great transgression and onlap of the continental interior; the subsequences share a number of characteristics and are conveniently discussed together.

In general terms, the facies belts of Sauk II and III

conform to the concentric pattern (inner detrital, middle carbonate, outer detrital) first described by Palmer (1960) and applied to western Canada by Aitken (1966). Thus, the Shield border, including the Wisconsin Arch and the Adirondack Dome, and the central spine of the Transcontinental Arch from Minnesota to Arizona appear to be blanketed in quartz arenites, much of the remainder of the craton is dominated by carbonates, and as the ancient continental slope is approached there is an increasing amount of shale. This gross concentric pattern on the craton proper is in part an artifact of facies mapping and results from post-Sauk erosion of a complex layercake with sandstones at the base and carbonates at the top; where the higher carbonates have been stripped, sands rise to predominance. However, it remains true that nearly all chronostratigraphic units, when traced from shelf break to Shield margin, pass into sandstones. The position of the middle carbonate--inner detrital facies boundary shifts with numerous offlapping reversals but generally encroaches on the Shield with advancing time during Sauk II and III deposition. The result is seen in complex intertonguing of carbonates with representatives of a spectrum of detrital environments (dunes, beaches and bars, intertidal and subtidal regimes). Shale units form important components of the intertonguing complex to the west of the Transcontinental Arch and in the southern Appalachians, particularly in Sauk II, whereas equivalent strata east of the arch and in the northern Appalachians are relatively shale-free.

At varying distances from the core of the craton and from the influences of Shield-derived coarse clastics, Sauk III comprises deposits formed on a great carbonate bank (Beekmantown, Knox, Arbuckle, Ellenburger, El Paso, Manitou, Pogonip, etc.). The bank is a museum of carbonate petrologies—stromatolites, oolites, rip-up conglomerates, bioturbated micrites, pelletoid and skeletal packstones, and more—commonly affected by pervasive dolomitization in central areas of the craton, preserving calcitic rocks in regions of more rapid subsidence at cratonic borders and in aulacogenic basins.

Tippecanoe Sequence

The maximum chronostratigraphic range of the Tippecanoe Sequence is from low in the Middle Ordovician (Chazy) to low in the Lower Devonian (Gedinnian). Because of the time involved in Tippecanoe onlap of the cratonic interior the base of the sequence is commonly as young as late Middle Ordovician and early Upper Ordovician (Caradocian); pre-Kaskaskia erosion is responsible for widespread stripping of younger Tippecanoe units such that these are preserved only on the shelf wedges, the centers of subsiding basins, and, more rarely, as blocks in breccia pipes.

Sub-Tippecanoe Surface

Subcrop geology of the sub-Tippecanoe surface is relatively simple; basal Tippecanoe strata rest on Early Ordovician Sauk carbonates over very large areas or are found overstepping the Sauk to lie on basement rocks. In spite of seeming absence of a major unconformity, it can be demonstrated that erosional relief of as much as 150 m was developed in areas of little structural discordance. At scattered places in the cratonic interior there appears to have been reactivation of basement faults and local exposure of lower Sauk units (e.g., Sandwich fault, northern Illinois). At the Cordilleran cratonic margin, as in west-central Idaho, there is evidence of pre-Tippecanoe deformation (and metamorphism?); here, basal Tippecanoe arenites were deposited on Belt rocks or on basal Sauk arenites. In general, however, the sub-Tippecanoe lacuna declines in time value with approach toward the Ordovician shelf break.

Tectonic Mode

The mid-Paleozoic pattern of intracratonic basins and arches that became a dominating factor in the distribution of facies and thicknesses developed soon after the initiation of Tippecanoe onlap in Middle Ordovician time. The picture is one of general cratonic subsidence marked by higher rates of downwarp at the present positions of such features as the Williston and Michigan basins and slower but appreciable subsidence on "positive" elements (Wisconsin Arch, Ozark Dome, Findlay Arch, Llano Uplift, Transcontinental Arch, etc.) (Fig. 2). Aulacogen-like downwarps, inherited from the time of Sauk deposition (Anadarko and Mississippi Embayment–Illinois Basin), resumed their subsiding tendencies but at lesser rates than previously and without the pronounced influence on facies distribution exerted by such "closed" basins as the Michigan Basin.

Tippecanoe I

The Tippecanoe I subsequence spans stratigraphic units from the basal unconformity to the top of the Ordovician. The basal deposits of the subsequence are intertidal to subtidal carbonates and calcareous mudstones without the intervention of significant basal sandstones in the Appalachian Basin, extending northwest to the trend of the Cincinnati Arch–Nashville Dome. Elsewhere on the cratonic interior and on the Cordilleran margin the most widely observed lithology is quartz arenite (St. Peter, Winnipeg, Harding, Eureka, etc.). Exceptions occur in Montana and Wyoming where Bighorn carbonates lie on Sauk rocks with only thin and patchy occurrences of basal sands (Lander), and in the depths of the Anadarko Basin where carbonates of the Joins Formation were deposited before the advent of sand accumulation.

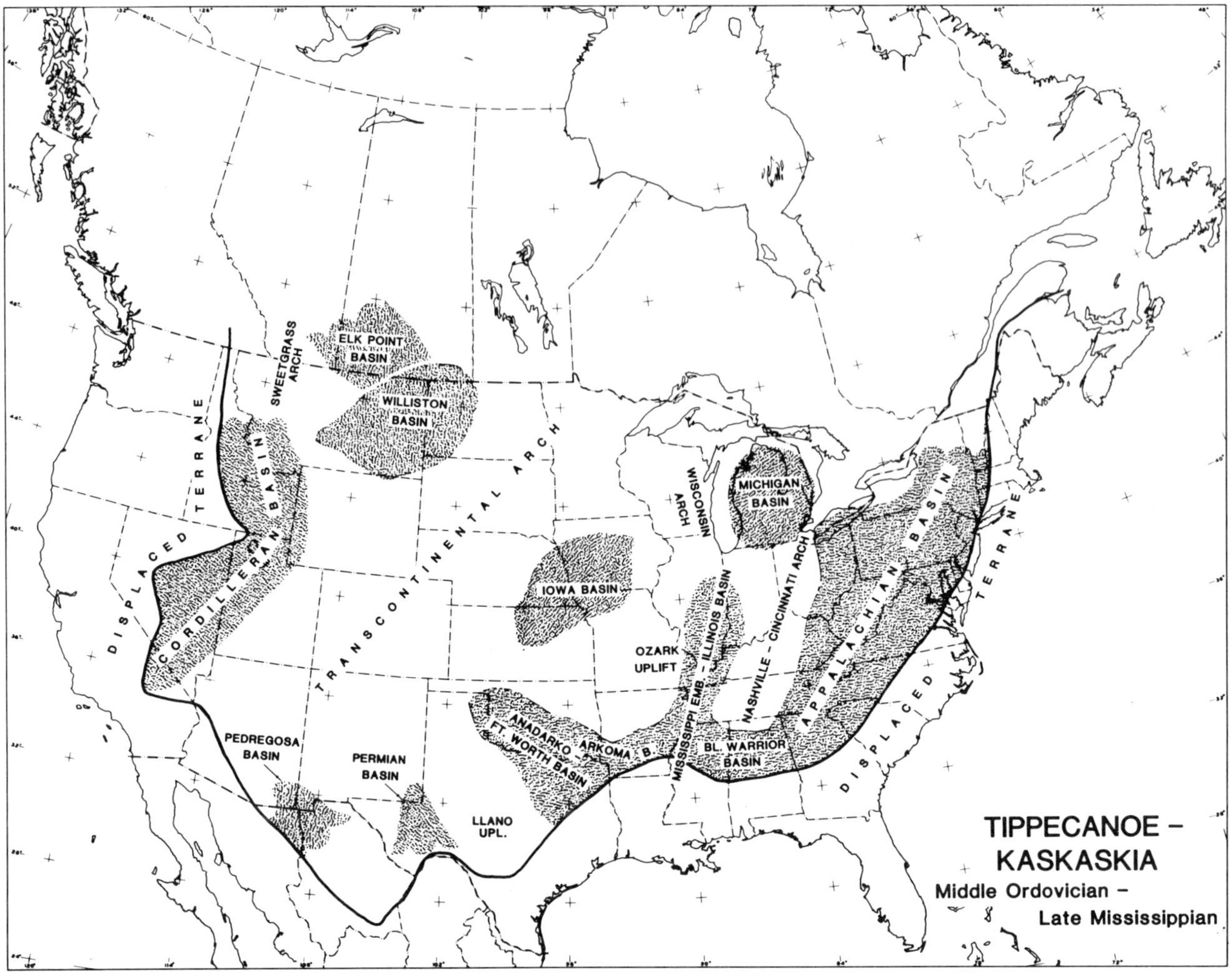

Figure 2. Major tectonic elements of the craton: Tippecanoe and Kaskaskia sequences.

Tippecanoe I quartz arenites are at an extreme level of textural and mineralogic maturity suggesting multicycle derivation from older sand units, especially from Sauk sands exposed on the sub-Tippecanoe surface. This hypothesis becomes difficult to support when consideration is given to the magnitude of exposure of feldspathic rocks at that surface. The sandstones in question are oldest at the Cordilleran margin and at the axis of the Anadarko Basin, climbing diachronously toward the Shield. In the Anadarko basin, sand units of the Simpson Group intertongue with shales and carbonates, indicating repeated onlaps followed by progradations of littoral deposits until transgression and onlap spread from the northern margin of the basin over the adjoining cratonic interior. Shale is also a significant component of lower Tippecanoe stratigraphy in the Williston Basin; elsewhere, basal sandstones are overlain by no more than a few meters of shale before giving way to significant thicknesses of intertidal and subtidal carbonates. Carbonates of the Williston Basin are punctuated by nodular displacement anhydrite members; these are widely interpreted as supratidal sabkha deposits but, if so, they represent supratidal progradation over remarkable distances (300 km) without stillstands or reversals. Tippecanoe I evaporites in other areas are relatively rare; extensive dolomitization is common on the flanks of "positive" elements of thinner accumulation, reflecting more numerous intrasequence interruptions of deposition and, presumably, greater frequency of subaerial exposure and resulting mixed-zone ("Dorag") pore waters. Although Tippecanoe I carbonates were obviously deposited in well-populated epeiric seas, there are few examples of concentrations of frame-building organisms and reef morphology.

The great carbonate blankets of Middle and Late Ordovician time extend without significant detrital influx across the Transcontinental Arch and far westward onto the Cordilleran shelf. Traced farther to the west, as in central Nevada, there is a gradual addition of clay and the appearance of shaly interbeds interpreted as representing deeper-water environments; evidence of extracratonic sources of detritals is lacking. Late Ordovician strata of the Illinois Basin and adjoining areas south and west to the Anadarko Basin include significant shale units. Minor proportions of these shales may be attributed to the distal effects of Appalachian clastic influx, but it is apparent that most is of intracratonic derivation.

Tippecanoe I sedimentation in the Appalachian area, west across the position of the Cincinnati Arch and into the Michigan Basin, was profoundly influenced by orogenic events of the Iapetus margin. Here, the long established passive shelf wedge was abruptly replaced in late Middle Ordovician time by piedmont fluviatile coarse clastics which grade northwest to delta-platform and prodelta, marine-border deposits. These, the Queenston "delta" system of clastic wedges, thicken rapidly to the southeast and intertongue distally (northwestward) with the persistant marine carbonates of the continental interior.

Tippecanoe II

The base of the upper Tippecanoe subsequence is placed near the chronostratigraphic boundary between Ordovician and Silurian. At this position a minor lacuna is commonly detectable, represented by the erosion of some tens of meters of older strata and a hiatal (nondepositional) gap in Early Silurian stratigraphy. There are large areas, however, where evidence for such a lacuna is lacking or where the interruptions of deposition are no more significant than many others that characteristically punctuate cratonic stratigraphy. Whether or not sub-Tippecanoe II lacunas are glacio-eustatic responses, there is reason to believe that there was a temporary change in the tectonic behavior of the craton and its passive margins as seen in a lessening of the subsidence rates of basins and in offlap relationships between Early Silurian and Late Ordovician units. The pause in subsidence was apparently brief and was followed by renewed basinal downwarp in a tectonic geography identical to that of Tippecanoe I. The distribution of Tippecanoe II deposits outside of basinal areas is profoundly affected by sub-Kaskaskia erosion; Silurian and earliest Devonian strata are largely absent, for example, over much of the Colorado Plateau and the Central and Northern Rockies, but there is no solid evidence to suggest that their depositional limits were more confined than Tippecanoe I.

The tectonic and environmental framework of deposition on the craton established during Tippecanoe I continued in control in Tippecanoe II following the apparent stabilization marked by the subsequence boundary. That is, carbonate domains prevailed over major areas of the craton and extended across the greater part of the Cordilleran shelf while clastic wedges impinged on the Appalachian margin until late in Silurian time. There were major differences, however. Frame-building and sediment-binding organisms became important elements of sediment accumulation, especially in equatorial regions east and south of the Transcontinental Arch. This is reflected in the profusion of reef platforms and isolated biohermal masses that ornament stable shelf areas and the "pinnacle" reefs of the flanks of the Michigan Basin. The latter basin and lesser downwarps in New York, Ohio, West Virginia, and Pennsylvania were the sites of massive accumulations of salt of Middle(?) and Late Silurian age. Controversy surrounds the relationships between salt deposition and the surrounding reef tracts, but it is clear that basinal subsidence accelerated markedly in the latter part of Silurian time, creating pronounced depth-controlled(?) environmental zonation of the basins and leading to restriction of circulation to and from the open sea. The Anadarko and Illinois basins also experienced relatively rapid Late Silurian subsidence but, presumably because of their open communication to continental margins, developed "deeper"-water marly and chert-rich facies rather than massive salt.

The extremes of tectonically induced environmental differentiation gradually relaxed in latest Silurian into earliest Devonian time, accompanied by broad-scale offlap leading to supra-baselevel exposure of the entire cratonic interior and major areas of the Cordilleran shelf by the end of Gedinnian time.

Kaskaskia Sequence

Chronostratigraphically the Kaskaskia Sequence includes all cratonic strata from mid-Lower Devonian (Siegenian) to near the top of the Mississippian (Namurian A). The great episodes of onlap were Middle Devonian, Late Devonian, and Early Mississippian; rocks of these ages are the commonly encountered cover of the sub-Kaskaskia unconformity. The sequence was eroded to varying levels during sub-Absaroka exposure; beyond basinal areas and the continental shelf wedges where Chester (Late Viséan–Early Namurian) beds are preserved, earlier Mississippian strata are the most commonly observed upper limits.

Sub-Kaskaskia Surface

The stratigraphy below the sub-Kaskaskia unconformity, reflecting the pattern of basins and arches developed during Tippecanoe deposition as well as new elements that emerged prior to Kaskaskia overlap, is much more complex than that revealed by sub-Tippecanoe subcrop maps. Basins are ringed by concentric patterns formed by the erosional limits of Tippecanoe and Sauk units, the latter being the subcrop stratigraphy over major areas of the Colorado Plateau and western Montana; there are more limited areas of Kaskaskia overstep on Precambrian basement.

Tectonic Mode

The tectonic mode and geography controlling Kaskaskia deposition was largely that inherited from Tippecanoe patterns with regionally significant differences. The Williston Basin during Tippecanoe deposition, for example, was an open embayment with unrestricted access to seas at the western (and northern?) continental margin; emergence of the Sweetgrass and Western Alberta arches deformed the western flank of the basin such that, early in Kaskaskia deposition, communication was limited to a passage from the Cordilleran shelf near the common corner of British Columbia, Alberta, and Yukon. The axis of greatest subsidence lay in central Alberta and southern Saskatchewan (Elk Point Basin) (Fig. 2) during most of Middle Devonian time; the Williston Basin was reconstituted at its former position late in Early Mississippian time.

The Cordilleran shelf continued as a gently subsiding appendage to a passive margin until latest Devonian time; at that time, possibly heralding an imminent collision with an easterly migrating oceanic arc, the shelf edge was markedly depressed. Allochthonous plates reached the Cordilleran shelf by late Early Mississippian time, a foreland basin developed, and the long passive history of the eastern Great Basin region came to an abrupt close, as did its long tectonic-environmental relationship to the cratonic interior. The Appalachian margin, restored to a passive state before the end of Tippecanoe deposition, enjoyed a brief quiescent episode before renewed late Middle Devonian reactivation, the emergence of orogenic source elements, and the rapid subsidence of a foreland-basin complex on the former position of the inner reaches of the shelf.

Kaskaskia I

The lower of two subsequences recognized as divisions of the Kaskaskia extends from the basal unconformity to a stratigraphic discontinuity near the top of Devonian strata. Basal sediments covering the unconformity are highly variable, ranging from arkosic detritus where basement rocks were exposed on local pre-Kaskaskia uplifts, to Shield-derived quartz arenites reminiscent of the Ordovician St. Peter Sandstone, to red beds, evaporites, and carbonates. The latter are perhaps the most widespread basal lithology, and where they rest on Sauk or Tippecanoe carbonates the position of the unconformity is not easily discerned. The age of earliest Kaskaskia I strata is also highly variable, reflecting widely differing times of submergence of the unconformity surface below depositional baselevel. In general, the oldest chronostratigraphic units (Siegenian and Emsian) are found only in the continental-shelf wedges and the Michigan and Illinois basins. Middle Devonian (Eifelian) appears to have been the time of greatest onlap, but there are major areas, as in much of the Rocky Mountain region, where Late Devonian units (Frasnian) form the base of the subsequence. In places lacking a continuous cover of lower Kaskaskia strata there are scattered remnants of nonmarine beds, apparently occupying channels incised into the sub-Kaskaskia erosion surface.

Generalizations as to the character of Kaskaskia I sediments are not possible except to state that pre-Famennian carbonates and evaporites are dominant over detritals in much of the cratonic interior west of the influence of Appalachian clastic wedges. Sabkha and apparent subtidal anhydrites and associated dolomites are common in the Michigan and Williston basins areas, extending broadly to the south and west of the latter. Middle Devonian salt extends southeast from its depocenter in Alberta and Saskatchewan toward the Williston Basin area; a briefer (Eifelian) salt-deposition episode affected the Michigan Basin. By mid-Late Devonian time (Late Frasnian) carbonate-evaporite deposition began to give way to a greater proportion of shale, including black shale representing restricted environments; concurrently, a marked diminution in the subsidence of basinal areas with respect to surrounding stable terrains is apparent.

The influence of orogenic highlands on the Appalachian margin becomes evident in Givetian(?) and younger Devonian with the renewed northwestward spread of clastic wedges over the northern half of the Appalachian Basin and beyond into the adjoining reaches of the cratonic interior. A complete suite of environments, from piedmont fluvial to deltaic, is represented, including manifold indications of complex transgressive-regressive relationships with marine-border conditions.

Meanwhile, back on the cratonic interior, offlap patterns appear in Late Famennian strata, heralding widespread supra-baselevel emergence and the close of Kaskaskia I deposition before the end of Devonian time.

Kaskaskia II

The offlap episode noted above, followed by renewed latest Devonian submergence and the onlap of Kaskaskia II sediments, produces a discontinuity in the succession that is obvious in some areas, obscure in others. Nowhere, except over the very considerable regions where Kaskaskia II broadly overlaps and oversteps the lower subsequence, is there evidence of more than a brief hiatus and minimal stripping of older strata. Commonly, basal Kaskaskia II units are petrologically similar to those at the top of Kaskaskia I but with a greater proportion of black shale and an admixture of impersistent sand and siltstone. These Devonian rocks cover very large areas beyond the depositional edge of Kaskaskia I and are overlapped in turn by limestones of indisputable Early Mississippian age. Great carbonate blankets, upper Kinderhook and Osage equivalents, often rich in replacement cherts, constitute the typical lower Kaskaskia II beds over much of the cratonic interior west of the Appalachian border, excepting the Michigan

Basin and adjoining parts of Ohio. Here, carbonate banks and shoals were repeatedly displaced by deltas fed by river systems from the north; at least one of these deltas penetrated well to the southwest, depositing distal-deltaic silts in the Illinois Basin.

The deltaic incursions of Shield-derived sediments in Kaskaskia II are indicative of both uplift of the previously very stable Shield and of renewed basinal subsidence of sufficient rate to influence stream courses and the loci of deltas. Other evidences of changing tectonic modes are seen in the sediments of the craton-margin shelf wedges. Here, the carbonate blankets of the interior interfinger with and ultimately pass into thick dark shales in apparent response to depression of the continental shelves below carbonate-productive depths.

Late Viséan (Meremac-equivalent) sediments of the Williston and Michigan basins show responses to continued basinal subsidence and tectonic guidance of facies and their distribution. Basin-margin anhydrites encircle the Williston Basin, and both basins are the sites of salt accumulation. Younger Kaskaskia II strata of Namurian A (Chesterian) age are poorly preserved outside of the subsiding basins and the shelf wedges because of erosion marking the sub-Absaroka unconformity. Chester rocks of the Illinois Basin clearly indicate renewed uplift within the Shield, leading to repeated progradation of deltaic cycles interfingered with marine carbonates. Deltaic influences spread south of the Illinois Basin (along the axis of the Mississippi Embayment?) to the Black Warrior Basin, passing into carbonate-dominated successions before intertonguing with dark shales in the southern part of the Appalachian Basin, the Anadarko Basin, and on the Ouachita shelf wedge. Shield-derived(?) clastics also dominate Chester sediments of the Williston Basin area, but deltaic cyclicity is less marked.

Clastic wedges of continent-margin orogenic provenance continued to prevail along the northern half of the Appalachian margin, intertonguing with cratonic sediments as before. Osagean (Early Viséan) time saw the first appearance of western-source detritals on the Cordilleran shelf; here, in central Idaho and eastern Nevada, the transformation of the miogeocline to a complex of foreland basins was initiated. By the end of Kaskaskia deposition on the cratonic interior, clastic wedges had prograded eastward to central Idaho and eastern Nevada, forming coarse intercalations with carbonate-bank deposits.

The youngest preserved Mississippian strata form the top of the Kaskaskia Sequence in the interior of the craton, and for all practical purposes the Mississippian-Pennsylvanian contact is the sequence boundary. However, as the continental shelf wedges and the axes of such subsiding elements as the Anadarko Basin are approached, the great sub-Pennsylvanian unconformity is seen to descend below strata assigned biostratigraphically to latest Mississippian ages. Here, the sequence-bounding unconformity, in contrast to its self-evident character over most of the craton, becomes very obscure, losing its identity in apparently conformable successions or among a welter of discontinuities related to continent-margin tectonics, at positions significantly inboard of the limits of recognition of older defining unconformities. It would seem, therefore, that the pause in sedimentation between times represented by Kaskaskia and Absaroka Sequences was brief and more confined in its area of influence than the inter-sequence lacunas previously discussed, although the changes in tectonic mode at this time and the consequent shift in depositional patterns are perhaps more profound than in any other episode of Phanerozoic history.

Absaroka Sequence

Chronostratigraphic units from latest Mississippian (post-Chesterian Namurian A) to Early Jurassic (Early Sinemurian?) fall within the bounds of the Absaroka Sequence. The time span (~250 Ma) is very long, about twice the length assigned to any other Phanerozoic sequence, and it is reasonable to anticipate the erection of formal sequence units to replace the designations Absaroka I, II, and III employed here. In part, as a concomitant of sub-Zuni erosion and in part because of net offlap during Absaroka deposition, the areal magnitudes of both deposition and preservation of Absaroka units decline markedly from the base upward, Absaroka II and III being virtually absent east of the Mississippi River except in isolated Atlantic and Gulf border grabens.

Sub-Absaroka Surface

Subcrop geology at the sub-Absaroka unconformity is more complex than that at any sequence boundary above the base of the Sauk, reflecting renewed stripping of older "uplifts" and arches plus the emergence of many new positive elements. The subcrop map is dominated by the Transcontinental Arch from which older Phanerozoic sequences were removed and on which Absaroka I strata (commonly mid-Pennsylvanian) rest on basement rocks and on the bevelled edges of Sauk through Kaskaskia units. Intersequence discordances typically are measured in meters per kilometer, but at the base of the Absaroka Sequence pronounced angular unconformity is often observable on and adjacent to a multitude of horst and half-graben yoked blocks and anticlinal structures across many areas of the cratonic interior.

Tectonic Mode

Tectonic behavior of the craton during Absaroka deposition stands in sharp contrast to that prevailing dur-

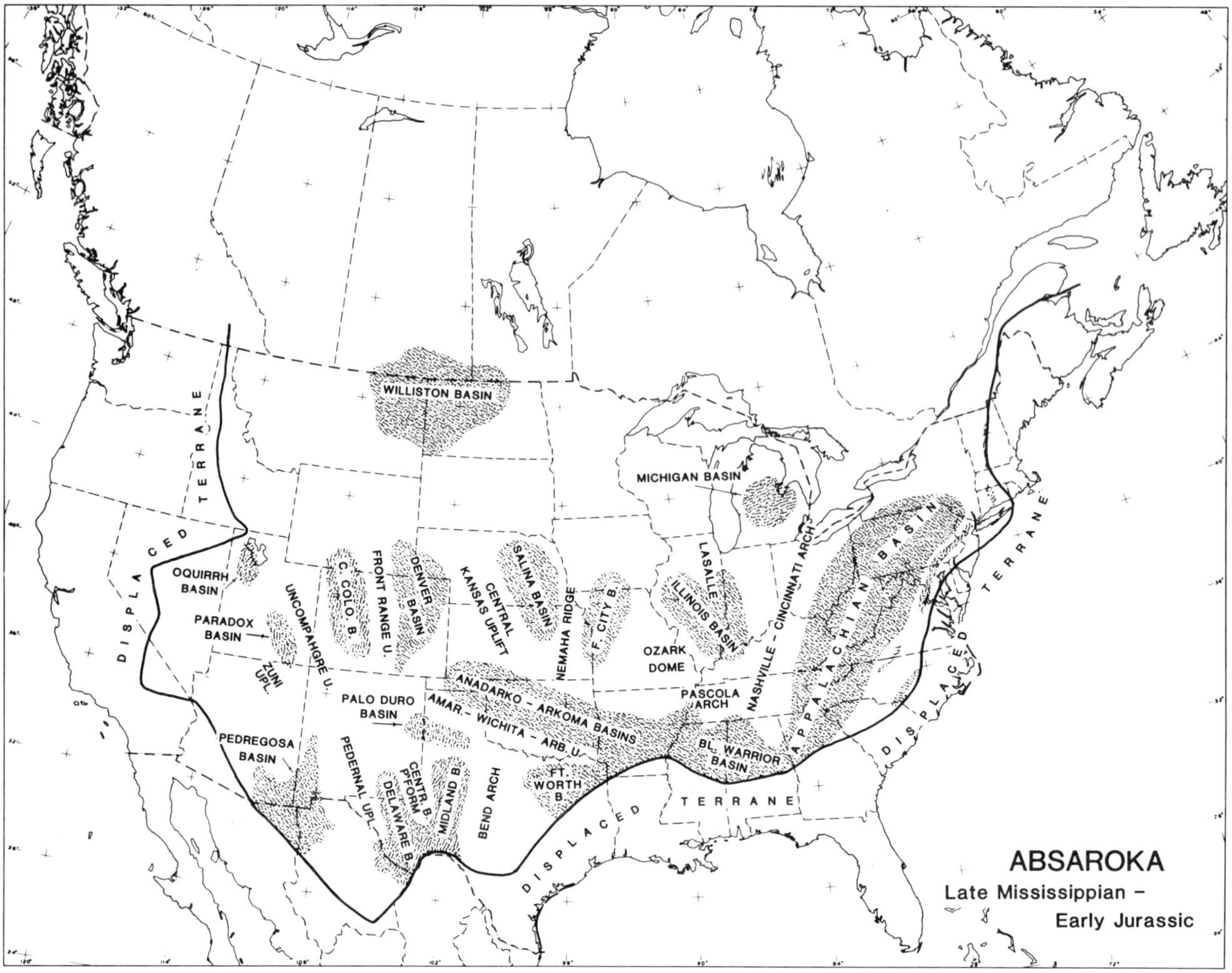

Figure 3. Major tectonic elements of the craton: Absaroka sequence.

ing earlier Paleozoic cycles when deformation of the interior was dominated by gentle flexure at wavelengths of hundreds of kilometers, abrupt vertical movements were confined to the continental margins, and major onlap-offlap cycles operated at a frequency of millions of years. Absaroka stratigraphy and structure (especially Absaroka I) indicate vertical movements on scales of kilometers, commonly involving fracturing and faulting at positions as central as Illinois and Kansas, and oscillations with respect to baselevel with periods measured in 10^5 years. A number of the older cratonic arches and basins renewed their differential rates of subsidence, and these were supplemented and, in places, grossly modified by greatly amplified vertical displacements and "new" additions to the tectonic geography of the craton (Fig. 3). Whereas the bulk of cratonic detritals were previously derived from the Shield or as clastic wedges at the margins, numerous local sources of Absaroka detritals are evident; at the same time large volumes of Shield-derived debris episodically flooded the craton in apparent response to the same kind of vertical tectonics that characterized the rest of the craton.

Continent-continent and arc-continent collisional events and the emplacement of exotic microplates during late Paleozoic–early Mesozoic time transformed the character of Appalachian, Ouachita(?), and Cordilleran margins, creating unresolved semantic questions as to the outboard limits of the craton but effectively removing these terrains from the purview of the continental interior.

Absaroka I

Latest Mississippian, Pennsylvanian, and Early Permian strata constitute Absaroka I; these make up a reasonably cohesive stratigraphic package distinct from younger Absaroka units. Absaroka I is conveniently discussed in terms of three stratigraphic-tectonic suites that dominate the subsequence, although all three are effective and interactive in many areas and cannot be considered as independent except for purposes of clarity. One of these is the succession of delta complexes that extends in a broad belt from the Appalachian area to Oklahoma and the Texas Panhandle. Each is characteristically supratidal and stream-dominated in the north and east, roughly half marine and half nonmarine in Illinois, largely intertidal to subtidal to the southwest. The reconstructed history is one of repeated marine transgression to the northeast followed by progradation, the establishment of distributary channels, interfluvial swamps in which masses of organic matter accumulated, and renewed marine onlap to form multiple cycles. Sandstones are a major component of the deltaic suite; in contrast to the dominance of clean arenites in earlier sequences, the deltaic channel, point-bar, bar-finger, and related barrier-island sands include significant proportions of wackes. As in the Kaskaskia precursors, major source areas appear to have been in the eastern Canadian Shield, supplemented by orogenic highlands on the Appalachian margin.

From southern Oklahoma to New Mexico to southern Wyoming, Absaroka I tectonics and sedimentation are dominated by intracratonic fault-bounded uplifts and adjoining yoked basins; the latter became the sinks for detritus eroded from the coupled rising blocks. Where Sauk through Kaskaskia sediments were thick, as along the pre-Pennsylvanian axis of the Anadarko Basin, alluvial-fan and proximal-deltaic detritals feature carbonate and chert clasts that are intercalated with quartz arenites and wackes related to the deltaic systems described above. Farther west and north, where pre-Absaroka sediments were thin or lacking on the Transcontinental Arch, basement crystallines were quickly exposed and large volumes of arkosic debris, now commonly red, flooded subsiding basin margins and are seen to intertongue with the products of marine environments, including, as in the Paradox and Central Colorado basins, considerable volumes of evaporites. In places (e.g., central Wyoming) where rapid basin subsidence did not occur, the horst-derived detritus matured to clean arenite (e.g., Tensleep) blankets prograding into the marine carbonates of the Cordilleran seaway.

The third major component of the Absaroka I record is the clastic-wedge corollary of collision events on the Appalachian and Ouachita margins. These largely extracratonic circumstances are better discussed in the context of the orogenic history of the areas in question.

Absaroka II and III

The upper two Absaroka subsequences are readily distinguished but are more rationally covered together in this presentation. Preservation, if not deposition, is limited to western and southern midcontinent areas, the Rocky Mountain–Colorado Plateau region and the Permian Basin area of West Texas and southeastern New Mexico; exceptions are found in the Early Jurassic faulted basins of the Appalachian Piedmont and the subsurface of the Atlantic and Gulf coastal plains.

Cratonic consequences of continental-margin tectonism decline through the time span of Absaroka II and III; similarly, there is progressive decay in the sediment source-sink activity of interior uplifts and their adjacent basins. What remains is the pattern of episodic marine transgression from a relatively continuous Cordilleran seaway and westward progradation of delta complexes and associated strandline environments and dune fields. By Early Jurassic time, net progradation carried nonmarine depositional conditions to the western continental margin.

There is at least one episode that sets Absaroka II apart from Absaroka III; this is the middle and late Permian evolution of the Permian Basin area. Here, an aulacogenic embayment (dating at least from Sauk depositional time) of the Ouachita-Marathon margin was subdivided into subsidiary basins by the emergence of Pennsylvanian blocks. Continuing rapid subsidence of the basins with respect to their margins created extremes of depositional environments including back-reef evaporitic lagoons and sabkhas, massive barrier-reef trends and fore-reef carbonate aprons, and starved, deep, basin interiors that became the sites of organic-rich deposits, turbidites, and eventually, thick evaporites. Concurrently, along the Idaho-Wyoming border, subsidence rates in excess of sediment supply appear to have caused conditions leading to the generation of phosphates and bedded cherts at positions previously and subsequently occupied by normal marine and deltaic environments.

Zuni Sequence

Rocks of the Zuni Sequence extend in time from Early Jurassic to Early Paleocene. The sequence is particularly well developed on the Atlantic and Gulf continental-shelf wedges, and these occurrences are appropriately treated elsewhere in this volume; here, discussion is limited to the Colorado Plateau, Rocky Mountains, and Western Interior.

Sub-Zuni Surface

The base of the Zuni is marked by yet another major unconformity on whose surface of erosion there were exposed rocks of all older Phanerozoic sequences and basement crystallines. In broad and general terms the degree of sub-Zuni erosion decreases along the Rocky Mountain belt from the Canadian border where Mississippian strata are truncated by the unconformity to the southern reaches of the Colorado Plateau where youngest Absaroka (Early Jurassic) beds are preserved. Again in general terms, the subcrop pattern is remarkably simple and systematic, betraying few evidences of localized uplift and erosion during the Absaroka-Zuni lacuna.

Tectonic Mode

Deposition of the Zuni Sequence represents a return to tectonic conditions on the craton as they existed during accumulation of the Tippecanoe and Kaskaskia sequences; that is, the cratonic interior was subject to broadly effective and long-period marine transgressions, basin subsidence was dominated by flexure rather than fracture, and the interior received large volumes of detritus from mountain belts developed by plate interaction at the continental margin.

Zuni I

The lowest subsequence extends from within Early Jurassic to high in Middle Jurassic time. Facies and their distribution mimic certain aspects of Absaroka II and III; marine conditions persisted to the west and northwest, red deltaic sediments formed the eastern border of the Western Interior, and a variety of marine-border sabkhas, beaches, and dune fields occupied the intermediate areas. Where Zuni quartz arenites (e.g., Entrada) rest on similar Absaroka rocks (e.g., Navajo) the unconformity is obscure to the point of invisibility; in most localities, however, this difficulty does not prevail, particularly over the very large areas in which marine Bajocian beds onlap to cover the unconformity surface. Zuni I deposition is closed by regional offlap accompanied by locally significant erosion.

Zuni II

Renewed marine onlap of the Western Interior (Oxfordian) is characterized by basal sandstones bearing chert clasts and marking the first appearance of western-source detritals derived from Cordilleran highlands produced by the first pulse of the Sevier orogeny. Clastic wedges, dominated by subaerial delta-platform deposits carried by northeasterly and easterly flowing distributary systems, had by the end of Jurassic time prograded to exclude marine environments from almost all of the region. In earliest Cretaceous time the entire area was again above depositional baselevel.

Zuni III

The third Zuni renewal of deposition is represented in the Western Interior by further spread of western-source clastic wedges, largely fluviatile and lacustrine in early phases. Marine transgression from Gulf and Arctic seaways expanded, perhaps to their maximum in Aptian-Albian time, at the same time that orogenic highlands of the Sevier belt advanced toward their ultimate eastern limit, creating strong stream gradients and the distribution of pebble conglomerates to what is now the position of Laramide uplifts and the western Great Plains. Concurrently, a continuous seaway formed at the prograding margin of the clastic wedges to become the site of craton-interior-derived deltaic and strandline sands at its eastern margin. At the height of Albian transgression marine and marine-border environments reached eastward to Minnesota and Illinois before the onset of yet another offlap episode in Early Cenomanian time.

Zuni IV

The youngest Zuni phase of onlap began in later Cenomanian time, and the Western Interior–Rocky Mountain province participated as before. Zuni IV facies and their distribution were strongly influenced by conditions accompanying the climax of Sevier orogeny: source highlands reached their maximum encroachment onto the craton and their maximum rates of uplift, accelerating the rates of sediment supply to clastic wedges and amplifying the eastward progradation of delta complexes at their distal margins. Here, extremes of marine-to-deltaic intertonguing occurred as sea level varied under eustatic and tectonic controls and as sediment supply rose and fell with pulses of orogenic activity in the Sevier provenance areas. Concurrently, outriders of the Cordilleran magmatic arc provided massive volumes of volcanics and volcanogenic sediments in increasing proportions with approach to the end of the Cretaceous, a major focus being in western Montana. Ultimately, reduction in the rate of sea-level rise or slowing in the rate of tectonic subsidence, or both, combined with floods of western-source detritus and delta-progradation from the craton-interior, brought about the final withdrawal of marine environments from the Western Interior region. A further feature of the closing stages of Zuni deposition was the intiation of isolated, basement-involved, uplifted blocks to form local sediment sources east of the frontal thrusts and folds of the Sevier belt.

Tejas Sequence

The youngest cratonic depositional succession extends from Late Paleocene to the present, representing a complex episode best documented on the Atlantic and Gulf continental shelves and slopes beyond the scope of this report. Additional and merciful limitations on the responsibilities of the Continental Interior synthesis are provided by assignment of the Cenozoic tectonics and sedimentation of the Rocky Mountains and of the history and consequences of Pleistocene continental glaciation to other syntheses. What remains to be considered is the relatively thin, but hydrologically important, Tertiary cover of the Great Plains, analysis of the several surfaces of erosion recorded in continental Tertiary stratigraphy and physiography, and the extensions of Laramide deformational, volcanic, and plutonic events documented in the Great Plains. Treatment of these topics is rationally deferred until mature consultation with other groups, particularly with those engaged in study of the Rocky Mountains and the passive margins, permits a unified approach to mutual problems.

Unresolved Questions

Careful consideration of 600 Ma of the history of a continental craton raises a multitude of questions; those that refer to cratonal tectonism and the relationship between the tectonic evolution of a continental interior and global tectonic events are of first-order relevance in terms of the goals of *The Geology of North America.* A few of the more obvious prime problems are noted below.

1. What is the significance of episodic emergences of the craton responsible for the sequence-bounding unconformities? The natural corollary is the meaning of the several episodes of cratonic submergence. In both cases the magnitude of vertical movement far exceeds that which could be ascribed to any combination of eustatics and loading-unloading by sediment and water. Is there a demonstrable correlation with plate interactions?

2. Why is deformation of the craton sometimes dominated by flexure, at other times by fracture? When the fractural mode is operating, is the continental crust in extension or compression? That is, are the down-dropped basins and block uplifts characteristic of Absaroka I deposition produced by thrusts of Wind River type or are they representative of distinctly different stress/strain patterns? In either case, what are the driving forces?

3. Is the tectonic mode prevailing on the craton during Sauk deposition as distinct from the younger Phanerozoic as is suggested here? If so, does this indicate that there are irreversible secular changes in cratonic behavior as a product of evolving crustal thickness, or rheologic change, or whatever?

4. Why do sedimentary basins and intervening arches remain fixed in positions for hundreds of millions of years? Is it demonstrable that basins are localized on or at the intersections of basement rifts and sutures?

5. What are the driving forces of basin subsidence? Again, eustatics plus surficial load cannot achieve the amplitudes recorded and, at distances from continental margins, thermal contraction or thermally induced phase changes are unsatisfactory hypotheses for many reasons.

References Cited

Aitken, J. D., 1966, Middle Cambrian to Middle Ordovician cyclic sedimentation, southern Rocky Mountains of Alberta: Bulletin of Canadian Petroleum Geology, v. 14, p. 405–441.

Kay, Marshall, 1951, North American Geosynclines: Geological Society of America Memoir 48, 143 p.

Palmer, A. R., 1960, Early Late Cambrian stratigraphy of the United States: Journal of the Washington Academy of Sciences, v. 50, p. 8–9.

Sloss, L. L., 1963, Sequences in the cratonic interior of North America: Geological Society of America Bulletin, v. 74, p. 93–113.

Stewart, J. H., 1972, Initial deposits in the Cordilleran geosyncline: evidence of a Late Precambrian Continental separation: Geological Society of America Bulletin, v. 83, p. 1345–1360.

—— 1976, Late Precambrian evolution of North America: Plate tectonics implications: Geology, v. 4, p. 11-15.

Vail, P. R., Mitchum, R. M., and Thompson, S. III, 1977, Seismic stratigraphy and global changes of sea level, Part 4: Global cycles of relative changes of sea level, *in* C. E. Payton, ed., Seismic stratigraphy—Applications to hydrocarbon exploration: American Association of Petroleum Geologists Memoir 26, p. 83–98.

Quaternary geology of Canada

R. J. Fulton
Geological Survey of Canada
601 Booth Street
Ottawa K1A OE8, Canada

Introduction

The objective of this volume is to synthesize the Quaternary geology of Canada. In essence, it is an update of "Quaternary Geology of Canada" (Prest, 1970), which synthesized knowledge of the Quaternary geology of Canada up to 1965. This volume will present current knowledge, ideas, and concepts, with particular emphasis on limits of glaciation, concepts of ice-sheet dynamics, nature of surface materials, and the interaction between Quaternary geology and the activities of man. The volume will be in two parts, one dealing with fundamental Quaternary geology and stratigraphy and the other dealing with aspects of Quaternary geology related directly to man and land use.

Part 1. Regional Quaternary Geology

This part will present a picture of the Quaternary geology of Canada. The nature of Quaternary deposits, landforms, and physiography will be described, and Quaternary stratigraphy and history will be synthesized. The Quaternary geology description will be tied to a map of surface materials of Canada (scale 1:5,000,000). This map will differ from the glacial map of Prest et al. (1968) in that it will portray materials as well as features and events. Map units will be based on the dual concepts of materials and genesis so that, in addition to indicating the nature of the materials at the land surface, the pattern of distribution will portray the latest Quaternary events. Particular attention will be paid to the history of the last glaciation. The emphasis will be on regions of accumulation and outflow, limits at the last glacial maximum, and the pattern of deglaciation. In line with this, a revision of the ice-retreat map (Prest, 1969) will be prepared.

The regional Quaternary geology part of the volume will be written in six chapters, each covering a natural Quaternary region. Individual regions are characterized by relatively homogeneous Quaternary deposits and congruent glacial histories. Regions used in this breakdown are Cordillera, Plains, Shield, High Arctic, St. Lawrence Lowlands, and Atlantic (Fig. 1). Each regional chapter will emphasize Quaternary information and problems peculiar to that area and will include an introduction outlining aspects that produce its unique Quaternary characteristics.

Cordillera

The Cordilleran chapter will highlight the history of the last glacial-nonglacial cycle, which is well documented in that area. It will discuss the relationships between topography, climate, and glacier buildup and decay. Land-sea relationships during advance and retreat of the last ice sheet and interaction between the ice and sea will be significant topics. Special consideration will be given to the Quaternary history of the Yukon Territory north of the glacial limit, using evidence obtained from sedimentation in the large basins scattered throughout this area. The activity of both conventional and rock glaciers will be used to document changes in postglacial climate.

Plains

The Plains chapter will include a special section on the development of preglacial, glacial, interglacial, and postglacial valleys. It will shed new light on the relationship between Laurentide and Cordilleran ice and review the difficult problems of determining and dating ice limits in this area. The rich vertebrate record of southern Alberta and the extensive glacial and interglacial records of Banks Island will form important components of the description of Quaternary stratigraphy. The story of deglaciation of the Plains will attempt to unravel the complex lobations, which occurred during ice retreat, and will give special emphasis to the formation of glacial lakes (such as Agassiz).

Shield

The Shield region includes the heartland of the Laurentide Ice Sheet, and consequently this chapter will present current ideas on the development and disintegration of this complex ice mass. Regional patterns of ice flow, erosion, and deposition will be discussed, as well as the configuration of the ice sheet, regions of outflow, and styles of deglaciation. The deglacial and early postglacial discussion will emphasize marine submergence and the history of isostatic recovery. The Canadian Shield is not generally regarded as

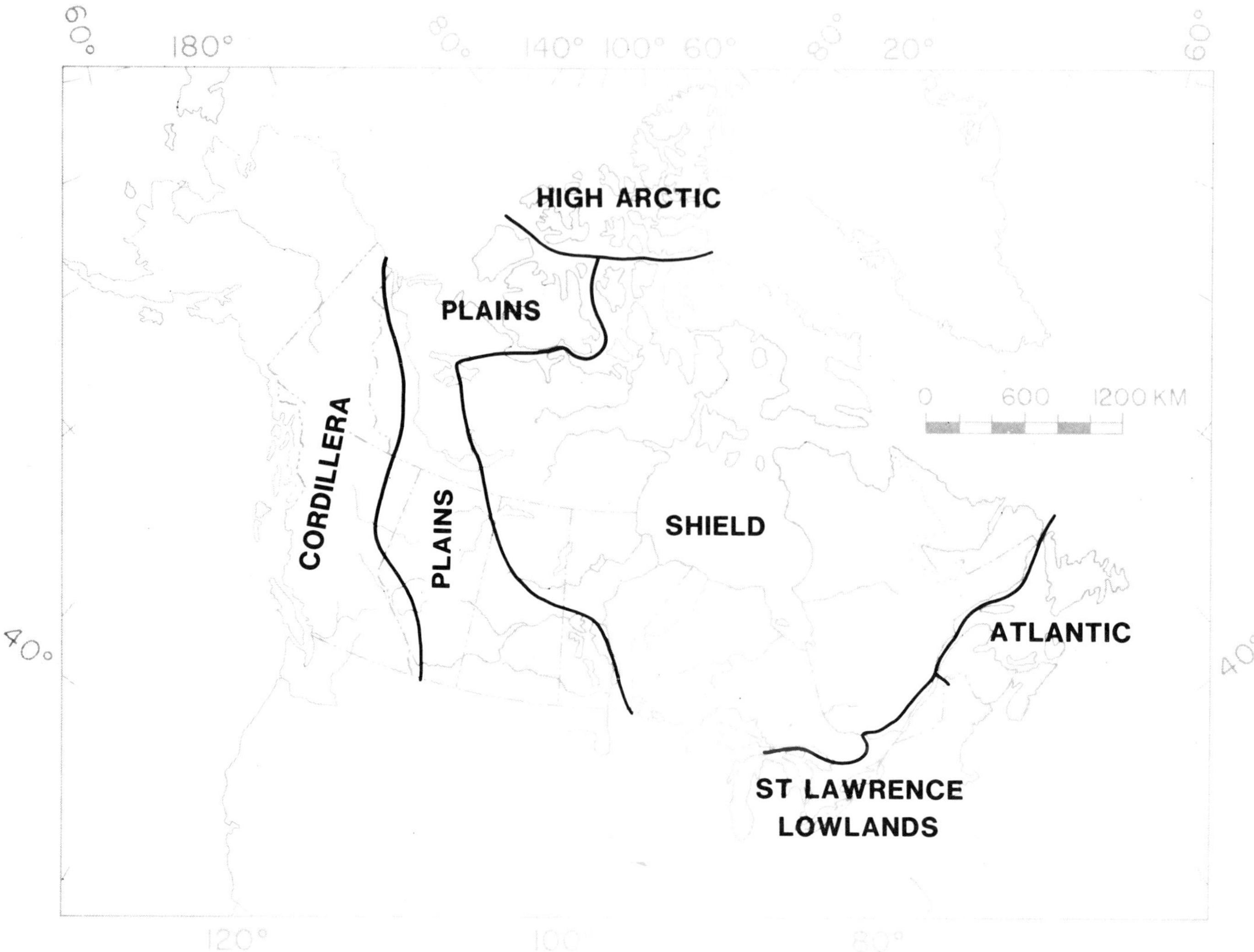

Figure 1: Areal subdivision of Canada used in preparing Part 1: Regional Quaternary Geology.

a place with well-developed Quaternary stratigraphy. However, in this volume the shield region includes both the James Bay–Hudson Bay area, with its excellent Quaternary record of the last interglaciation and glaciation, and Baffin Island where the Institute of Arctic and Alpine Research (INSTAAR) of the University of Colorado has conducted much stratigraphic research.

High Arctic

One special difficulty in conducting Quaternary geology studies in the High Arctic is the general lack of good exposures of Quaternary sediment. This chapter will include descriptions of some of the unique (to Canada) features of the area. There is considerable controversy over the nature of the last glaciation in this area, so this chapter will discuss different theories involved in the controversy and propose the best solution based on current knowledge. Particular emphasis will be placed on land-sea relationships in providing answers to the enigma of degree of ice cover during the last glaciation. One section will be devoted to the glaciology of present-day glaciers and information on Quaternary climate which has been obtained from ice cores.

St. Lawrence Lowlands

The St. Lawrence Lowlands are small when compared to the other regions of Canada but are important because they contain more than half the Canadian population. They are significant from a Quaternary point of view because the Quaternary framework developed in the area is a correlation standard for the rest of the country, and many of Canada's Quaternary scientists were trained and received their biases in this area. As a consequence, the nature of the Quaternary geology of this area has influenced Quaternary work throughout Canada. This regional chapter will include a brief review of the stratigraphy of classical sites such as the Don Valley Brickyard, Scarborough Bluffs, and Lake Erie coastal bluffs. The picture of development of the Great Lakes and the Champlain Sea will be updated, and the controversy over the relationship between Great Lakes drainage and formation of the Champlain Sea will be aired. In addition, the deglaciation picture of the area will be revised to take into account current ideas relative to the development of calving bays, lobation of ice sheets, and flow reversals.

Atlantic

The Atlantic region warrants separate treatment because it supported a complex of local ice caps, with ice shelves and ice rises, that was satellitic to the main Laurentide Ice Sheet that periodically invaded parts of the area. Quaternary physiographic evolution, based on the shelf sedimentary sequence as a reflection of uplift and degradation, will be adapted from the "Cenozoic Development" chapter of the *Continental Margin: Eastern Canada* volume, in order to set the stage for glacierization and the resultant ice-flow patterns. The climate-stratigraphic scheme that has been developed from studies of the Quaternary record on land will be extended to incorporate the marine record. The complex glacial history, characterized by the interaction of ice from several sources, will be illustrated by maps showing patterns of ice-flow indicators, provenance of superposed till sheets, and till lithology. The problem of glacier limits will be discussed in terms of the concept of weathering zones and the theory of cold-based ice. Reconstruction of postglacial events will feature glacio-isostatic movement and Holocene changes of sea level.

Part 2. Applied Quaternary Geology

The second part of the Quaternary volume will cover special aspects of Quaternary geology. These will consist of diverse subjects ranging from palynology to placers. Human concern and land and resource use are common themes used to link these various topics. The chapters will consist of sections describing geologic phenomena and related environmental and engineering information, followed by discussions of how these are related to human activity.

Quaternary Paleoenvironments

The first chapter of Part 2 will basically consist of documentation of the nature of environments that characterized Canada during the Quaternary through use of palynology, plant macro remains, tree-ring and diatom studies, and vertebrate and inverbebrate paleontology (including insects). Emphasis will be on the late Wisconsinan and Holocene because considerably more is known about this period than the rest of the Quaternary and because the Holocene is the time in which we live. Palynology will be the prime data base because pollen has been the most commonly used paleoenvironmental indicator. The basic tenets of this chapter are that Quaternary climate and environment have fluctuated widely and will continue to fluctuate. If we are to be ready to meet the challenges of future changes, we must know what degrees and rates of fluctuation to expect and understand how these might be reflected in environmental changes. The chapter will also touch on the ways in which man's activities could be affected by these changes, and on some of the causes of climatic change.

Geomorphological Processes

The chapter on geomorphological processes will emphasize the common, steady processes rather than the catastrophic ones, which will be highlighted as "hazards" in the final chapter. Canada is a northern country and consequently great diurnal and annual temperature fluctuations occur; most areas are in a state of frozen immobility part of the year, and high snow falls are generally preserved throughout the winter with a resulting large spring freshet. The manner in which these geographical-climatic factors dominate process activity will be the main theme of this chapter, and permafrost and related processes will be prominently featured. An important adjunct is a new permafrost map of Canada (scale 1:5, 000, 000). This will differ from previous maps (e.g., Brown, 1967) in that it will show the nature and content of ground ice and location of features associated with permafrost. The effect of process activities on man will be brought out as will also the effect of man's activities on processes. In addition, the role of climate in controlling geomorphological processes and the effects of climatic change on process activities will be discussed.

Terrain Geochemistry

This chapter will present general information on the geochemistry of surface materials of the different regions of Canada, with special emphasis on the composition of tills. It will include data on methodology, interpretation of data, and descriptions of the programs through which the terrain geochemistry data were gathered. The first main theme of the interpretive part of the chapter will be the impact of terrain geochemistry on human activities. This will include sections on drift prospecting, geochemistry and health, lake sediment sampling, and mineral exploration and geochemistry as used in Quaternary stratigraphy. The second main interpretive theme will be the effect of human activities on geochemistry. Topics in this section will include acid rain and sediment contamination by toxic elements.

Geodynamics

This chapter will be concerned with crustal movements largely due to forces that are well documented only in the Quaternary. Lacustrine and marine shoreline displacement data will form the basis for this review because they are the primary evidence of Quaternary crustal movements. Loading and unloading of ice and water, the main forces causing these displacements, will be discussed, and various geophysical models proposed to explain the tectonic adjustments to these forces will be examined. The impact on man will be detailed by describing the past and present effects on human activity. The depth to which this chapter discusses conventional tectonics and seismicity has not yet been determined, but it is felt that these subjects should be covered in the regional bedrock volumes to better bring out their relationship to regional forces and features and the tectonic evolution of the continent.

Resources

Sand and gravel will be the main focus of this chapter. In addition to looking at the genesis of these Quaternary resources and the effect of genesis on exploration and exploitation, the chapter will include a brief survey of the prime aggregate sources and future availability of aggregate in Canada's major urban areas. Other extractive resources such as clay, marl, and peat will also be discussed, and a separate section will deal with placer deposits. In addition, soils, in the agricultural sense, will be discussed as a Quaternary resource, and a brief look will be taken at how Quaternary events have influenced the distribution, texture, and utility of Canadian soils.

Planning

The central theme of this chapter will be the need for knowledge of Quaternary geology (surface materials distribution, process activity, etc.) in the planning process. Specific consideration will be given to Quaternary deposits as foundation materials and as fill, to the importance of a knowledge of Quaternary stratigraphy in groundwater exploitation and protection from contamination, and to siting and land use planning concerns relative to Quaternary geodynamics, hazards, and best use of land and resources. In addition, the chapter will examine geologic hazards and the application of a knowledge of geology in taking hazards into account in the land use planning process. A special section of this chapter will be devoted to a survey of major geologic hazards in Canada. Prime hazards will be catalogued, driving forces will be discussed, and their significance in the different regions of Canada will be addressed.

Acknowledgments

J. A. Heginbottom, who will be editor for Part 2 of the synthesis volume, critically read this manuscript. The following persons supplied outlines for the noted chapters: J. J. Clague—Cordillera; A. MacS. Stalker—Plains; W. W. Shilts—Shield; W. Blake, Jr., and D. A. Hodgson—High Arctic; D. R. Grant—Atlantic; T. W. Anderson and J. V. Matthews, Jr—Quaternary Paleoenvironments; J. A. Heginbottom—Geomorphological Processes; R.N.W. DiLabio—Terrain Geochemistry; and L. E. Jackson—Resources and Planning.

References

Brown, R.J.E., 1967, Permafrost in Canada: Geological Survey of Canada, Map 1246A.

Prest, V. K., Grant, D. R., and Rampton, V. N. (compilers), Glacial map of Canada: Geological Survey of Canada, Map 1253A, scale 1:5,000,000.

——1969, Retreat of Wisconsinan and Recent ice in North America: Geological Survey of Canada, Map 1247A.

——1970, Quaternary geology of Canada in Geology and economic minerals of Canada, Chapter XII: Geological Survey of Canada, Economic Report no. 1, 5th edition, 838 p.

Quaternary history of North America and adjacent oceans

H. E. Wright, Jr.
Department of Geology and Geophysics
University of Minnesota
Minneapolis, Minnesota 55455

W. F. Ruddiman
Lamont-Doherty Geological Observatory
Columbia University
Palisades, New York 10964

Introduction

This Quaternary volume in the D-NAG series is to be topical rather than regional. The Quaternary is characterized by major environmental changes, and an understanding of the dynamic interactions of physical, chemical, and biological processes during the Quaternary can illuminate the operation of the natural world and the formation of the geologic record.

The book will be in two parts, the first of which will be concerned with the last deglaciation, approximately 18,000 to 6,000 years ago. For this interval, a wealth of information exists, closely controlled in chronology by radiocarbon dating. The approach here will be to focus on the response of natural processes to the climatic forcing that initiated the deglaciation, including (1) the relative sequencing of the various climatic responses and (2) the interrelations among these processes. The last deglaciation is the most accessible geologically and is representative of earlier major deglaciations, each of which brought to an end 100,000-year-long cycles of global climatic deterioration. Thus determining the speed, timing, and mechanism of the most recent deglaciation is fundamental to an understanding of the origin of these 100,000-year cycles that constitute the dominant rhythm of Quaternary climate.

The second part of the volume will treat the record prior to the last deglaciation—the Early and Middle Wisconsinan and previous glacial and interglacial cycles. The major discovery that many more climatic cycles are recorded in ocean cores than are thought to be evident in the terrestrial record and that the glacial intervals were consistently longer than the interglacials has added a new dimension to concepts of Quaternary climatic change. This portion of the record for the most part lies beyond the reach of radiocarbon dating, necessitating other strategies for obtaining chronologies and correlations. These strategies rely on (1) other dating techniques, both radiometric and paleomagnetic, (2) correlation with the relatively well-dated and continuous oceanic record, and (3) correlation with models of the earth's response to the orbital forcing that is known to have significantly affected the last few million years of climate. Plans for this part of the volume are intended to encourage these efforts to correlate and interpret long-term records in the orbital framework.

The Last Deglaciation

Particular emphasis will be placed on the sensitivity of the natural processes to climatic variations and on the response time following the initial impetus toward climatic change. One can envision that the earth adjusts somewhat to the long regime of the glacial mode and that a rapid climatic shift sets in motion a complex of interdependent adjustments of the natural processes, each proceeding at its own pace, until the relative stability of an interglacial mode is attained. The adjustment period may witness the temporary existence of seemingly anomalous associations of plants or animals or of landscape features—at least anomalous with respect to the relative stability of the interglacial world of today. Discussions will concentrate on such anomalies.

The concept of rapid climatic changes must first be examined, especially in the context of causal mechanisms. The various natural processes will then be treated in turn, with attention not only to response time but also to interaction with other processes. Thus, the changes in fauna will be treated in the context of the changing vegetational scene, which in turn will consider the periglacial effects of the wasting ice sheet. Many chapters will involve the contributions of two or more specialists if necessary, but particular efforts will be made to integrate and emphasize interrelations among the natural processes involved. A map series depicting the nature of the North American landscape and ocean margins at different times during the deglacial cycle will accompany this part. The following chapters are planned.

1. The chronology of Earth-Sun orbital variations over the past 20,000 years and their potential effects on the land, atmosphere, and ocean in different geographic regions.

2. A review of paleoclimatic models of the last deglaciation and the mechanisms and rates of climatic change. Square-wave climatic change, regional interrelations, air/sea/ice/land feedbacks, interhemispheric relations, and modern analogues. Entrapment of polar air by the North American ice sheet (Bryson effect) and the effects on seasonality in the periglacial area; opening of the Rocky Mountain corridor. The treatment will be based on climatological theory and models rather than primarily on geologic evidence.

3. Direct measures of changing ice volume in deep-sea cores: oxygen-18 content of benthonic Foraminifera. Indirect measures: oxygen-18 content of planktonic Foraminifera as controlled by meltwater; productivity suppression of plankton by meltwater; ice-rafted detritus; supply of outwash sediment to ocean margins.

4. Retreat of North American glaciers. Significance and chronology of ice-margin fluctuations in different segments of the Laurentide ice sheet, the Cordilleran ice sheet, and various mountain glaciers. Dynamics of wasting ice sheets. Surging. Periglacial effects.

5. Changes in sea-surface temperatures inferred from assemblages of planktonic Foraminifera and other microfossil groups. The late-glacial interstadial in the North Atlantic (13,000–11,000 yr B.P.).

6. Coastal changes in relation to isostatic and eustatic sea-level changes.

7. Periglacial physical effects of wasting ice sheets. Loess deposition, soil formation, and frost features.

8. Hydrologic changes: rivers, proglacial lakes, temperate lakes, desert lakes, groundwater. Relative importance of precipitation and temperature. Relic landforms. Catastrophic events related to lake drainage. Effects of isostatic crustal deformation on rivers and lakes.

9. The vegetational sequence in different areas, not only close to the wasting ice sheets but also at low latitudes and other areas far from the ice sheets (e.g., Beringia) where periglacial influences may have been minimal. Anomalous assemblages. Differential migration rates of dominant tree species from glacial refuges. Relations to soil conditions.

10. Faunal changes during deglaciation. Extinctions and extirpations. Relations to the food resource and directly to climatic parameters.

11. Human cultures. Chronology of immigration and expansion. Megafaunal overkill. Adaptations to changing food resources and climatic conditions.

12. Paleomagnetic sequences. Possible terrestrial and extraterrestrial mechanisms. Geographic variations. Excursions. Relations to climatic change.

13. Tectonics and volcanism. Effects on geomorphic processes. Tephrochronology.

Earlier Quaternary Record

Quaternary events prior to the radiocarbon-dating limit of about 40,000 years ago suffer from inadequate chronological control, as well as from the spotty occurrence of deposits and landforms as a consequence of subsequent burial or erosion. Thus, the record is incomplete, at least in most continental environments, and the chronological sequence must depend on such techniques as paleomagnetic reversals, isotope dating of volcanic deposits or carbonate sediments, or other dating methods not thoroughly proven. Thus, special attention will be paid to dating techniques in this section.

The ocean-core record, on the other hand, may be complete, although the chronological sequence may be locally biased by gaps or by variations in sedimentation rates. In addition, the ocean is the final repository for much of the sediment eroded from the land during the long span of Quaternary glaciation. As a result, this section moves progressively from the more nearly complete deep-ocean and coastal records to the isolated occurrences of long pluvial-lake cores and finally to those land records constrained in some measure by dating and correlation. The following chapters are planned for this part of the volume.

1. The chronology of Earth-Sun orbital variations over the past 1.8 million years and their potential effects on the land, the atmosphere, and the oceans in different geographic regions. Regional (latitudinal) variations in the strength of the 23,000- and 41,000-year cycles of solar radiation arriving at the top of the atmosphere.

2. An overview of paleoclimatic modeling of the response of the ice, air, sea, and land to long-term orbital cycles. Response of ice models to changing insolation. Impact of isostatic loading. Models of oceanic moisture flux.

3. Correlation of the ocean-core record with the solar-radiation curve, particularly for areas close to North America. Chronological controls: isotope methods, paleomagnetic stratigraphy, variations in sedimentation rates, faunal extinctions, and other markers. Spikes in the isotope record.

4. Direct land/sea correlations: pollen, desert dust,

and ice-rafted debris in marine sediments. Volume, nature, and distribution of Quaternary sediments in the oceans near North America: estuaries, shelves, canyons, fans, and abyssal plains.

5. Coastal terraces and on-lap deposits. U-series and amino-acid dating. Effects of tectonics. Early and Middle Wisconsin high sea levels. Fauna and flora of continental-margin deposits.

6. Long records from pluvial lakes. Role of tectonics and volcanism.

7. Early and Middle Wisconsin glaciation of the Laurentide ice sheet and in the western mountains. Loess stratigraphy in the mid-continent.

8. Pre-Wisconsin glacial/interglacial cycles in the mid-continent region and the western cordillera. Interglacial soils and plant remains. Loess stratigraphy and alluvial terraces. K/Ar dating of Quaternary lavas in areas of mountain glaciation. Tephrochronology.

9. Speleothems and dated isotopic records. U-series dating. Correlation to atmospheric conditions.

10. Faunal changes. Land-mammal ages, migrations, their relations to paleoclimate.

Innuitian Orogen and Arctic Platform

H. P. Trettin
Geological Survey of Canada
3303 - 33rd Street N. W.
Calgary, Alberta
T2L 2A7
Canada

Introduction

The outline presented and discussed in this paper is preceded by a brief introduction to the geology of the region that explains the proposed organization of geological topics. This introduction is restricted to Canada because several important reports on the geology of northern Greenland are currently in press or preparation. The summary is based mainly on the framework established by Thorsteinsson and Tozer (1970) with modifications necessitated by a large amount of more recent work.

Innuitian Orogen and Arctic Platform in Canada: Tectonic Framework, History, and Some Major Problems

The Innuitian Orogen comprises deformed rocks of Precambrian to early Tertiary age that are unconformably overlain by upper Tertiary to Holocene sediments. The Phanerozoic component is divisible into two major successions, Early Cambrian to Late Devonian and late Early Carboniferous to early Tertiary in age, respectively, that are separated by a major angular unconformity. Towards the craton, the Phanerozoic strata of the orogen pass into undeformed beds of the Arctic Platform that lie unconformably on upper Proterozoic to Archean rocks of the Canadian Shield (Fig. 1B). The boundary between these two major provinces is taken as the limit of continuous deformation, which coincides mainly with the limit of mid-Paleozoic folding but locally with the limit of Tertiary thrust faulting or folding. From a stratigraphic point of view, it corresponds roughly to an ill-defined hinge separating the rapidly subsiding outer part of the early Paleozoic shelf (the previous miogeosyncline or miogeocline) from the continental interior (craton), but there are no persistent facies changes associated with it. The most important facies boundary, the early Paleozoic shelf margin, lies well within the orogen.

The Arctic Platform is separated from the Hudson and Interior platforms by water or extensive Precambrian terranes. It is divisible into unstable domains of different ages, such as uplifts and basins, and intervening stable regions. Only three major positive elements, the Minto Arch, Boothia Uplift, and Bache Peninsula Arch, are shown in Figure 1B.

Proterozoic of the Innuitian Orogen

A belt of Proterozoic shelf carbonate rocks with some mudrock and sandstone in east-central Ellesmere Island is overlain with low angular unconformity by Lower Cambrian sandstone of the early Paleozoic shelf province (see below). The strata are assumed to be related to middle or upper Proterozoic supracrustal successions of the Shield, but their specific age and correlation are unknown.

Extensive Proterozoic terranes in northern Ellesmere and Axel Heiberg islands consist of gneiss and a thick succession of sedimentary and less abundant volcanic rocks, metamorphosed to greenschist and amphibolite facies. One gneiss belt has Rb/Sr isochron and zircon ages around 1 Ga, indicating formation during a Grenville-age orogeny. Some metasediments are older, but most seem to be younger than the gneisses. Their stratigraphy is poorly understood because of excessive deformation and lack of age determinations. These terranes are mostly in fault-contact with strata of the early Paleozoic magmatic belt but are locally uncomformably overlain by upper Middle and Upper Ordovician sediments. The problem whether they are part of the Canadian Shield or exotic can only be resolved on the basis of the early Paleozoic geology.

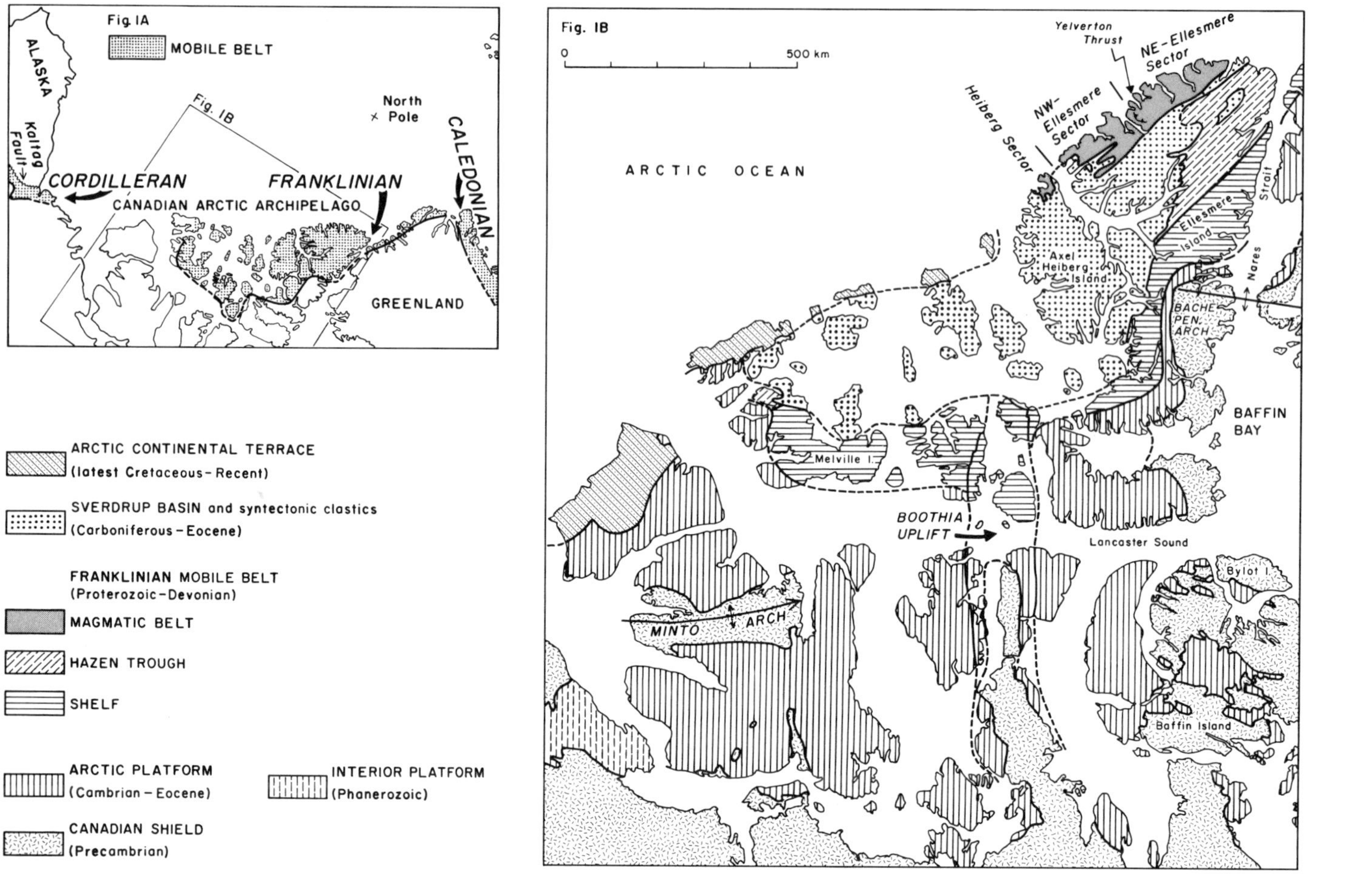

Figure 1. A. Cambrian-Devonian mobile belts and index. (Limit of Late Devonian–Early Carboniferous deformation in the Cordillera and Kaltag Fault after D. K. Norris (personal communication, 1981). The same deformation probably extends beyond the Kaltag Fault into Alaska but is not shown there because of various uncertainties. Limit of Franklinian and Caledonian mobile belts in Greenland adapted from Escher, 1970). B. The Innuitian Orogen and most of the Arctic Platform in Canada (The Arctic Platform also includes lower Paleozoic outliers on southern Baffin Island and adjacent islands not covered by this map).

Cambrian to Early Carboniferous

The Cambrian to Devonian succession of the orogen represents a continental margin–type mobile belt, originally named the Franklinian Geosyncline and divided into a miogeosyncline and a eugeosyncline. These terms are still commonly used in the literature but are here replaced by informal, descriptive names because of some changes in concept. The name "Franklinian," however, is retained to emphasize continuity with the earlier concepts.

Deposition. Three major depositional belts are apparent in Cambrian to Early Devonian time: a southern shelf; an axial deep-water basin, known as the Hazen Trough; and a northern sedimentary and volcanic belt, here referred to as the magmatic belt because it also contains granitic to ultramafic intrusions. The shelf includes all of the Arctic Platform and adjacent parts of the orogen; the trough and magmatic belt are limited to the orogen. The shelf and trough extend into northern Greenland, but the magmatic belt is exposed only in northern Ellesmere and Axel Heiberg islands.

The *shelf province* is characterized by a southward thinning succession mainly of carbonate rocks with lesser amounts of clastic sediments and evaporites. Facies are extensive laterally, and three major settings—an outer belt of carbonate buildups, a lagoon, and an inner belt of shallow subtidal or peritidal environments—are apparent during some intervals. This pattern was disturbed by pronounced subsidence of the outer shelf from Late Ordovician time onward, which resulted in accelerated growth of buildups in some areas and encroachment of deep-water facies in others, and by Late Silurian–Early Devonian movements of the Boothia Uplift and Early Devonian movements of the Bache Peninsula Arch.

Three highly diachronous phases are distinguished in the depositional history of the *Hazen Trough* in the Arctic Islands: a rift-related(?) phase of coarse clastic deposition in deltaic and deeper settings (Cambrian); a starved-basin phase, characterized by subsidence to a depth of about 1 km, accompanied by very slow accumulation of mudrock, radiolarian chert, and resedimented carbonates (late Early Cambrian to Early Devonian); and a flysch phase during which the trough was rapidly filled to shelf level with sediment gravity flows (latest Ordovician to early Middle Devonian). The flysch sediments were derived from metamorphic and sedimentary sources that lay north of northeastern Ellesmere Island and at the northeastern termination of the trough, presumably in the Caledonian Orogen. Apart from transverse flows entering the Hazen Trough from the north, the predominant direction of current flow and progradation was from northeast to southwest.

The *magmatic belt* contains a great variety of volcanic and sedimentary rocks. Most volcanics are of arc type and suggestive of subduction, but some seem to have originated during extensional tectonic regimes. The sediments comprise: (1) very coarse to fine clastic sediments, associated with unconformities and deposited in nonmarine and shallow marine settings; (2) shelf carbonates; and (3) tongues of all three phases of the Hazen Trough. Facies patterns are complex and individual lithofacies of small areal extent. Three lateral divisions are distinguished that differ in phases of their structural and depositional history; they are here referred to as the NE-Ellesmere, NW-Ellesmere, and Heiberg sectors (Fig. 1B).

A *molasse basin* was superimposed on shelf and trough in the Middle and Late Devonian. It was occupied by rivers, deltas, and shallow marine environments that received clastic sediments from easterly and northerly sources and prograded in southwesterly and westerly directions.

Deformation, Plutonism, and Metamorphism in the Magmatic belt. In the NE-Ellesmere sector, early Middle Ordovician (or earlier) ophiolitic plutonism was followed by an orogeny in the Middle and Late Ordovician. A final orogeny of Late Silurian to Early Carboniferous age encompassed the entire region. Both orogenic intervals are divisible into several phases, each of which contributed to the progressive stabilization of the region.

Early Middle Ordovician (or Earlier) Ophiolitic Plutonism, NE-Ellesmere Sector. A large pluton in this sector, the M'Clintock West massif, consists mainly of serpentinite and altered gabbro with remnants of wehrlite and small sheets and plugs of leucocratic granitic rocks. It is separated from Proterozoic to Upper Carboniferous rocks by steeply dipping faults, and is unconformably overlain by Upper Carboniferous strata. The pluton is comparable to lower and middle parts of the ophiolite suite; the absence of upper parts of the suite, comprising sheeted dykes, pillow basalt, and sediments, can be explained by faulting or erosion. Together with four smaller bodies of comparable lithology the pluton delineates a suture belt, traceable for about 100 km, although distorted and displaced by Ordovician and later orogenies.

The age of the ophiolites is fundamental for the tectonic history of the region. An earlier K/Ar (hornblende) age of 390 ± 20 Ma (late Early Devonian) on a granitic sheet within the M'Clintock West massif posed problems for plate-tectonic interpretation because some upper Middle Ordovician to Middle Silurian units straddle the suture belt without significant changes. This question has now been resolved by a zircon analysis on the same granitic sheet, indicating crystallization no later than early Middle Ordovician (unpublished analysis by D. Loveridge, Geological Survey of Canada, Ottawa). Interpretation of crustal separation as being prior to the late Middle Ordovician is permissible, mainly because of lack of stratigraphic control.

Middle to Late Ordovician Orogeny, NE-Ellesmere Sector. This event is inferred primarily from a high angular unconformity separating strata of Proterozoic to Ordovician age from overlying sediments of late Middle Ordovician age. Angular unconformities also are observed locally between rocks of Proterozoic and early Late Ordovician age, and within the Upper Ordovician succession. The fact that upper Middle Ordovician and younger sediments overlap the ophiolitic belt suggests that the orogeny involved collision of terranes separated in early Middle Ordovician (or earlier) time. It terminated arc-type volcanism in this region and raised mountains that produced great thicknesses of sediments. Metamorphism seems to have been of subgreenschist to greenschist facies. Proterozoic rocks locally have a Middle to Late Ordovician overprint on their K/Ar age.

Late Silurian–Early Carboniferous Orogeny. Three phases of structural development are apparent in the Heiberg sector where the stratigraphic record of the mobile belt probably terminates in the late Early Devonian.

Phase 1, characterized by folding and uplift, occurred mainly in the early Late Silurian (early Ludlovian), but uplift may have continued into the earliest Devonian.

Phase 2 was a rifting event, probably of late Early Devonian age. It is inferred from a sudden change from deltaic to coarse clastic deep-water sedimentation, accompanied by increasingly mafic volcanism.

Phase 3 comprised folding, uplift, and the emplacement of small, high-level, post-tectonic granitic intrusions with a Late Devonian or Early Carboniferous K/Ar age. The sedimentary record of the molasse basin suggests that uplift commenced in Middle or Late Devonian time.

The structural history of Ellesmere Island is difficult to decipher because the sedimentary and volcanic record of the magmatic belt mostly terminates in intensely deformed strata of early Late Silurian (Ludlovian) age that are locally unconformably overlain by upper Paleozoic beds. The tectonic interpretation of this region will have to be based mainly on the intrusive record but has to await completion of current age determinations on zircon; preliminary results indicate that the K/Ar ages used so far are unreliable.

The metamorphism in both Ellesmere and Axel Heiberg islands is mainly of subgreenschist grade and to a lesser extent of greenschist grade. The amphibolite facies is attained only in a small area adjacent to a transcurrent(?) fault.

Deformation of Shelf and Hazen Trough. The structural history of the shelf and adjacent parts of the Hazen Trough is better known. An early phase of epeirogenic movements and a terminal orogeny are distinguished.

Late Silurian–Early Devonian Epeirogenic Movements. The Boothia Uplift, which extends from the Arctic Platform into the mobile belt and is about 800 km long and about 150 km wide, was elevated intermittently from Late Silurian (late Ludlovian) to late Early Devonian (late Zlichovian) time. Its trend is at right angles to the depositional and structural trends of the mobile belt and evidently was controlled by the grain of the Precambrian crystalline basement. The Bache Peninsula Arch, a northwestward-plunging salient of the Shield, was positive in mid-Early Devonian (late Pragian) time.

Ellesmerian Orogeny s.s. The shelf and southern parts of the trough were deformed in latest Devonian and/or earliest Carboniferous time. This deformation probably was diachronous, migrating from the magmatic belt to the craton, as northern parts of the trough probably were folded prior to the Famennian. Structural trends are sinuous in southwesterly directions in the eastern part of the mobile belt and arcuate in the western part, where they turn from southwesterly through westerly into northwesterly directions. In the central part they are interrupted by the north-trending Boothia Uplift. Tectonic transport was towards the craton. In the shelf province the deformation is bounded at depth by a decollement, which lies in Ordovician evaporites in the western part and probably close to the Precambrian crystalline basement in the eastern part.

Here, as in the magmatic belt, the folding was followed by uplifts that are interpreted as isostatic adjustments to the crustal thickening caused by the preceding compression.

Preliminary Plate-Tectonic Considerations. The clues to the plate-tectonics history of the Franklinian mobile belt seem to lie mainly in the magmatic belt, but our understanding of this region is inadequate because of various factors—structural complexity; the fragmentary state of the stratigraphic record; absence of fossils older than late Middle Ordovician; scarcity of reliable age determinations; and, last but not least, the reconnaissance nature of the field work. Some progress can still be made prior to the completion of the Innuitian volume, but many important problems undoubtedly will remain.

Present inferences may be summarized as follows.

1. Parts of the NE-Ellesmere sector were involved in a Wilson-cycle of oceanic evolution that included ophiolite formation in early Middle Ordovician (or earlier) time and a collision in the Middle to Late Ordovician. The suture left by this cycle lies well within the magmatic belt and to the north and northeast of an extensive Proterozoic terrane.

2. The Cambrian to Early Silurian history of the Hazen Trough and adjacent parts of the magmatic belt can be interpreted as marginal response to this cycle and provides additional information about it. For example, the phase of coarse clastic sedimentation in the Cambrian points to an epeirogeny caused by mantle upwelling that initiated the cycle. The subsequent deep foundering appears to be due to both lithospheric cooling and extension concomitant with seafloor spreading farther to the north. The wide extent of the trough suggests that the latter process was far more extensive than apparent from the distribution of ophiolites. The Hazen Trough itself, however, does not seem to have been a site of crustal separation; wherever the base of the highly diachronous deep-water succession is exposed it lies on shallow marine sediments and not on ophiolites. The flysch sediments of latest Ordovician and Early Silurian age were produced by the terminal orogeny and support the inference that this event was limited to northeasternmost parts of the mobile belt.

3. After the Middle to Late Ordovician collision, the margin of the North American Continent must have lain in the present offshore area. This configuration and the scarcity of strata younger than early Late Silurian and older than Late Carboniferous make it impossible to develop an adequate interpretation for the Late Silurian–Early Carboniferous orogeny and its various phases. Perhaps the intense compressional deformations, followed by a radical change in tectonic and depositional regime, indicate another collision, more extensive than the Middle to Late Ordovician event. In addition, there also is evidence for strike slip in the magmatic belt that may have been caused by transform motions at the continental margin, but age and importance of these movements are uncertain. Structural evidence indicates that the NE-Ellesmere sector was transported southwest on the Yelverton thrust (Fig. 1B) and on transcurrent faults bounding it on the northeast and southeast but gives no indication of the magnitude of offset. Sedimentary relationships between the Hazen Trough and the NE-Ellesmere sector do not require such offset but permit up to perhaps 150 km of displacement. This inference is based on the facts that paleocurrent directions in northeastern parts of the trough indicate a source area to the north, and that adjacent parts of the NE-Ellesmere sector indeed contain unconformities and coarse clastic sediments of corresponding age.

Relationships with Caledonian-Appalachian and Cordilleran Mobile Belts. The Franklinian mobile belt closely approaches the Caledonian-Appalachian belt in northeasternmost Greenland and probably merges with it in the adjacent offshore area. The most obvious characteristic they have in common is the timing of orogenies. The Middle to Late Ordovician orogeny of the NE-Ellesmere sector coincides approximately with the Taconic Orogeny, and both have been interpreted as collisions. The early Ludlovian orogeny of the magmatic belt (Heiberg sector, phase 1) probably is approximately coeval with extensive terminal deformations in the Caledonides that were followed by molasse-type sedimentation in latest Silurian to Early Devonian time (Old Red Sandstone). It is generally assumed that the Caledonides of northeastern Greenland were affected by this event although field evidence indicates only that the deformation is no older than middle Wenlockian (medial Silurian).

Outcrops of the Franklinian mobile belt are separated from outcrops of the Cordilleran belt by a stretch of about 1,000 km covered with water or younger sediments. Common features are (1) the presence of large, ensialic deep-water basins such as Selwyn Basin, Richardson Trough, and Hazen Trough, and (2) evidence for Late Devonian–Early Carboniferous deformation. Seismic surveys have not given any indication of a connection through the present continental shelf area. If it exists—and the overall trend of the mid-Paleozoic orogens suggests that this may be so—it must have lain farther offshore.

In recent years, the hypothesis that the missing link is represented by northern Alaska has gained wide but not universal acceptance. It proposes that northern Alaska originally lay back-to-back with the southwestern Arctic Islands and that it obtained its present position by counterclockwise rotation in the Mesozoic. This hypothesis calls for a systematic comparison of the two regions that could be included either in the present volume or in the one dealing with Alaska.

Carboniferous to Late Cretaceous

During this interval a large depression of extensional origin, the Sverdrup Basin, developed upon the bevelled structures of the mid-Paleozoic orogen—mainly in the region of the Hazen Trough. Its outcrop area is about 1,300 km long and up to 400 km wide and contains about 12 km of strata in the depocentre. Geophysical evidence indicates that the crust beneath the basin thins to 32 km in the axial region. The basin was bordered on the northwest by the Sverdrup Rim, a fault-bounded domain that was intermittently positive. Clastic sediments in the basin were derived mainly from terranes to the south and east and only to a minor extent from the northwest.

After basin-wide initial clastic deposition in the Late Carboniferous, concentric facies patterns developed in Late Carboniferous to Late Permian time. Subtidal evaporites, deep-water mudrocks, spicular chert, and limestone dominated the axial region while coarse clastic sediments or mixed carbonate and clastic sediments were laid down at the margins. The southwestern parts of the basin were weakly deformed in latest Carboniferous to Early Permian time (Melvillian Disturbance).

Mesozoic stratigraphy is presently in a state of revision, necessitated by a wealth of new subsurface and surface information. Suffice it to say that sedimentation was almost entirely clastic and that it demonstrates a complex interaction of eustatic changes in sea level, subsidence, and basin filling.

Basaltic volcanism occurred, in different parts of the basin, in Carboniferous, Permian, and Cretaceous time. Volcanism and intrusion of dykes and sills mark phases of extension. In the southwestern parts of the basin, evaporite diapirs seem to have ascended through the same extensional fractures as the igneous rocks.

Latest Cretaceous and Tertiary

The Sverdrup Rim ceased to be positive and subsided from latest Cretaceous time onward as part of the more extensive Arctic Continental Terrace. The latter received an oceanward thickening, homoclinal succession of Cretaceous to Recent sediments that is up to 3 km thick.

In the Sverdrup Basin three phases of deformation occurred that were most pronounced in Axel Heiberg and Ellesmere islands.

1. Broad basins and uplifts formed during the latest Cretaceous and early Tertiary. Syntectonic clastic sediments, locally more than 3 km thick, accumulated in the basins in lacustrine, fluvial, and shallow marine settings.

2. A phase of compressive deformation, the Eurekan Orogeny, took place in mid-Tertiary (post–middle Eocene, pre–late Miocene) time. In east-central Ellesmere Island it was followed by strike-slip faulting that probably was related to inferred sinistral strike-slip along Nares Strait. Strike-slip also seems to have occurred in the environs of the magmatic belt in Ellesmere Island, but sense and magnitude of displacement there are uncertain.

3. Vertical movements that continued in late Tertiary time produced the present topographic relief. The present major mountain ranges have negative Bouguer anomalies, indicating that they are in approximate isostatic equilibrium. The contours of a large anomaly in northeastern Ellesmere Island correlate roughly with topography but cut across major thrust faults of mid-Tertiary age at high angles. This seems to indicate that the anomaly postdates the mid-Tertiary compressive deformation and is the result of relatively young, deep-seated processes.

The structure of the southeastern parts of the Arctic Platform, including Baffin and Bylot islands and Lancaster Sound and also of some southwestern parts is characterized by extensive normal faults, forming grabens and half-grabens. The faults cut strata as young as early Tertiary and probably are post–middle Eocene in age. It is commonly assumed that the faulting began in Late Cretaceous time, but the evidence for this is very indirect.

A problem of importance, not only for the Innuitian volume but for tectonic theory in general, is the so-called Nares strait dilemma, an apparent conflict between geological and geophysical inferences concerning the amount of strike-slip along the Nares Strait. Early exploration indicated no significant offset. Subsequently, geophysical studies demonstrated that Baffin Bay has oceanic crust. Assuming that it was produced by Tertiary seafloor spreading, a large amount of offset (160 to 400 km according to different authors) was required. However, recent compilations, made jointly by geologists working in Greenland and Ellesmere Island, have shown that no more than a few tens of kilometres of offset are possible. This seems to rule out seafloor spreading and seems to imply that Baffin Bay is another mid-Tertiary graben, but raises the problem how crustal thinning from perhaps more than 30 km to about 4 km was accomplished with only a moderate amount of lateral extension.

Preliminary Outline

Introduction
Physiographic provinces
History of geological exploration
Tectonic framework
Geophysical characteristics
  Gravity field
  Aeromagnetic field
  Conductivity anomalies
  Seismology and deep structure
  Seismicity
  Heat flow
Precambrian
  Basement of Arctic Platform
  Proterozoic of Innuitian Orogen
Cambrian to Devonian deposition, Franklinian mobile belt, and Arctic Platform
  Shelf
  Hazen Trough
  Magmatic belt
  Middle to Late Devonian molasse basin
Ordovician to Early Carboniferous deformation, metamorphism, and plutonism, Franklinian mobile belt
  Deformation
    Magmatic belt
      Middle to Late Ordovician orogeny
      Late Silurian–Early Carboniferous orogeny
    Hazen Trough and shelf
      Late Silurian–Early Devonian epeirogenic movements
        Boothia Uplift
        Bache Peninsula Arch
      Ellesmerian Orogeny (Late Devonian or earlier–Early Carboniferous)
        Arctic islands
        Greenland
  Metamorphism
  Plutonism
Carboniferous and Permian, Sverdrup Basin
  Deposition
  Melvillian Disturbance (Late Carboniferous–Early Permian)
Mesozoic, Sverdrup Basin, and Arctic Platform
  Deposition
  Dykes and sills
  Evaporite diapirism
Tertiary deposition, Canada
Carboniferous to Tertiary deposition, Wandel Sea Basin (Greenland)
Tertiary deformations
  Early Tertiary epeirogeny
  Mid-Tertiary orogeny (Eurekan Orogeny of Arctic islands and corresponding events in Greenland)
  Late Tertiary epeirogeny and physiographic development
Quaternary
Summary of tectonic history and interpretation
  Cambrian to Early Carboniferous
  Carboniferous to Late Cretaceous
  Latest Cretaceous to Holocene
Resources
  Petroleum
    Geology
    Geochemical studies
  Coal
  Mineral deposits
In pocket: Geological map, scale 1:2 million
  Cross-sections
  Geotectonic correlation charts

Comments on Organization of Subject Matter

The preliminary outline follows a tentative generalized model developed for the Canadian regional volumes. Most headings are either self-explanatory or clear from the geological introduction, but a few require comment.

Geophysical Characteristics. The purpose of this chapter would be (1) to give an overview of the kind of geophysical information available for this region and (2) to provide factual background information for later chapters. The geophysical information would be presented as maps or diagrams, accompanied by brief, not too technical explanations of the methods used, the density and reliability of the data, and the meaning of the results in general terms. Farther-reaching interpretations could be incorporated into later chapters dealing with deformation, tectonic history, etc.

General Treatment of Cambrian to Tertiary Geology. The purpose of these chapters is to describe the geological history of the region in the form of separate accounts, for chosen time intervals, of such major processes as deposition, deformation, metamorphism, and intrusion. The subject of deposition will occupy the largest amount of space and will be dealt with by eight or more authors and therefore requires a somewhat uniform approach. It should consist of two sections of unequal length, as follows.

1. Time-stratigraphic framework and faunal zonations: The epochs, stages, and perhaps substages used and

the relevant faunal zonations are presented in tabular form, together with a brief explanation and with references to all pertinent biostratigraphic literature, but there will be no space for discussion of biostratigraphic detail.

2. Lithofacies and depositional history: This section should provide (a) comprehensive descriptions of major lithofacies; (b) a clear picture of their relationship in space and time; and (c) an overview of paleogeographic and basin development as inferred from the depositional record. Most of this information will have to be conveyed by illustrations.

The lithofacies mentioned are informal units defined by the authors for the purpose of the description of chosen time intervals and designated by descriptive names (e.g., starved-basin phase of Hazen Trough). Formal stratigraphic units, especially formations and groups, will appear on the geotectonic correlation charts accompanying all Canadian volumes; accordingly, they should also appear on columnar sections, and their relationship with the informal lithofacies should be mentioned in the text (if convenient, in tabular form).

Deformations are discussed by age and region and the following order may be followed: (1) age of deformation with relevant stratigraphic or isotopic evidence; (2) general character and domains; (3) structural style and major structural elements within domains. In many areas structures of different ages are superimposed upon each other and so the same cross-sections will have to be referred to in different chapters.

Coordination of Canadian and Greenland Geology. This subject will have to be discussed in detail with the geologists working in Greenland. Generally, the early Paleozoic mobile belt trends across the Nares Strait without major changes and therefore requires a unified overall treatment, although some subjects (e.g., deformation) can be split. The present outline is based on the geology of the Arctic islands but can readily be modified to accommodate the forthcoming new information from Greenland. On the other hand, it probably will be more convenient to discuss the Carboniferous to Tertiary geology separately because the Sverdrup Basin of the Arctic islands and the Wandel Sea Basin of northeasternmost Greenland are far apart and differ in depositional and structural history.

Summary of Tectonic History and Interpretation. The tectonic history essentially will have been told in the preceding chapters, but in a disrupted manner and from different viewpoints; here it is summarized in a concise, coherent fashion and interpreted—insofar as this is possible. Because of the fragmentary and indirect nature of the evidence, a variety of plate-tectonic models will have to be discussed and the remaining problems will have to be defined. This chapter should also include a discussion of the relationships with the Appalachian and Cordilleran mobile belts and with Arctic Alaska.

Resources. This chapter will provide a useful overview of the fuel and mineral resources of this frontier region but will be confined to known deposits and proven reserves. Emphasis should not be on quantitative estimates but on geological controls.

Acknowledgments

I am indebted to A. F. Embry, R. Thorsteinsson, and J. O. Wheeler for critical reviews and improvement of the manuscript.

References Cited

Escher, A., compiler, 1970, Tectonic/geological map of Greenland by the Geological Survey of Greenland: Copenhagen, Denmark, Geological Survey of Greenland.

Thorsteinsson, R., and Tozer, E. T., 1970, Geology of the Arctic Archipelago, *in* R.J.W. Douglas, ed. Geology and economic minerals of Canada: Geological Survey of Canada, Economic Geology Report no. 1, p. 547–590.

Geology of the Canadian Appalachians

Harold Williams
Department of Geology
Memorial University of Newfoundland
St. John's, Newfoundland
Canada A1B 3X5

Introduction

The Appalachian orogen in Canada is a northeast-trending belt of upper Precambrian and Paleozoic rocks that were deformed in late Precambrian (Avalonian), Middle Ordovician (Taconian), Devonian (Acadian), and Permian-Carboniferous (Alleghanian) times. The Canadian Appalachians lie southeast of the Canadian Shield and Interior Platform of North America, and the system is bounded to the east by the Atlantic Ocean (Figs. 1 and 2). Toward the southwest, its rocks can be traced through the southeast United States to Alabama. Toward the northeast, its submerged offshore extensions are covered by Mesozoic and Cenozoic sediments of the Atlantic continental margin. Syntheses of the orogen and its offshore extensions in Canada and the United States are keyed to a similar format, using many of the same maps and illustrations for purposes of continuity and clarity.

Rocks and structures like those of the Canadian Appalachians reappear on the opposite side of the Atlantic Ocean in the British Caledonides and extend northward to East Greenland and Scandinavia. An overview of the East Greenland Caledonides will be included in the Canadian Appalachian volume, because these rocks are a natural extension of those along the west flank of the Appalachian orogen and because of contiguity with the North American continent.

The present stratigraphic and tectonic analysis of rocks of the Canadian Appalachians is based on both a temporal and spatial breakdown. Temporally, its rocks can be subdivided into four broad categories: those of late Precambrian–early Paleozoic age, those of middle Paleozoic age, those of late Paleozoic age, and those of early Mesozoic (Triassic) age. These temporal divisions are separated by three major orogenic episodes in the development of the orogen, namely, the Taconian, Acadian, and Alleghanian orogenies (Fig. 3).

Rocks of late Precambrian–early Paleozoic age show sharp contrasts in thickness, facies, and structural style across the Canadian Appalachians, and this realization has led to several stratigraphic-tectonic zonations. Most recent geologic syntheses and models are based upon such zonations. Contrasting rocks of this earliest time frame define five major geologic provinces (also referred to as zones or belts, some of which are suspect terranes). Several are composite. From west to east these are designated Humber, Dunnage, Gander, Avalon, and Meguma (Fig. 2). Only the Humber and Avalon zones are traceable along the full length of the orogen in Canada and the United States.

Middle Paleozoic rocks of the Appalachian orogen do not reflect the contrasts so evident in older rocks. A sub-Silurian unconformity across most of the Humber, Dunnage, and Gander zones indicates juxtapositioning and a loss of stratigraphic distinctiveness following Taconian orogeny. A lack of clear evidence for Taconian deformation in the Avalon and Meguma zones suggests that their

Figure 1. Regional setting of the Canadian Appalachians.

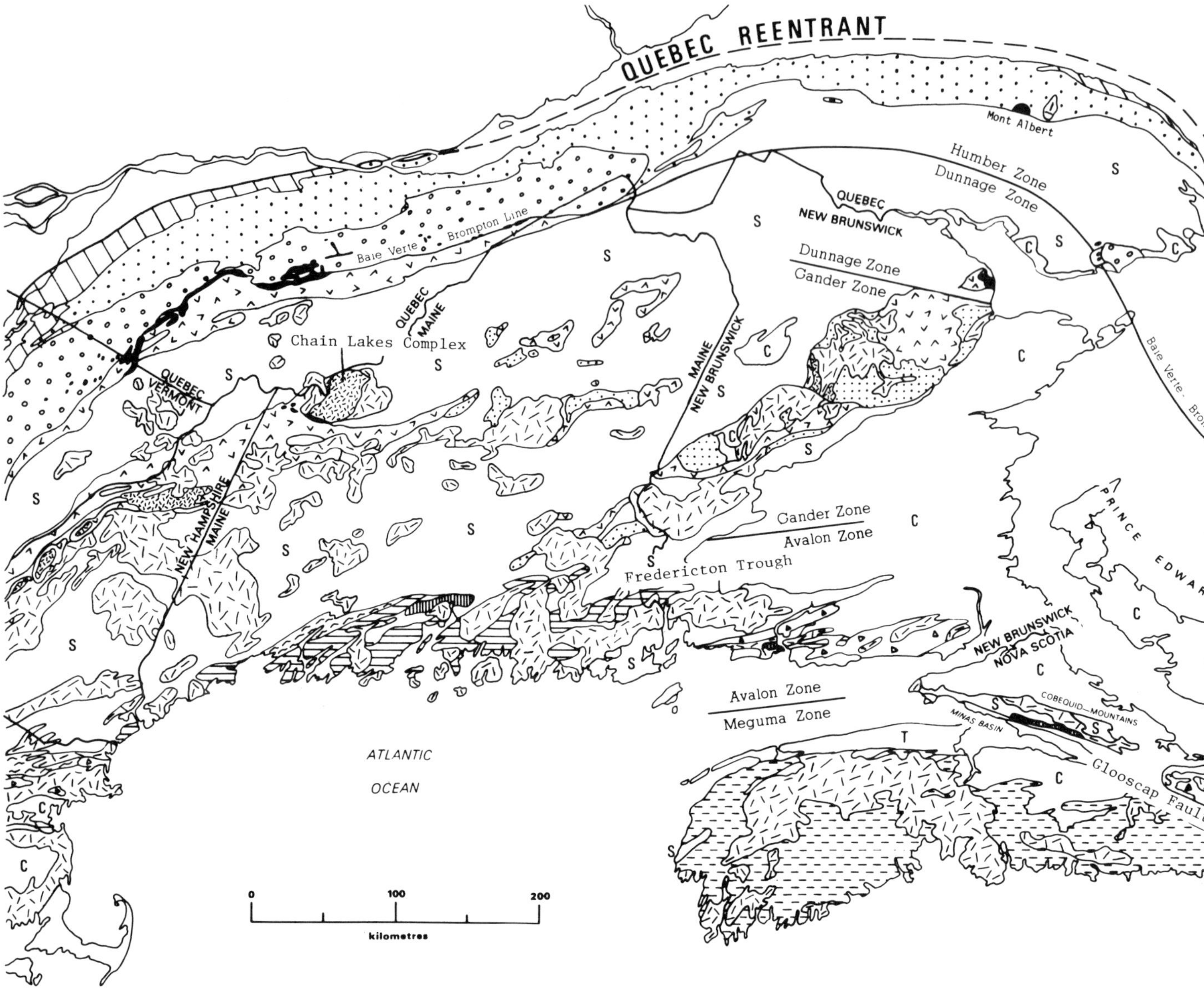

Granitic to dioritic and gabbroic intrusions (mainly Mid-Paleozoic but ranging in age from Late Precambrian to Jurassic)

Metamorphic rocks of unknown affinity (mainly Mid-Paleozoic)

SUCCESSOR BASINS OF APPALACHIAN OROGEN

T — Terrestrial sedimentary and volcanic rocks related to opening of present Atlantic Ocean (Triassic-Jurassic)

C — Late orogenic to post orogenic mainly terrestrial sedimentary and minor volcanic rocks (Carboniferous-Permian)

S — Mainly clastic sediments and terrestrial volcanic rocks deposited in troughs and basins across vestiges and deformed margins of Iapetus (Middle Ordovician to Devonian)

Figure 2. Tectonic stratigraphic zonation of the Canadian Appalachians (after Williams, 1979).

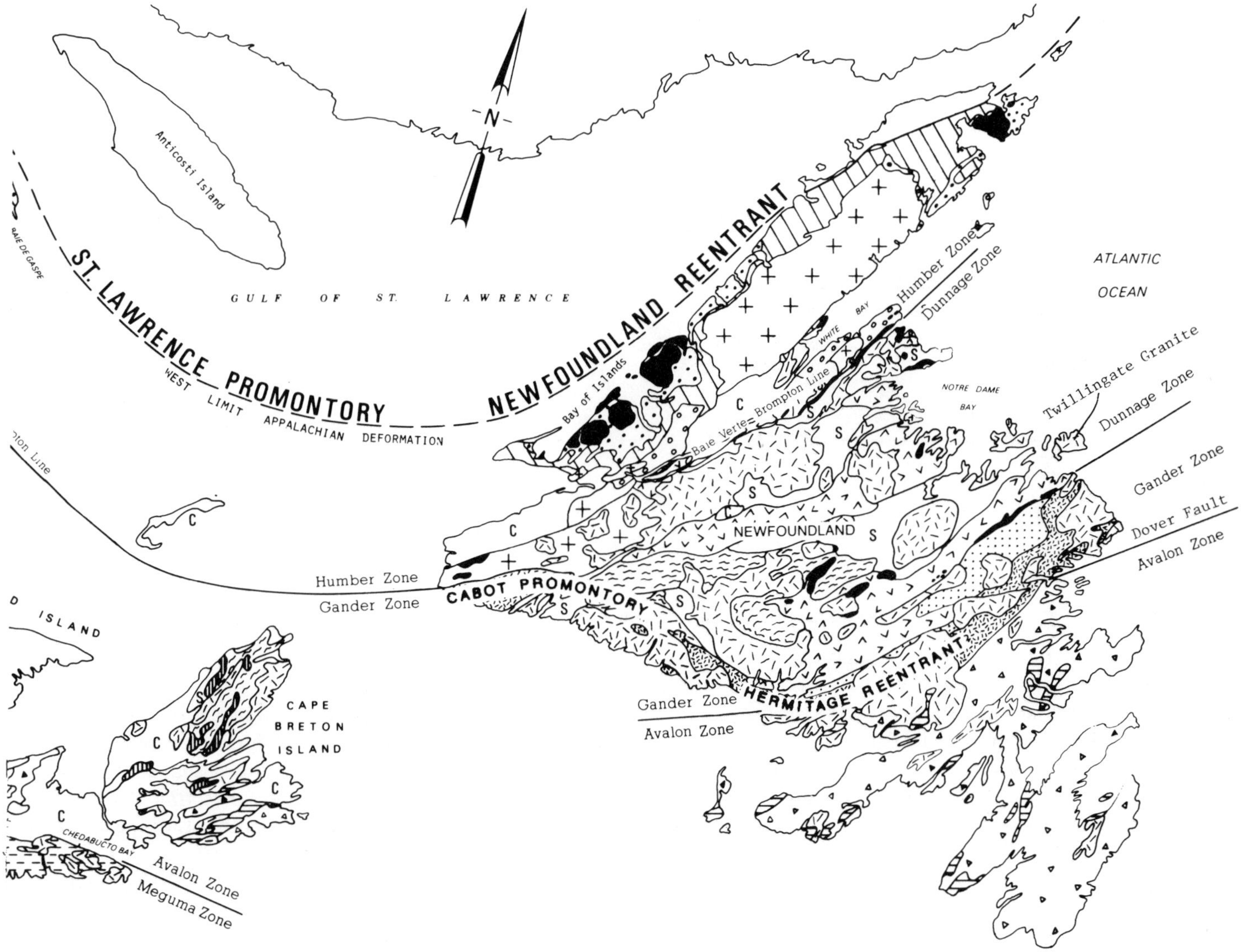

HUMBER ZONE	DUNNAGE ZONE	GANDER ZONE	AVALON ZONE	MEGUMA ZONE
ANCIENT CONTINENTAL MARGIN OF EASTERN NORTH AMERICA	VESTIGES OF IAPETUS OCEAN	EASTERN MARGIN OF IAPETUS	AFRICAN EUROPEAN ELEMENTS OF APPALACHIAN OROGEN	

Humber Zone

- Transported mainly ophiolitic rocks (Late Cambrian to Early Ordovician)
- Transported mainly sedimentary rocks (Late Precambrian to Middle Cambrian)
- Mainly carbonate and overlying clastic sediments (Cambrian to M. Ordovician)
- Clastic sediments and bimodal volcanic rocks (Late PC to Early Cambrian)
- Grenvillian basement

Dunnage Zone

- Marine volcanic rocks, slate, chert, greywacke and melange (Mainly Ordovician)
- Parautochthonous ophiolitic rocks (mainly Early Ord.)

Gander Zone

- Clastic sedimentary rocks (Early Ord. and Older)
- Metamorphosed cover rocks and sialic basement

Avalon Zone

- Shale, sandstone, minor limestone (Cambrian to Early Ordovician)
- Sedimentary and volcanic rocks (Late Precambrian-Hadrynian)
- Marbles and quartzites with assoc. gneisses and migmatites (Late PC and older)

Meguma Zone

- Greywacke and slate (Early Ord. and older)

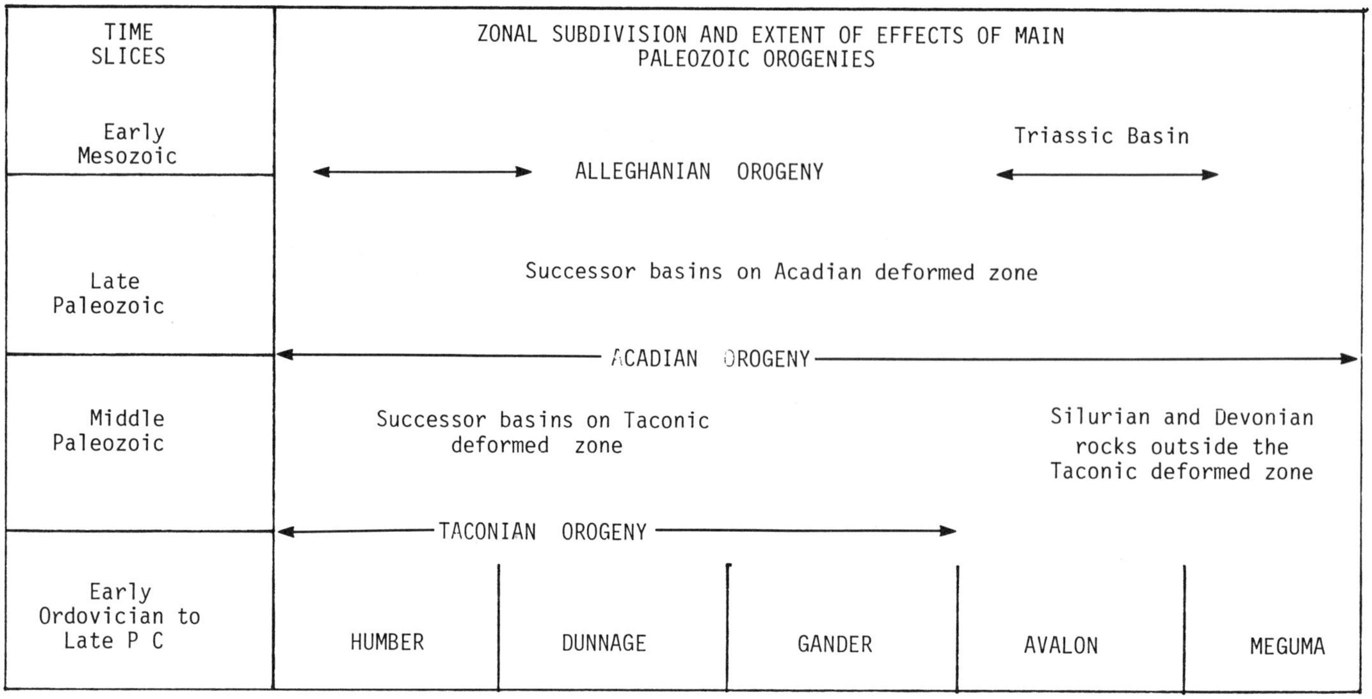

Figure 3. Time-slice and spatial subdivisions of rocks of the Canadian Appalachians.

amalgamation into the orogen may have been later. However, their Silurian and Devonian rocks are more akin to rocks of the same age elsewhere across the orogen, rather than to the facies of underlying rocks.

Carboniferous-Permian rocks of the Canadian Appalachians are mainly terrestrial cover sequences upon the Acadian deformed zone. They are flat lying over large areas, but locally they are moderately to intensely deformed.

Early Mesozoic, mainly Triassic rocks, are terrestrial red beds and volcanics deposited upon the Alleghanian deformed zone.

Recent models for the evolution of the Appalachian orogen involve the opening and closing of a late Precambrian–early Paleozoic Iapetus Ocean. The western margin of Iapetus is the well-defined Humber Zone, and its stratigraphic and sedimentologic analyses are among the most sophisticated for any ancient continental margin. Ophiolitic vestiges of Iapetus define the Dunnage Zone, and the Gander Zone has been viewed traditionally as the eastern margin of Iapetus. The Avalon and Meguma zones are more difficult to rationalize in a simple Iapetus model.

There are no late Precambrian–early Paleozoic linkages between the Humber and more outboard zones. There is no assurance therefore that they were ever previously connected to or evolved in sight of the North American continent or one another. These paleogeographic uncertainties are stressed in this analysis, along with the stratigraphic and sedimentologic evidence for the accretionary history of the orogen. Finally, an attempt is made to relate structural, metamorphic, plutonic, and other analyses to the stratigraphic analysis.

Thematic maps, all compatible at 1:2,000,000, are planned to accompany separate chapters devoted to special features of the orogen. These will allow visual contrasts and comparisons between stratigraphic zones, time of deformation, plutonism, metamorphism, and other parameters important in the development of the orogen. It is anticipated that compilations of this type will do much to improve understanding of the existing data base.

This report is based upon earlier syntheses of the Canadian Appalachians (Williams, 1979; Williams and Hatcher, 1982). All material presented here is taken directly from these articles and the reader is directed to them for complete referencing.

Late Precambrian–Early Paleozoic Development of the Appalachian Orogen

Humber Zone: Ancient Continental Margin of Eastern North America

The Humber Zone consists of a Grenvillian basement overlain by mixed volcanic and clastic sedimentary rocks, in turn overlain by a prominent Cambrian-Ordovician carbonate sequence. Its western margin is defined by the limit of Appalachian deformation. Its eastern margin with the Dunnage Zone is the Baie Verte–Brompton Line, a steep structural belt characterized by ophiolite occurrences (Fig. 2).

The evolution of the Humber Zone involved rifting of Grenvillian basement with concomitant mafic dike intrusion, mafic volcanism, and the accumulation of thick clastic sequences. A thinner sequence of sedimentary rocks, locally supratidal, formed at the stable continental shelf. These sediments thicken eastward and record an upward transition from immature arkosic sandstones to mature quartzites, overlain by a prominent carbonate sequence. Coeval deposits at the continental slope and rise consist of coarse limestone breccias of westerly provenance and turbidites and shales with distinctive faunas.

Destruction of the ancient continental margin began toward the end of Early Ordovician time. The first indication of instability is shown by an ancient karst topography developed across an upwarped carbonate shelf. Later subsidence is recorded first by the deposition of deeper-water carbonates across the disturbed shelf, then by a flood of clastic rocks from the east. These are structurally overridden by a sequence of contrasting rock assemblages in separate slices. The structurally lowest slices consist of sedimentary rocks from the nearby continental margin, and the highest ophiolitic slices represent farther traveled oceanic crust and mantle. The geometry of the slices and facies relationships indicate that the structural pile was assembled from the east, possibly through peeling of successively landward sections. Ordovician deformation and prograde metamorphism affected eastern parts of the Humber Zone. These effects probably accompanied ophiolite obduction and an attempt to submerge the North American margin beneath an overriding oceanic plate, as depicted in Figure 4.

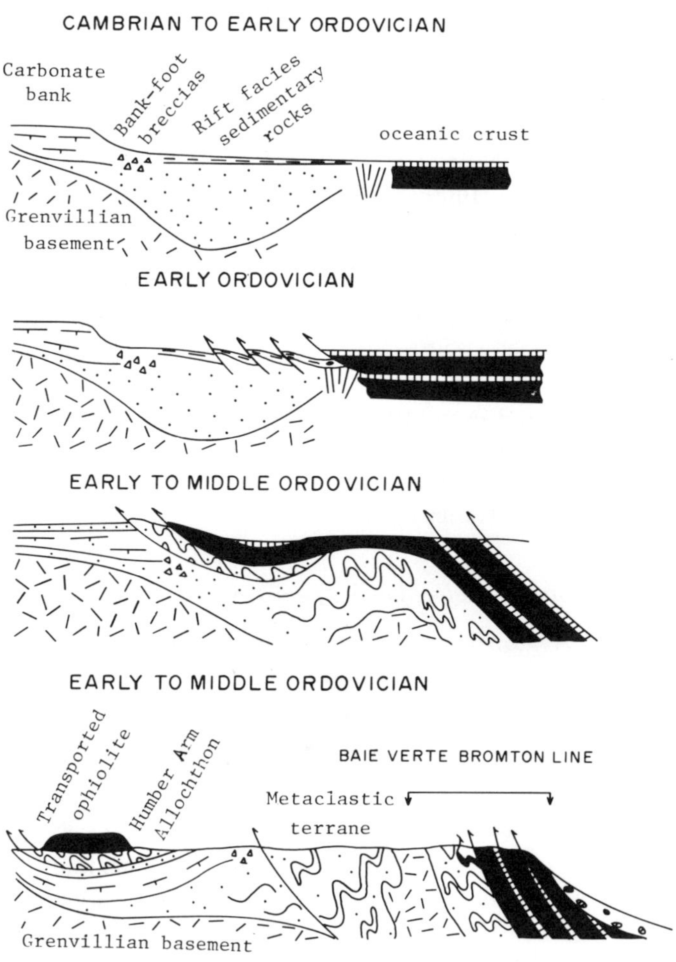

Figure 4. Model for Taconian orogeny and the destruction of the ancient continental margin of eastern North America (modified from Williams, 1979).

The sinuous course of the Humber Zone is interpreted as the result of an initially orthogonal continental margin along which rectilinear rifts were linked orthogonally by transform faults. The form of the modern Atlantic margin mimics the form of the Humber Zone, implying modern continental breakup by reactivation of ancient rifts and transforms. Parallelism of the Humber Zone and the Grenville structural province further implies that the form of the Paleozoic margin reflects even older structures.

Palinspastic restoration of rocks in the Humber Zone indicates an ancient continental margin with horizontal dimensions comparable to the modern Atlantic margin.

Several features of the Canadian Appalachians suggest that the present subsurface edge of the Humber Zone is at or near the Baie Verte–Brompton Line, although palinspastic considerations suggest that it lay much farther east before Paleozoic orogenesis. Thus, the form of the Humber Zone, the course of its lithofacies belts, and the position of the Taconian structural front are all roughly parallel and nearly coincident with the gravity gradient across the orogen. These features, coupled with the lack of a wide Alleghanian thrust belt along the west flank of the Canadian Appalachians, suggest much less telescoping compared to the Southern Appalachians of the United States.

Dunnage Zone: Vestiges of Iapetus

Rocks of the Dunnage Zone represent vestiges of an oceanic domain, as they are marine volcanic and sedimentary rocks that overlie the ophiolite suite of rock units. The Dunnage Zone is bounded by the Baie Verte–Brompton Line to the west. Its eastern boundary is marked by ophiolite occurrences of the Gander River Ultramafic Belt in Newfoundland, and it coincides with the Rocky Brook–Millstream Fault of New Brunswick. It is hidden by Silurian and younger rocks in southeast Quebec. Where volcanic rocks similar to those of the Dunnage Zone occur above sialic basement of the New Brunswick Gander Zone, the true Dunnage-Gander distinction can only be made by ascertaining whether or not the volcanic rocks overlie and ophiolitic or sialic basement.

Most ophiolitic complexes of the Dunnage Zone are overlain by thick volcanic and volcaniclastic sequences that are interpreted as ancient analogues of modern island arcs. Melanges are common throughout the length of the zone. Most consist of volcanic and greywacke blocks in an Ordovician shale matrix. Some are interpreted to mark the sites of ancient oceanic trenches and subduction zones, but few exhibit the structural and metamorphic features typical of subduction melanges.

The Arnold River gneisses of southeastern Quebec and the correlative Chain Lakes Complex of Maine appear to be out of structural and metamorphic context with respect to surrounding ophiolite complexes of the Dunnage Zone. They are viewed, therefore, as a suspect terrane within the oceanic tract. The Twillingate Granite and associated rocks in northeast Newfoundland may represent a similar alien terrane.

At the Baie Verte–Brompton Line, ophiolite complexes of the Dunnage Zone are structurally and metamorphically gradational with ophiolitic melange and metaclastic rocks that are an integrl part of the Humber Zone. Locally in northeast Newfoundland, Dunnage ophiolitic rocks are thrust eastward upon clastic rocks of the Gander Zone. Where rocks of the Dunnage Zone are less deformed and metamorphosed than those of nearby parts of the Humber and Gander Zones, they may be protected by the form of opposing continental margins. Thus, in northeast Newfoundland, the widest remnant and best preserved rocks of the Dunnage Zone lie between matching reentrants in the bordering Humber and Gander zones. Furthermore, the Dunnage Zone is absent at matching promontories (Fig. 2).

The age of volcanic rocks in the Dunnage Zone indicates that island arcs were active there during the passive development of the Humber Zone to the west. A cover of Middle Ordovician shales above the volcanic rocks indicates that the evolution of the volcanic islands ceased at the time of destruction of the North American continental margin and final emplacement of transported assemblages across the Humber Zone. The relationships imply also that the North American margin was destroyed by collision with an island arc, and the Dunnage Zone was juxtaposed with the Humber Zone in Early to Middle Ordovician time.

Seismic refraction studies indicate that the Newfoundland Dunnage Zone is underlain by an overthickened dense crust. This is further suggested by a positive Bouguer gravity anomaly across the zone. Its rocks therefore appear to be rooted, rather than to constitute a thin structural slice above sialic crust of the Humber Zone. Rocks of the Dunnage Zone seem to have been caught in an Ordovician vise whose jaws are represented by the North American continental margin (Humber Zone) to the west and the Gander Zone to the east. Subsequent Acadian (Devonian) deformation reflects this local Ordovician model with westward polarity of structures across the Humber Zone and eastward polarity across the Gander Zone. Palinspastic restoration of the widest segment of the Dunnage Zone in northeast Newfoundland indicates an original oceanic width of at least 1,000 km.

Gander Zone: Possible Eastern Margin of Iapetus

Rocks of the Gander Zone are mainly pre-Middle Ordovician metaclastics that are in most places polydeformed and metamorphosed, thus resembling metaclastic rocks at the east margin of the Humber Zone. Migmatites, granitic gneisses, and some foliated megacrystic granites along the eastern margin of the zone are interpreted either as basement to the clastic cover rocks or as deeper migmatitic levels of the cover sequence. Paleozoic megacrystic biotite granites and garnetiferous muscovite leucogranites and associated pegmatites are everywhere common throughout this zone..

In Newfoundland, the eastern margin of the Gander Zone is defined by the Dover and Hermitage Bay faults. In New Brunswick, the Gander-Avalon zone boundary is covered by Silurian greywackes of the Fredericton Trough and by Carboniferous strata.

The thick polydeformed arenaceous sequence of the Gander Zone is interpreted as a prism of sediment built up parallel to a shoreline, possibly the eastern margin of Iapetus. However, the stratigraphic analysis is weak compared to that for the Humber Zone. Where thick volcanic sequences overlie the arenaceous rocks of New Brunswick, the relationships suggest eastward subduction and the evolution of an Andean margin.

Unassailable evidence for an Ordovician linkage between the Gander and more westerly zones is provided by a continuous belt of Upper Ordovician and Silurian rocks in Quebec and New Brunswick (Aroostook-Matapedia groups) that forms an overlap assemblage extending from the Humber Zone of Quebec to the Gander Zone of New Brunswick and Maine.

Avalon Zone: Appalachian Suspect Terrane

The Avalon Zone consists mainly of upper Precambrian volcanic and sedimentary rocks that are relatively undeformed and unmetamorphosed compared to rocks of nearby parts of the Gander Zone. Locally, the rocks are cut by Precambrian intrusions with important sub-Cambrian unconformities. In other places the upper Precambrian rocks pass upward with structural conformity into Cambrian shales with distinctive Atlantic realm trilobite faunas. The oldest Precambrian rocks of the Avalon zone are marbles and quartzites, which underlie the upper Precambrian volcanic and sedimentary rocks in Nova Scotia (George River Group) and New Brunswick (Green Head Group).

Rocks like those of the Avalon Zone occur offshore at the Grand Banks and Flemish Cap, indicating that in the north this zone is wider than the combined widths of other zones in the Canadian Appalachians. In New Brunswick and Nova Scotia, the Avalon Zone is much narrower and it is separated from the more easterly Meguma Zone by the Glooscap Fault. Offshore extensions of the Glooscap zone boundary are marked by prominent magnetic and gravity anomalies.

The Avalon Zone contains a variety of diverse Precambrian geologic units. These include tillites, marine sequences with late Precambrian faunas, and volcanic suites that range chemically from tholeiitic to calc-alkaline and bimodal. Precambrian ophiolitic melange and ophiolite suites occur in Avalonian correlatives of Rhode Island and Northwest Africa, respectively. The Avalon Zone may therefore include a number of smaller provinces. If composite, it was assembled in the late Precambrian as its Cambrian rocks and faunas are similar throughout its extent.

Although the geology of the Avalon Zone is well known, the controlling features of its late Precambrian evolution are still doubtful. Its late Precambrian development may relate to rifting and the initiation of the Iapetus cycle. Alternatively, it may relate to a subduction episode and a cycle that preceded the opening of Iapetus.

There are no confirmed pre-Silurian linkages between the Avalon Zone and nearby zones. In addition, the Avalon Zone is unaffected by Taconian orogeny, the effects of which are evident in more westerly zones. This negative evidence suggests that its time of accretion may be somewhat later than that of more westerly zones. In New Brunswick, the Gander-Avalon zone boundary coincides with a Silurian marine trough. In Newfoundland, Silurian-Devonian sedimentologic linkages and Devonian intrusions establish a mid-Paleozoic connection between the Gander and Avalon zones.

Obvious truncation of the Avalon Zone along its faulted boundaries between Newfoundland and New Brunswick and the common occurrence of wide and steep mylonite zones along its boundary with the Gander Zone suggest major transcurrent displacements.

Meguma Zone: Continental Embankment and Appalachian Suspect Terrane

The Meguma Zone is restricted to Nova Scotia, but it may extend offshore to the southeast end of the Grand Banks. Its oldest rocks, the Meguma Group, are Cambrian-Ordovician strata up to 13 km thick. The succession consists of a lower greywacke unit overlain by a shaly unit that contains Early Ordovician graptolites in its upper parts. Magnetic patterns suggest that the Meguma Group terminates offshore before reaching the edge of the Nova Scotia continental shelf.

Paleocurrent analyses indicate that the Meguma Group was deposited by currents that flowed from the southeast toward the northwest. The makeup and volume of sediments further suggest a low-lying metamorphic provenance of continental dimensions. Accordingly, the Meguma Group is interpreted as a continental embankment. Recent geochronologic studies on gneisses that are interpreted as possible basement to the Meguma Group indicate Pb/U ages up to 2,500 million years (J. D. Keppie, 1981, personal commun.).

If the Meguma Zone represents another ancient continental margin now contained within the Appalachian orogen, then this implies a Paleozoic ocean that once existed between the Avalon Zone and Meguma Zone. No vestiges of such an oceanic tract remain, and evidence of its western margin, that is, along the east side of the Avalon Zone, is entirely lacking.

The Meguma Zone was first deformed by Acadian orogeny, and its folded rocks are cut by large Devonian intrusions. These are distinct from intrusions of the nearby Avalon zone, and they nowhere straddle the Avalon-Meguma zone boundary. The earliest firm linkage between the Avalon and Meguma zones is provided by a Carboniferous overlap assemblage. Alleghanian deformation is localized along the Avalon-Meguma zone boundary, and its steep, straight character implies major transcurrent movements.

Middle Paleozoic Development of the Appalachian Orogen

Silurian and Devonian rocks of the Appalachian orogen are mixed marine and terrestrial sedimentary and volcanic rocks, with local facies distribution that is entirely unrelated to the earlier Paleozoic zonation of the system. These rocks are less conducive to meaningful zonal separation compared to older rocks of the orogen. The mid-Paleozoic development of the orogen was entirely different, therefore, from its Cambrian-Ordovician development. A sub-Silurian or sub-Middle Ordovician unconformity across the Humber, Dunnage, and Gander zones indicates destruction of earlier continental margins and the Iapetus oceanic tract. Even the outboard Avalon and Meguma zones, of unknown mid-Paleozoic location with respect to western zones, contain mid-Paleozoic rocks that are more akin to those deposited farther west than to distinctive underlying rocks that define the zones in which they occur.

The mid-Paleozoic development of western parts of the system is viewed, therefore, as the history of deposition and deformation of successor basins, which formed across the already destroyed margins and oceanic tract of Iapetus. Possibly, the Avalon and Meguma zones were still far removed from North America. If so they also experienced major environmental changes compared to conditions that prevailed during their Cambrian and Ordovician development.

The Cambrian-Ordovician evidence for the development of the Humber Zone continental margin and adjacent oceanic crust of Iapetus is virtually unassailable. However, no comparable evidence exists for the roles of continental margins and an oceanic tract in the Silurian-Devonian development of the Appalachians, and most tectonic elements essential to such roles, that is, deposits typical of continental margins, ophiolite suites and melanges of appropriate age, volcanic arc sequences, or marine deposits upon undisturbed ophiolite suites are unknown in the Silurian-Devonian record. Possibly major oceans were destroyed at the Gander-Avalon and Avalon-Meguma zone boundaries. However, there is no evidence of their presence in the stratigraphic record, and all that can be said is that these outboard terranes were truncated by major faults and emplaced finally by transcurrent movements.

Acadian orogeny affected the entire Canadian Appalachians, including the earlier Taconian deformed zones. Its characteristic upright structures contrast with the subhorizontal structures and major horizontal displacements associated with Taconic movements. Almost all of the major granitic plutons were emplaced in this later phase of orogenic development, and regional metamorphism was intense in many places. Yet the controlling features of this phase of orogenesis remain enigmatic.

Late Paleozoic Development of the Appalachian Orogen

Carboniferous rocks of the Canadian Appalachians are mainly terrestrial sandstones and conglomerates with local thick marine sandstone-shale sequences and minor volcanic rocks. These form a cover upon the Acadian deformed zone that involved the entire exposed orogen. The Carboniferous rocks are mainly undeformed, except along the Avalon-Meguma zone boundary in southeast New Brunswick and Nova Scotia, and along a northeast-trending narrow zone in western Newfoundland.

Penetrative Carboniferous deformation in New Brunswick and amphibolite facies metamorphism and plutonism in Carboniferous rocks along strike in eastern Massachusetts appear to be confined to areas underlain by Avalonian basement. The same is true for certain deformed zones farther south in the Carolina Slate Belt of the United States and in Hercynian deformed zones of Europe. This circumstance may reflect a collisional event in eastern and/or offshore parts of the orogen.

Early Mesozoic (Triassic) Development of the Appalachian Orogen

Triassic rocks of the exposed Canadian Appalachians are restricted to a fault-bounded half-graben in Nova Scotia. These are red beds and basalts that relate to rifting and the onset of the modern episode of Atlantic spreading. The Triassic graben coincides with a faulted zone of deformed Carboniferous rocks and the still earlier Avalon-Meguma zone boundary. The situation is yet another example of the control of earlier structural boundaries on the position of more recent features.

Accretionary History of the Appalachian Orogen compared to Times of Deformation, Plutonism, Metamorphism, and Metallogeny

The stratigraphic analysis of accretionary history of the Canadian Appalachians compared to time of deformation, plutonism, and metamorphism is summarized in Figure 5. The Dunnage and Gander zones were juxtaposed with the Humber Zone and with one another during the Middle Ordovician Taconian orogeny (Fig. 5, A and B). The presence of the Avalon Zone cannot be clearly demonstrated until Silurian or Devonian time, and the Meguma Zone appears to have maintained a distinct plutonic history during much of Devonian time. Thus, the accretion of the orogen seems to have progressed from west to east, or from the ancient continental margin of North America outward.

The boundaries of the earliest accreted zones (Humber, Dunnage, and Gander) are marked by broad soft melanges and ophiolite complexes. In contrast, the Gander-Avalon boundary and the Avalon-Meguma boundary are marked by steep faults with local wide zones of ductile deformation and the development of mylonites. Thus, the mechanism of earliest accretion involved obduction and/or subduction, whereas later accretion was controlled by major transcurrent movements. Paleomagnetic disparities further imply major transcurrent movements affecting eastern terranes in the late development the orogen.

Time of deformation and the distribution of the effects of the Appalachian orogenic episodes are consistent with the stratigraphic analysis for the accretionary history of the orogen (Fig. 5). Thus, the more western zones that were attached to the North American margin, or to one another, during Middle Ordovician time are the areas that best exhibit the effects of Taconian orogeny It seems reasonable, therefore, to equate Taconian deformation and the earliest accretionary phase of the orogen. The lack of Taconian deformation in the Avalon and more eastern zones further supports the stratigraphic evidence for later accretion of more easterly zones.

Mid-Paleozoic Acadian orogeny affected the entire orogen. Its effects are most intense east of the Taconian deformed zone. Since its timing corresponds to the best estimate of the time of Avalonian accretion, it is tempting to equate the two.

The effects of late Paleozoic Alleghanian orogeny are restricted to a narrow zone in southeast New Brunswick and along the Avalon-Meguma boundary in Nova Scotia. Its effects are also evident in a narrow zone of relatively mild deformation along the east margin of the Humber

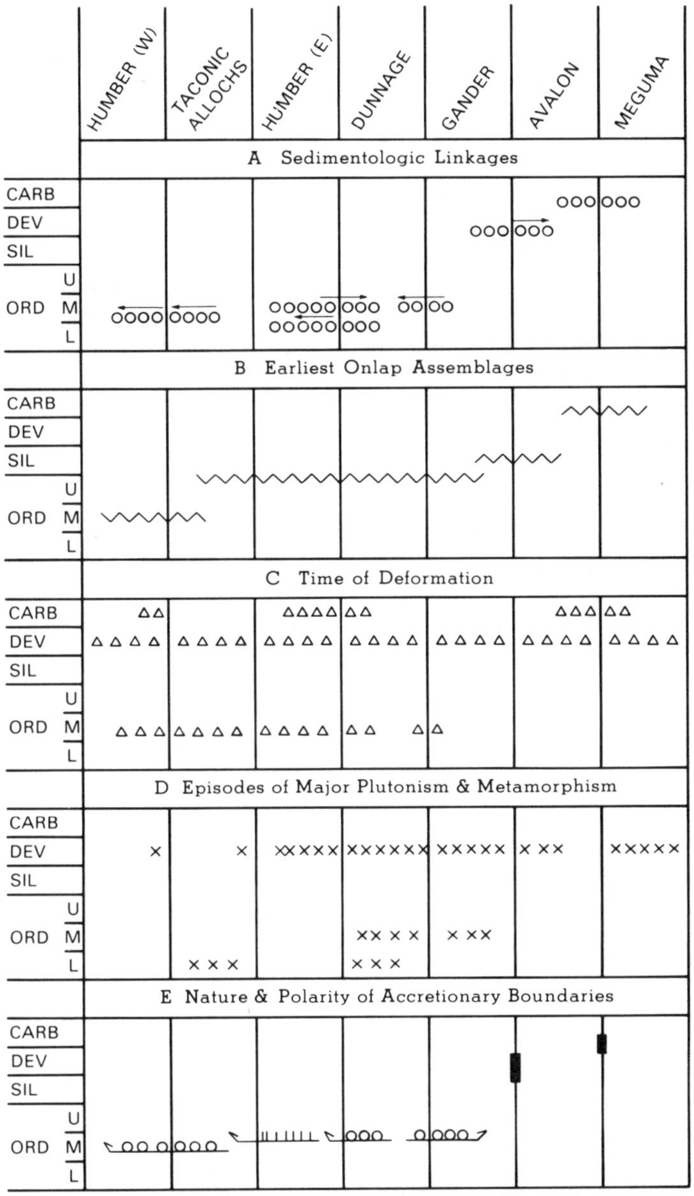

Figure 5. Sedimentologic and stratigraphic analyses of the accretionary history of the Canadian Appalachians compared to time of deformation, plutonism, and metamorphism. A. Sedimentologic linkages (arrow indicates direction of sedimentary transport); B. Earliest onlap assemblages; C. Time of deformation; D. Episodes of major plutonism and metamorphism; E. Nature and polarity of accretionary boundaries: indicates a hard thrust contact; indicates a thrust with melange and ophiolitic rocks; indicates a steep fault zone commonly with mylonites (after Williams and Hatcher, in press).

Zone in Newfoundland. Alleghanian plutonism and regional metamorphism are restricted to the Avalon Zone at the Avalon-Meguma boundary. The timing and spatial relationships of Alleghanian deformation imply a controlling mechanism related to continued deformation at the Avalon-Meguma zone boundary.

Thus, time of deformation in the Appalachian orogen shows some coincidence with the main phases of accretion of the orogen. The effects of Taconian orogeny are restricted to early accreted westerly zones. Acadian orogeny was most intense in more easterly zones but also involved the Taconian deformed zones. The effects of Alleghanian orogeny, especially its intrusive and metamorphic effects, are localized in the Avalon Zone and at the Avalon-Meguma zone boundary.

This crude analysis clearly implies a relationship between accretionary events and times of deformation. Furthermore, it may explain the episodic nature of orogeny and rationalize the sequential aspect of the orogenic cycle with the concept that subduction and plate convergence are essentially continuous processes.

The timing of *plutonism* and *metamorphism* within the orogen compares well with the timing of major deformation (Fig. 5). The intrusion of late Precambrian granitic batholiths of the Avalon Zone relates to the early history of this zone so that they were an integral component before its mid-Paleozoic accretion. Ordovician batholithic intrusions are minor in the orogen and are confined mainly to the Humber, Dunnage, and Gander zones. Tonalites, diorites, and granodiorites are the common rock types, compared to a preponderance of potassic granites emplaced during later plutonic episodes.

Large Acadian granitic batholiths are widespread throughout the Canadian Appalachians, and they are common in all zones east of the Humber. The granitic intrusions bear little similarity to those of modern island arcs and Andean continental margins, and this view has been expressed independently for plutons in Newfoundland, New Brunswick, and Maine. Thus, a subduction mechanism of generation is not favoured. Furthermore, the mid-Paleozoic intrusions cut Silurian-Devonian rocks that are overlap assemblages across the earlier zones. The latest views relate Paleozoic plutonism in the Canadian Appalachians to melting of a compressed and overthickened crust, or to generation along major shear zones. A paucity of Ordovician granitic plutons, emplaced at a time when subduction was most active, also supports an alternative mechanism of generation.

Late Paleozoic granitic plutons are rare in the Canadian Appalachians. A few small plutons, dated isotopically, occur in the Avalon Zone.

Plutonism and metamorphism are related both spatially and temporally throughout the orogen. The main climactic episodes relate to early Paleozoic Taconian orogenesis and mid-Paleozoic Acadian orogenesis. Noteworthy is the lack of plutonic rocks within the deformed Humber Zone and their preponderance in more eastern zones. This pattern agrees with a west-to-east accretionary model for the orogen. Furthermore, a shearing mechanism for the generation of the middle to late Paleozoic plutonic rocks supports the concept of outboard accretion by major transcurrent movements.

Mineral deposits that formed during the late Precambrian and early Paleozoic development of the orogen are as distinct as the rocks that define its zonal subdivision at this time. These vary from Mississippi Valley–type lead-zinc deposits in carbonates of the Humber Zone, Cyprus-type copper-pyrite deposits and asbestos occurrences in ophiolitic rocks of the Dunnage Zone, polymetallic deposits of the Gander Zone, local oddities such as pyrophyllite and manganese in the Avalon Zone, and gold in the Meguma Zone. Following mid-Paleozoic and late Paleozoic accretionary events, mineral deposits that relate to plutonism and hydrothermal systems and that occur in overlap assemblages as well as older rocks obviously bear no relationship to zonal subdivisions based on older rocks.

Concluding Remarks

The North American connection, or that part of eastern North America that was removed during the initiation of the Appalachian orogenic cycle, is poorly defined within the orogen. Zones outboard of the North American Humber Zone, all structurally uncoupled, are therefore of unknown paleogeography.

Before the advent of plate tectonics, a failure to recognize and stress major tectonic junctions between contrasting geologic provinces led to gross oversimplifications and major misconceptions in the interpretations of the Canadian Appalachians. Even now, some plate models assume or imply paleogeographic proximity of contrasting provinces or zones. This analysis stresses paleogeographic uncertainties and focuses on the accretionary history of the orogen as a unifying theme to present and compare a multitude of features that bear upon the evolution of the Canadian Appalachians.

References Cited

Williams, Harold, 1979, Appalachian orogen in Canada: Canadian Journal of Earth Sciences, v. 16, p. 792–807.

Williams, Harold, and Hatcher, R. D, Jr., Suspect terranes and accretionary history of the Appalachian orogen: Geology (in press).

The Appalachian/Ouachita Orogens: United States and Mexico

Robert D. Hatcher, Jr.
Department of Geology
University of South Carolina
Columbia, South Carolina 29208

George W. Viele
Department of Geology
University of Missouri
Columbia, Missouri 65201

Introduction

The Appalachians and Ouachitas are Paleozoic orogens developed along the eastern and southern Precambrian continental margins of North America. Their histories record considerable diversity in stratigraphic, deformational, metamorphic, and plutonic events. Our purpose here is to outline the various subdivisions of the orogens and comment upon what we know of their history at the present time.

The Appalachian orogen, as presently exposed, extends from Newfoundland to Alabama in eastern North America (Fig. 1). The Ouachita orogen extends from Arkansas to north-central Mexico (Fig. 2). The connections between these two orogens are buried beneath the Mississippi Embayment of the Gulf Coastal Plain; consequently, details of that region may never be precisely known.

The subdivisions outlined below for the Appalachians, and to some degree for the Ouachitas, represent a time-space categorization that we hope will serve to break down the classic geographic subdivisions that have historically dominated our descriptions of the Appalachians. These subdivisions provide a means by which the history of the orogens may be more easily described and related to the constructive processes that built them on the margins of North America during Paleozoic time. These subdivisions reflect the likely plate tectonics history of the orogens and the manner in which the present southern and southeastern continental margins of North America evolved throughout Paleozoic time.

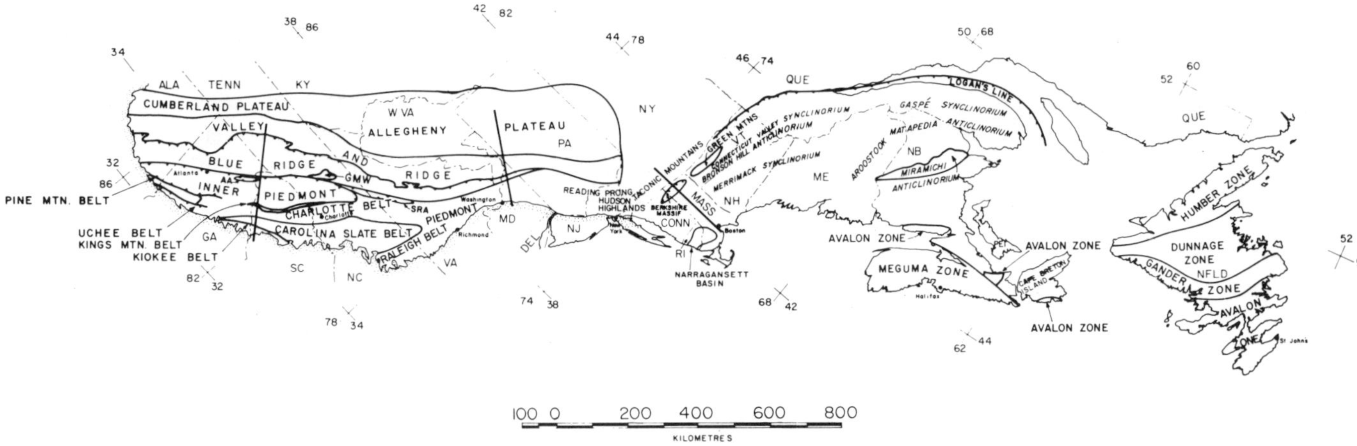

Figure 1. Distribution of subdivisions in the Appalachian orogen.

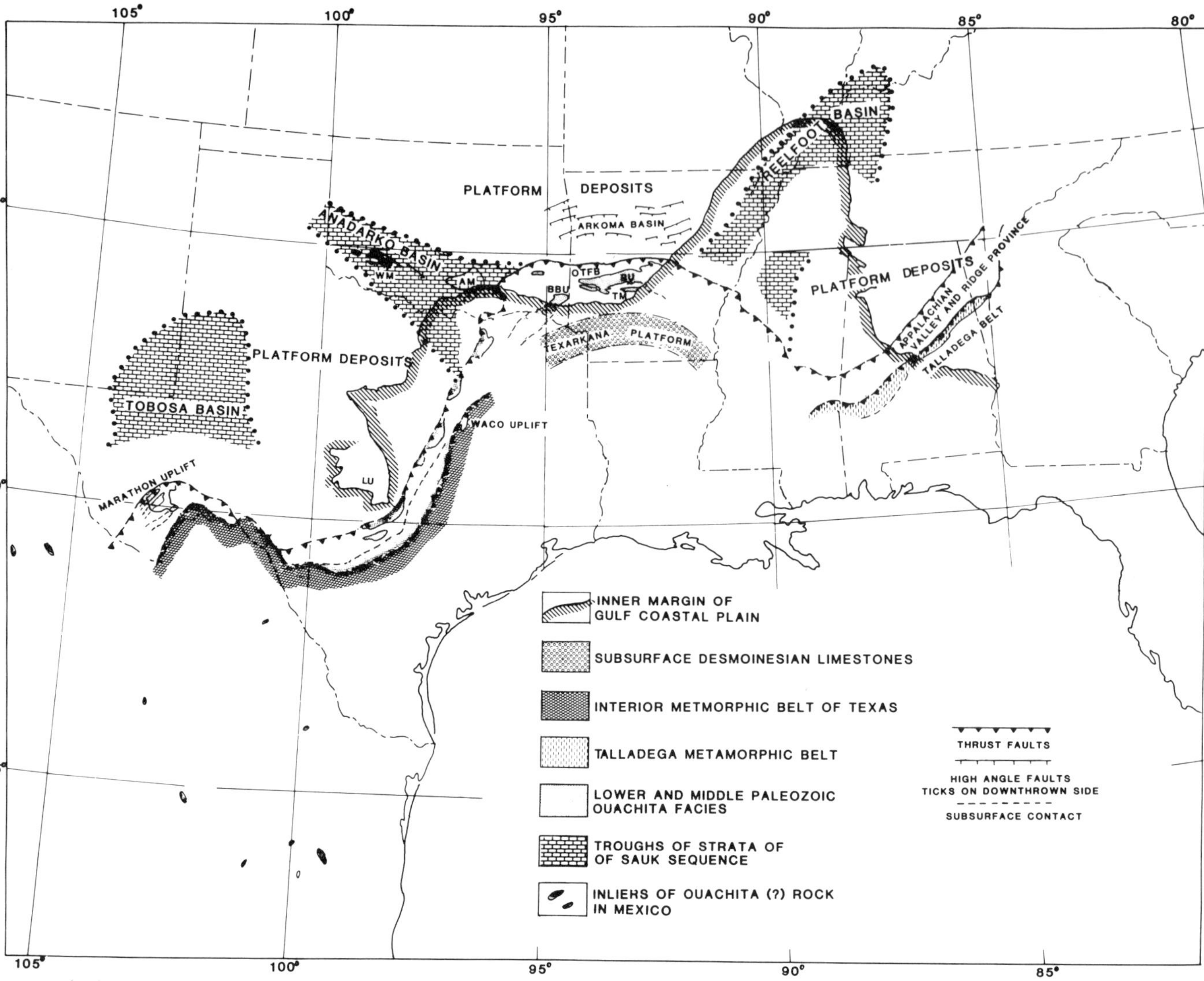

Figure 2. Surface and subsurface distribution of the Ouachita orogen. (AM) Arbuckle Mountains. (BU) Benton Uplift. (BBU) Broken Bow Uplift. (LU) Llano Uplift. (OTFB) Ouachita ThrustFold Belt. (TM) Trap Mountains. (WM) Wichita Mountains.

Appalachian Orogen

Late Precambrian–Early Paleozoic Passive Margin

Following the Proterozoic (1 Ga) Grenville orogeny in eastern North America, a passive continental margin developed in the late Precambrian and began to receive sediments from the eroding Grenville terrane. These sediments constituted a continental rise, slope, and shelf sequence that evolved in time into a stable trailing margin sequence. Sediment thicknesses range from zero to twelve to fifteen thousand meters. In addition, extensive associated felsic and mafic volcanic rocks occur in several areas throughout the southern and central Appalachians (Rankin, 1976). Initially, in part of the southern Appalachians, the sediments were relatively mature. Subsequently, a great thickness of poorly sorted sediments appeared, reaching its greatest thickness as the Great Smoky Group in the southern Appalachians. Later, the character of sediments changed again to a more mature assemblage before the beginning of Cambrian time.

The basal Cambrian clastic succession in the Appalachians is more uniform and dominantly more mature than the underlying succession of Precambrian sedimentary rocks. Overlying the basal Cambrian clastic sequence is a succession of carbonates, then clastics and more carbonates, that comprise the stable margin carbonate bank and associated sediments that existed in eastern North America from Early Cambrian time until the end of Early Ordovician time. The clastics that occur in this sequence are derived entirely from the craton. The Cambrian(?) Unicoi clastic-mafic volcanic assemblage that occurs in the central and southern Appalachians is the last vestige of volcanism in the evolving continental-margin sequence prior to Middle Ordovician time.

Avalon Zone

The Avalon Zone (Williams, 1978) consists of sequences of late Precambrian to early Paleozoic volcanic rocks and clastic to volcaniclastic sediments that appear unrelated to the North American margin at the time the platform sequence was being deposited along the late Precambrian coastline. This zone is represented by the Carolina slate belt in the eastern Piedmont of the southern Appalachians and in the Narragansett Basin–Boston platform area of southeastern New England. A part of the coastal Maine carbonates and metaclastics may likewise represent the Avalon Zone. High-grade equivalents of the Carolina slate belt rocks probably comprise most of the country rocks of the Charlotte and Kiokee belts and possibly the Raleigh belt in the southern Appalachians (Snoke and others, 1980). There are no outcrops of rocks assignable to the Avalon Zone in the central Appalachians. They are either covered by the Atlantic Coastal Plain or were removed or not deposited in this sector. Volcanic rocks in the Avalon Zone exhibit a variety of compositions ranging from mafic to felsic. They have been variously interpreted as a system of volcanic rocks belonging to an island arc, a rifted margin, or a sedimentary–volcanic arc sequence deposited on continental crust (Hatcher, 1972). Gravity data suggest that the rocks of the Avalon Zone are underlain by a mafic or intermediate crust because the gravity field beneath the southern Appalachians part of the zone is high (Hatcher and Zietz, 1980).

Taconic Orogen

In Middle Ordovician time, following regional uplift of eastern North America and development of an extensive unconformity on top of the marginal carbonate bank, shallow seas encroached upon the continent once more and the carbonate bank was briefly reestablished. Shortly thereafter, however, the eastern part of the margin began to subside rapidly forming a foredeep basin in which black shale was deposited. This rapid subsidence probably represented loading of a part of the crust farther east by advancing thrust sheets of the Taconic orogen. In New England, the Taconic klippes, preceded by debris, were emplaced into the basin from the east with concomitant ophiolite emplacement farther north. Farther south in Pennsylvania, the Hamburg klippe was emplaced, and large amounts of clastic sediments were deposited from Middle Ordovician until Early Silurian time in the southern and central Appalachians. Accumulations reached a maximum thickness of a few thousand meters in areas of east Tennessee and central Pennsylvania. These reflect uplift taking place in the central part of the orogen probably related to contemporaneous thrusting and penetrative deformation, Barrovian metamorphism (high pressure–low temperature in Vermont), and plutonism. Plutonic activity in the form of granitic, intermediate, and mafic plutons is found throughout the Appalachian orogen in the New England region (Robinson and Hall, 1980) and occurs over wide areas, but primarily in the central part of the orogen in the southern and central Appalachians. In northern Maine there is an Arenigian unconformity that probably represents the oldest response to Taconic activity in the internal parts of the orogen. An ocean of unknown size was closed at this time resulting in the emplacement of ultramafic bodies along a prominent trend in the western part of the orogen east of the primary continental margin sediments and basement rocks of the craton, that is, in the central to eastern Blue Ridge of the southern Appalachians, the Piedmont of the central Appalachians, and east of the basement massifs in New England. Relict blueschist metamorphism occurs in north-central Vermont.

Basement rocks belonging to the Grenville cycle were incorporated into the deforming metamorphic core during the Taconic orogeny. These were remobilized, sometimes detached, penetratively deformed to a greater or lesser degree, and generally thrust toward the craton. Many were reactivated in the Acadian orogeny and moved farther westward on large faults (Hatcher, 1978, 1981).

Although the Taconic event has been considered to be a relatively minor event in the central and eastern New England Appalachians, it is a major event in western New England (Robinson and Hall, 1980). It is also considered to be the major thermal event affecting the rocks of the Blue Ridge and western Piedmont, as well as parts of the eastern Piedmont farther south (Butler, 1972). However, the evidence concerning the role of later orogenies and the degree to which they affected the southern Appalachians is presently inconclusive, although more evidence favoring major Acadian and Alleghenian metamorphism and plutonism is appearing.

Acadian Orogen

During Silurian and Early Devonian times, there was a return to relatively quiet continental margin or supracrustal conditions over most of the Appalachians. In the continental interior, the Silurian-Devonian rocks are represented by carbonates and relatively clean clastics. However, volcanic and deep-water sedimentary rocks are found in the Silurian and Devonian of New England, possibly indicating a return to continental margin or deep basin conditions. Some of the sediments formed at this time could be the successors to those of the earlier continental margin that was overridden during the Taconic event. Some of the Silurian volcanic rocks were formed under subaerial conditions.

During Devonian time, the rocks of central and southern New England were subjected to high-temperature relatively high-pressure metamorphism, and many of the earlier plutons and older structures were reactivated and to some degree overprinted by structures representing this Acadian event. Abundant plutons are associated with this event in New England and most are S-type granites. It is uncertain to what degree Acadian metamorphism affected the crystalline rocks of the Blue Ridge and Piedmont in the southern Appalachians. It has been suggested that the Acadian event was an event of major faulting, based upon Rb/Sr whole-rock ages obtained from dating mylonites (Odom and Fullagar, 1973).

South of Virginia, Silurian and Devonian sediments are practically nonexistent on the craton in southern Kentucky, Tennessee, and north Georgia. Farther south there are some sediments of this age in Alabama, but for most areas of the southern Appalachians, the Devonian and Silurian on the foreland were times of little deposition. There is no evidence in the South of a Devonian clastic wedge that might reflect the Acadian orogeny. The Talladega sequence of Alabama and Georgia must be a pre-Acadian clastic-volcanic sequence. It contains Lower Devonian fossils and was affected by apparent Acadian dynamothermal events.

Alleghanian Orogen

The Alleghanian event is a Late Carboniferous–Permian event roughly correlative to the Variscan-Hercynian events of Western Europe and northwestern Africa. Associated with the Alleghanian events in North America are the deformation of the foreland; deposition of molasse-type sediments on the craton west of and within the Appalachian orogen; deformation and metamorphism along the eastern margin of the Appalachians in the southern Appalachian Piedmont, southeastern New England; and possibly along the coast of the Maritime Appalachians in Canada; and the intrusion of plutons, mostly granites, in association with the deformational-metamorphic events.

The Alleghanian deformation in the foreland of the Appalachians postdates the extensive clastic deposition in the so-called Appalachian basin from Alabama to Pennsylvania. It is characterized by thin-skinned thrusting directed toward the continent in which thrust sheets are detached from the underlying basement and follow incompetent zones in the stratigraphic section. Toward the west they ramp across the more competent zones to high incompetent detachments. Deformation involves rocks as young as early Permian in West Virginia, Pennsylvania, and Ohio and as young as Pennsylvanian in the Valley and Ridge of Pennsylvania and several places farther south. Thus, the timing of this event in the foreland of the Appalachians is reasonably well established. Thrusts of the Valley and Ridge are overridden in the southern Appalachians by the Blue Ridge thrust sheet, which along its toe in east Tennessee and Georgia involves Mississippian rocks. These rocks serve to indicate that movement on the Blue Ridge thrust is likely Alleghanian throughout most of its extent. Farther to the southeast, vestiges of movement during the Alleghanian are more difficult to resolve within the crystalline complex. It is likely that brittle deformation along the Brevard fault which brings up slices of carbonate from beneath the thrust sheet is Alleghanian (Hatcher, 1978). The metamorphism and penetrative deformation occurring in the Kiokee and Raleigh belts of the eastern Piedmont have been dated as Alleghanian by cross-cutting relationships of deformed plutons. The Modoc and Augusta faults are likewise Alleghanian structures, since they are intimately involved in the deformation occuring in this area (Snoke and others, 1980). In the New England area there is some suggestion on the foreland of Alleghanian deformation in northeastern New York and Vermont (P. A. Geiser, 1981, personal commun.). However, this is not well established. In southeastern New England, metamorphism, plutonic activity, and penetrative deformation associated with deformation of the Pennsylvanian rocks of the Narragansett basin are well established. Some of the large brittle faults that occur in southeastern Maine and extend into the Canadian maritime region may likewise be Alleghanian features.

Ouachita Orogen

Subdivisions appropriate for description of the Appalachian orogen are appropriate only in part for discussions of the Ouachita orogen. Fundamentally, both folded belts record the late Paleozoic closing of oceans and the consolidation of continents rimming the Atlantic. Within this general theme several variations of tectonic history are apparent, but significant differences in tectonic history distinguish the Ouachitas and Appalachians. However, it is certain that the Appalachian and Ouachita orogens join in the subsurface of Alabama and Mississippi. Therefore, similarities and dissimilarities especially will be noted in this discussion.

Precambrian

Precambrian rocks are not exposed within the Ouachita orogen but do occur in scattered localities on the foreland side of the folded belt. The oldest known rocks crop out in Oklahoma at the southeastern end of the Arbuckle Mountains where the principal unit of the basement terrain is the 1,374-Ma Tishomingo Granite (Bickford and Lewis, 1979). Layered cumulates in the Wichita province are about the same age and may indicate the existence of an old intracratonic rift system (Powell and Phelps, 1977). Moreover, on the southern side of the Wichita Mountains of Oklahoma, a series of layered seismic reflections has been interpreted as a thick wedge of Proterozoic strata (Brewer, 1981), although their total areal extent is unknown. Farther south in the Llano uplift in Texas, rocks of the 1 Ga Grenville province appear, and in the Waco uplift, Texas, the Shell No. 1 Barret well penetrated quartz diorite, which is probably Precambrian, notwithstanding a last cooling date of 336 Ma (Rozendal and Erskine, 1971). The Grenville rocks of the Llano uplift probably connect with the Grenville province of the east-central United States, although the trace of the front is buried in the subsurface. It seems probable also that, as in the Appalachians, a continental margin formed after the Grenville orogeny, but its position across the southern United States has not been discerned.

Early–Middle Paleozoic

The first clear evidence of rifting along the southern margin of the North American craton comes from the region of the Wichita Mountains of Oklahoma. Here, in fault-bounded troughs, 525-Ma spilitic basalt, granophyre, and rhyolite rest unconformably on the layered cumulates and are overlain unconformably by sandstone, limestone, and dolomite of the Sauk Sequence. These latter rocks were deposited over a broad region of the Ouachita foreland, extending from West Texas to the southern Appalachians, with essentially the same scenario of deposition. The rocks are largely platform limestone and dolomites and exhibit well the unconformities that form the sequence boundaries of Sloss (1963). For the most part the strata form a thin blanket less than 800 m thick, but three basins opened and deepened toward the present-day Ouachita Mountains: the Reelfoot Basin in the Mississippi Embayment, the Anadarko-Arbuckle Basin in southeastern Oklahoma, and the Tobosa Basin in West Texas. In the first two, thicknesses of the Sauk Sequence alone are in excess of 2,000 m. It is possible but not proven that the Reelfoot and Tobosa Basins, like the Anadarko-Arbuckle Basin, also formed above failed rift arms or aulacogens.

During Devonian and Mississippian times, numerous beds of chert were deposited in the foreland area. Sedimentation rates evidently were slower and the water probably was deeper. The silica sedimentation may record the slow foundering of the southern margin of the North American craton.

In the Arkoma Basin and along the Ouachita thrust fault belt, thick deposits of Carboniferous strata lie between the lower-middle Paleozoic rocks of the platform and rocks of equivalent age but distinctly different facies (Ouachita) in the Benton, Broken Bow, and Marathon uplifts. In addition, the rocks of the uplifts are allochthonous, both factors hiding the transition from rocks of the platform to rocks of the Ouachita facies.

In all three uplifts the Lower-Middle Paleozoic Ouachita facies may be divided into two parts: The lower, starting in late Cambrian time in the Marathon region and in Early Ordovician time in the Ouachita Mountains, is made up predominantly of black, graptolitic shales interstratified with subordinate amounts of micrite and clean sandstone, both exhibiting graded bedding, sole marks, and other characteristics of deposition by gravity-driven flows. The definitive characteristic of the lower part, however, is the presence of lenses and beds of exotic material. In the Marathon region, the bouldery debris is made up largely of shallow-water carbonate rocks derived from the platform margin, whereas in the Benton uplift in Arkansas, the debris includes many boulders of granitic rock. These bouldery strata are viewed as continental rise-and-slope deposits. During Ordovician time, the shelf break at the southern margin of the North American craton must have been steep, rugged, and accented by high-angle faults.

The upper part of the Ouachita facies ranges from Late Ordovician to Early Mississippian in age, and comprises siliceous shales and thin- and thick-bedded cherts or novaculites. Where not metamorphosed, the novaculites contain abundant radiolaria, graptolites, and conodonts and are generally viewed as deep-water deposits, although their environmental setting remains a matter of debate.

The Lower-Middle Paleozoic Ouachita facies show no evidence of the unconformities defining the sequence boundaries of the craton. Indeed, the rocks exhibit no evidence of sediment introduced from the craton. They were probably deep-water sediments that were deposited slowly over a time span of 120 Ma.

The Taconic orogeny is not recognized in the Ouachita Mountains, and evidence is scanty for Acadian events. In several places, K/Ar dates suggest a Devonian metamorphic event, but no structures have been mapped associated with these ages. In the Talladega block of Alabama, the Devonian(?) rocks comprise metamorphosed conglomeratic quartzites, cherts, and greenstones, all suggesting tectonic activity, but in the Ouachita Mountains this interval of time was marked by the slow accumulation of siliceous shale and novaculite. The Ouachita Mountains seem to mimic a Lower-Middle Paleozoic continental margin outlined by rifting and opening of an ocean basin (Thomas, 1976). Thus, the Lower-Middle Paleozoic history constitutes the first half of a Wilson cycle: rifting, spreading, and subsidence uninterrupted by compressional tectonic events.

Late Paleozoic

In Mississippian time, the sedimentary-tectonic regime of the Ouachita Mountains changed abruptly as rapid deposition of clastic rocks succeeded the slow accumulation of bedded cherts. In Mississippi and Alabama, lithic and quartz arenites, derived from an orogenic source terrain, prograde from southwest to northeast. In Arkansas and Oklahoma, submarine fan deposits prograde north-northwest to a distal flysch facies and contain in their basal part beds of rhyodacitic tuff that thicken southward. In the Marathon region similar deposits of flysch came from the east and southeast. By Early Pennsylvanian time, a thick trough of deep-water flysch extended from Mississippi to at least as far as the Marathon region of Texas. Still farther to the south in Mexico, scattered inliers of Paleozoic strata probably represent a southern extension of the Ouachitas. Flysch-like deposits containing a strong admixture of volcanic fragments are the most common, and their age is Late Pennsylvanian to Permian. The data are scanty, but the outcrops constitute a logical southward extension of the flysch trough of the Ouachitas. Possibly the rocks correlate with the flysch deposits of the Val Verde Basin lying north of the Marathon uplift. Here, a late orogenic pulse formed a new flysch trough that filled quickly with late Pennsylvanian and early Wolfcampian deposits.

Many workers have described the sedimentology of the flysch deposits. It is established that on the north side of the Ouachita orogen, after Mississippian time, major sediment feeds came from the north and northeast, but feeds also came from the south in Arkansas, from the east in the Fort Worth Basin, and from the southeast in the Marathon region. Once in the trough, the general direction of prograding was along it toward the west and southwest, although concurrently the axis of deposition migrated toward the foreland. On the north side of the trough the flysch thins rapidly toward the foreland, in part across syndepositional, down-to-the-south, high-angle faults, and grades into shallow marine and fluviatile platform deposits. This flysch trough, containing an aggregate thickness of as much as 12 km of strata, is distinctly unlike the molasse-like deposits of the Appalachian foreland.

Clastic wedge sediments appeared as early as Mississippian time in the Alabama and Mississippi portions of the Ouachita orogen. Westward, from Arkansas to Oklahoma to the Marathon region along the northern margin of the folded belt, the successor deposits to the flysch were Des Moinesian and were deposited in shallow marine and fluviatile environments. Sediment feeds as before were from the north, east, and south and prograded toward the west along the trend of the old flysch basin. South of the outcrops of the folded belt, in the subsurface of Arkansas and Texas, numerous wells have penetrated shallow-water, fossiliferous Des Moinesian limestones. They constitute the Texarkana platform and rest unconformably across broadly folded Atokan strata, thereby indicating that the Ouachita orogeny was over by this time in this region. Toward the north within the Benton uplift, multiple phases of deformation are evident, but all fabric elements are penetrative from Ordovician to Lower Pennsylvanian rocks. Still farther north, in the Arkoma Basin, both high-angle faults and thrust faults offset Des Moinesian strata, indicating that here the last pulse of deformation was of Late Pennsylvanian or possibly Early Permian age.

Appalachian-Ouachita Connections

The nature of the connections between the Ouachita and Appalachian orogens has long been a subject of discussion and concern. No precise answer to the question is yet available because the area where the connection occurs is buried beneath the Cretaceous and Tertiary sediments of the Gulf Coastal Plain in the Mississippi Embayment. Subsurface and geophysical studies indicate that a belt of deformed rocks extends between the exposed Appalachian and Ouachita structures. New gravity and magnetic studies (I. Zietz, 1981, personal commun.) show that geophysical trends reflecting basement rocks are truncated on the southwest against the Ouachita structural front. Major differences in contemporaneous facies and thickness are demonstrated by stratigraphic sections exposed in the Appalachians and Ouachitas in Alabama and Arkansas. The stratigraphic transition is documented by subsurface data.

Constructive History

The Appalachian and Ouachita orogens were built along the Precambrian continental margin of North America following the rifting and opening of an ancient ocean. The size of this ocean is presently indeterminant. The evidence confirming that the continents were rifted to produce this ocean is present in the late Precambrian–Cambrian sedimentary and volcanic rocks of eastern North America. Following the rifting and spreading events in late Precambrian time, the eastern margin of North America developed into a carbonate bank platform that existed until the end of Early Ordovician time. At this time the platform uplifted, eroded, and subsided once more. Then a foredeep basin formed along the eastern flank of the North American Continent prior to influx of clastic sediments. The arrival of the first thrust sheets onto this flank apparently occurred in Ordovician time. These carried fragments of oceanic crust and mantle as ophiolites shedding debris. Whether the entire ocean basin or only a part of it was closed at this time remains to be determined, but a segment of oceanic crust was destroyed. Accompanying the closing of this part of the ocean were pulses of extensive penetrative deformation, metamorphism, and plutonic activity from Early to Middle Ordovician until Early Silurian times. These pulses constitute the Taconic orogeny.

Sedimentation, which was limited in the South but more extensive in the North, followed a period of quiesence and was accompanied by some volcanic activity. This was followed by another compressional event perhaps related to the closing of an ocean in the North. It produced high-temperature metamorphism in New England, faulting and possibly metamorphism in the southern Appalachians. Extensive emplacement of plutons of various compositions occurred throughout the Appalachians at this time. These events constitute the Acadian orogeny.

The Alleghanian orogeny occurred near the end of Paleozoic time and terminated compressive tectonic activity in the Appalachian orogen. This event may be related to the final closing of the ancient large ocean that may have existed off eastern North America during Paleozoic time. The discrepancy in stratigraphic, deformational, metamorphic, and other evidence along the orogen indicates that perhaps the collision between New England and another continent may have involved Africa and South America, or possibly both. This would help account for the diachronous nature of the collisions and the apparent discrepancy in the nature of the deformations that occur up and down the orogen.

The variable that must be considered, in addition to across-strike transport, in any reconstructive hypothesis for the Appalachian orogen is the possibility that large segments of the orogen have been transported along strike for tens or even hundreds of kilometers to be emplaced in their present positions within the orogen. The only part of the Appalachian orogen that can be considered reasonably intact and relatable to the North American craton is the continental margin. The internal parts were developed offshore and may well have been moved for considerable distances along strike prior to the final closing and collision events that terminated orogenic activity throughout the mountain chain. Therefore, it is necessary to consider that a number of segments of the Appalachian orogen may be suspect terraines similar in characteristics to those in the North American Cordillera.

Deformation in the foreland is presently thought to be related to transmittal of stresses through the deformed crystalline mass with only a local thermal high on the east side of the deformed mass, resulting in the thin-skinned deformation of the foreland and transportation of some of the crystalline rocks over the foreland along the western edge of the orogen. Emplacement of the Valley and Ridge thrusts by gravity has also been suggested, but this mechanism encounters significant difficulties where problems of the transition from the external foreland parts of the orogen to the internal parts must be considered. Gravity may have played a part in the readjustment process of final emplacement.

Most Appalachian geologists agree that parts of the Appalachian orogen are allochthonous. The distance from a major gravity gradient to the outermost thrusts and the possible amount of overthrusting decrease northward. Seismic reflection data have led to the proposal that the entire southern sector of the Appalachian orogen was thrust onto the continental margin along a master decollement which was active throughout most of Paleozoic time (Harris and Bayer, 1979; Cook and others, 1979). The idea that major parts of the crystalline core were thrust over the foreland was suggested earlier based upon occurrence of lower grade carbonate slices in the Brevard zone (Hatcher, 1971, 1978). Magnetic and gravity data, in conjunction with surface geology, also suggest that all of the southern Blue Ridge and much of the Piedmont are allochthonous (Hatcher and Zietz, 1980).

The constructive history of the Ouachita orogen is difficult to assemble because so much of the orogen is buried by younger rocks. Nevertheless, two cross-sections illustrate both the general tectonic style and differing viewpoints of the tectonic history (Rozendal and Erskine, 1971; Viele, 1979). Within the foreland of the orogen in Arkansas and Oklahoma, down-to-the-basin faults were active during deposition of the Pennsylvanian flysch and later, for the northernmost high-angle faults offset molassic Des Moinesian strata; these same beds were involved in thrust faulting and associated folding. Therefore, compressional and extensional events either existed at the same time or were very narrowly separated from one another in space and time. South of the belt of high-angle faulting, thrust faults translate the Carboniferous flysch toward the craton and constitute the dominant structures of the orogen, although buried high-angle faults are shown on many proprietary, seismic profiles. Belts of strata containing exotic clasts and zones of chaotic deformation are relatively common in the Carboniferous strata, and their origin as tectonic melange or stratigraphically bounded olistostromes is debated. It is generally known that good-quality reflection profiles are difficult to obtain in areas of the Ouachita thrust-fault belt, but is it because the flysch scatters and absorbs seismic energy, or is it because the profiles accurately show the chaotic geometry of the rocks? In the Benton uplift of Arkansas, the most intense deformation occurs along its northern margin where polyphase folding, cleavage, and other features of high ratios of strain are common. Isolated pods of serpentinite also occur here. The deformation decreases southward, for in the Trap Mountains and southward the folds are relatively open and uncleaved, and indeed, the southernmost exposures of Ouachita rocks dip homoclinally southward. In western Arkansas a recent COCORP profile indicates south-dipping reflectors to a depth of five seconds north of the Benton uplift, a zone of few coherent reflectors on its northern margin, a broad arch rising to about three seconds directly beneath it, and dipping reflectors extending to depths of eight seconds about 25 km south of the Coastal Plain onlap (K. D. Nelson, personal commun.).

The tectonic configuration is quite different in the subsurface of Texas where the interior metamorphic belt lies toward the Gulf of Mexico or on the outboard side of the Ouachita folded belt. Cross-sections of this region show the Carboniferous Ouachita facies and the interior zone as strongly allochthonous sheets thrust toward the craton over basement highs. One such high is the Waco uplift where the drill penetrated phyllites, slates, quartzites, and then 1,800 m of marble and 37 m of quartz-biotite schist before entering quartz diorite. The marble and underlying foliated quartzite have been correlated with the Ellenberger-Reagan sequence of the Texas platform, and therefore it is implied that North American platform rocks extend under the Ouachita folded belt. The early middle Paleozoic shelf break in such an interpretation necessarily lies beneath or even east of the Ouachita orogen.

Alternatively, the inboard side of the Ouachita orogen marks the surface trace of a major suture formed by the south-directed subduction of a remnant ocean basin and the collision of the North American continent with a southern microplate. The early middle Paleozoic slope and abyssal strata were scraped off the subducting oceanic crust and incorporated into accretionary wedges. In this view the interior metamorphic belt of Texas appears to be an exotic terrain docked against the Ouachita folded belt, but not as an integral part of it, thus the basement rocks of the Waco uplift are believed to be non–North American. The early middle Paleozoic continental margin in this interpretation is placed farther inboard along the trend of the Ouachita Mountains. An aeromagnetic map of the United States (Zietz, 1981, personal commun.) supports this hypothesis to the extent that none of the magnetic anomalies of the craton and foreland transect the Ouachita folded belt. Moreover, the anomaly pattern on the Gulf side of the Ouachita orogen is very different from North American patterns, suggesting a fundamental difference in basement morphology.

Major Problems

A number of major problems in the Appalachian and Ouachita orogens remain unsolved at the present time. Many may never be solved to the satisfaction of everyone. Some have partial solutions; others that have only recently been recognized will doubtlessly be modified and solved in the future as still other problems become known.

1. Resolution of the question of large-scale continentward transport of major segments or all of the Appalachian orogen is a major problem. Most geologists will agree that the Appalachian orogen in the south is to a great degree allochthonous and the magnitude of transport decreases northward. However, the point of eastward termination of the large thrust sheet is a matter of conjecture. Some maintain that it extends to points well east of the Coastal Plain overlap, whereas others maintain that it may be terminated within the exposed portions of the orogen.

2. The extent and nature of the Acadian orogen in the southern and central Appalachians are major problems that particularly relate to the question of whether the regional metamorphism is in large part Acadian or whether it is all Taconic, as has been suggested previously.

3. The question of why we do not see abundant evidence of Alleghanian deformation in the core of the Appalachian orogen although abundant evidence is present along the flanks of the southern and central Appalachians needs further study.

4. The extent and nature of the boundaries between possible suspect terrains in the Appalachians are problems that will receive considerable attention in the future. Problems of the subsurface extent of Appalachian and Appalachian-Ouachita elements will doubtlessly receive more attention as additional geophysical and drilling data become available.

5. The question of the relationship between the narrowing of the Appalachian orogen both southward from New England and northward from the North Carolina–Virginia Piedmont into the central Appalachians is something that has not been addressed. This problem will doubtlessly be explored further as more information becomes available from buried parts of the orogen beneath the Coastal Plain in this region.

6. The problem of the timing of closing of small ocean basins or major oceans in New England is one that is clouded by the succession of Silurian and Devonian sediments and Acadian thermal activity. The geophysical data and additional detailed geologic mapping on the surface may help solve this problem.

7. Dating of brittle deformation in eastern New England is as yet unresolved.

8. The recognition and resolution of a number of sutures within the orogen is a problem that is receiving a great deal of attention today. This may be a topic that may help resolve some of the questions related to the early history of construction of the Appalachian orogen along the continental margin and the resolution of suspect terrains within the orogen.

9. Within the Ouachita orogen the environment of deposition of the Lower-Middle Paleozoic rocks represents a critical boundary condition for tectonic reconstructions. If the rocks are truly of shallow-water origin, the entire tectonic history will have to be rewritten. If they are truly abyssal, the absence of oceanic crustal rocks is puzzling.

10. The petrology of the Carboniferous flysch needs additional study because studies of its provenance represent the easiest path to knowledge of whatever may have lain to the south: a microcontinent, a volcanic arc, or an open ocean.

11. The association of multiple folding and chaotic belts and the overall tectonic setting of the cratonal side of the Ouachita Mountains is a problem. The thrust-fault

belts of the Ouachitas and their adjacent uplifts should not be equated to other foreland thrust-fault belts without caution and additional investigation.

12. The position of the early middle Paleozoic shelf break and, therefore, the edge of the North American craton remains enigmatic. If it lies close to the Gulf of Mexico, the rocks of the Ouachitas have been thrust for hundreds of miles. If it was on the cratonal side of the folded belt, it marks the approximate position of a suture joining two plates.

Acknowledgments

Early drafts of this chapter benefited from comments by D. R. Wones, Peter Robinson, P. H. Osberg, D. W. Rankin, W. A. Thomas, and A. R. Palmer.

References Cited

Bickford, M. E., and Lewis, R. D., 1979, U-Pb geochronology of exposed basement rocks in Oklahoma: Geological Society of America Bulletin, v. 90, p. 1540–1544.

Brewer, J. A., 1981, COCORP deep seismic reflection traverse across southern Oklahoma aulocogen [Abs.]: American Association of Petroleum Geologists Bulletin, v. 65, p. 1496.

Butler, J. R., 1972, Age of Paleozoic regional metamorphism in the Carolinas, Georgia and Tennessee southern Appalachians: American Journal of Science, v. 272, p. 319–333.

Cook, F. A., Albaugh, D. S., Brown, L. D., Kaufman, Sidney, Oliver, J. E., and Hatcher, R. D., Jr., 1979, Thin-skinned tectonics in the crystalline southern Appalachians: COCORP seismic-reflection profiling of the Blue Ridge and Piedmont: Geology, v. 7, p. 563–567.

Harris, L. D., and Bayer, K. C., 1979, Sequential development of the Appalachian orogen above a master decollement—A hypothesis: Geology, v. 7, p. 568–572.

Hatcher, R. D., Jr., 1971, Structural, petrologic and stratigraphic evidence favoring a thrust solution to the Brevard problem: American Journal of Science, v. 270, p. 177–202.

——1972, Development model for the southern Appalachians: Geological Society of America Bulletin, v. 83, p. 2735–2760.

——1978, Tectonics of the western Piedmont and Blue Ridge southern Appalachians; Review and speculation: American Journal of Science, v. 278, p. 276–304.

——1981, Thrusts and nappes in the North American Appalachian Orogen, *in* K. R. McClay and N. J. Price, eds., Thrust and nappe tectonics: Geological Society of London Special Publication 9, p. 491–499.

Hatcher, R. D., Jr., and Zietz, Isidore, 1980, Tectonic implications of oregional aeromagnetic and gravity data from the southern Appalachians, *in* D. R. Wones, ed., The Caledonides in the U.S.A.: IGCP Project 27, Virginia Polytechnic Institute, Memoir 2, p. 235–244.

Odom, A. L., and Fullagar, P. D., 1973, Geochronologic and tectonic relationships between the Inner Piedmont, Brevard zone and Blue Ridge belts, North Carolina: American Journal of Science, v. 273–A, p. 133–149.

Powell, B. N., and Phelps, D. W., 1977, Igneous cumulates of the Wichita province and their tectonic implications: Geology, v. 5, p. 52–56.

Rankin, D. W., 1976, Appalachian salients and recesses: Late Precambrian continental breakup and the opening of the Iapetus Ocean: Journal of Geophysical Research, v. 81, p. 5606–5619.

Robinson, Peter, and Hall, L. M., 1980, Tectonic synthesis of southern New England, *in* D. R. Wones, ed., Caledonides in the U.S.A., IGCP Project 27, Virginia Polytechnic Institute Memoir 2, p. 73–82.

Rozendal, R. A., and Erskine, W. S., 1971, Deep test in Ouachita structural belt of central Texas: American Association of Petroleum Geologists Bulletin, v. 55, p. 2008–2017.

Sloss, L. L., 1963, Sequences in the cratonic interior of North America: Geological Society of America Bulletin, v. 74, p. 93–113.

Snoke, A. W., Kish, S. A., Secor, D. T., Jr., 1980, Deformed Hercynian granitic rocks from the Piedmont of South Carolina: American Journal of Science, v. 280, p. 1019–1034.

Thomas, W. A., 1976, Evolution of Ouachita-Appalachian continental margin: Journal of Geology, v. 84, p. 323–342.

Viele, G. W., 1979, Geologic map and cross-section of eastern Ouachita Mountains, Arkansas: Geological Society of America Map and Chart Series MC-28F, scale 1:250,000.

Williams, Harold, 1978, Tectonic-lithofacies map of the Appalachians orogen: Memorial University, Newfoundland, scale 1:1,000,000.

Geology and tectonic evolution of Alaska

George Plafker
U.S. Geological Survey
Alaskan Geology Branch
345 Middlefield Road
Menlo Park, California 94025

D. L. Jones
U.S. Geological Survey
Paleontology and Stratigraphy Branch
345 Middlefield Road
Menlo Park, California 94025

Introduction

The last major synthesis of the geology of Alaska was *Areal Geology of Alaska* (P. S. Smith, 1939, U.S. Geological Survey Professional Paper 192). Up to that time, less than half of what was then the Territory of Alaska had been surveyed by geologists, even by the crudest exploratory standards. As a consequence, Professional Paper 192 was concerned primarily with the general spatial distribution, lithology, and relative ages of rock units and Quaternary deposits in Alaska with incidental mention of structure, metallic mineral deposits, and coal resources.

Since 1939 there has been a vast increase in the amount and diversity of earth science research in Alaska. The initial impetus was during World War II when expanded studies of mineral resources, surficial deposits, and Aleutian Arc volcanoes were undertaken throughout Alaska. In the postwar years there has been an explosive expansion of geologic and geophysical mapping and research both onshore and offshore. This accelerated effort was spurred on by the emergence of Alaska as a major petroleum producer, by the discovery of major metallic mineral deposits at various localities in the state, and by the extensive systematic helicopter-supported mapping and resource appraisals that have been carried out mainly by government agencies as part of the requirements for Alaska land classifications. In addition, the great 1964 earthquake stimulated intense interest in neotectonic activity and geologic hazards statewide. As a result of work since 1939, modern geophysical and geologic data are now available for much of Alaska, and few areas remain unexplored.

The primary objectives of the Alaskan volume will be to provide a concise up-to-date synthesis of the geology, geophysics, and mineral resources of the state and to trace its geologic evolution. Emphasis will be placed on evaluating these data in the light of models of plate tectonics and the concept that much of Alaska is comprised of a collage of accreted tectonostratigraphic terranes.

History of Geologic and Geophysical Research

This chapter will trace the history of geologic investigations in Alaska from the early days of exploration during the late 1800s and early 1900s, through the gold rush period, the time of great expansion of geologic investigations in response to the national emergency of World War II, followed by the intense interest in neotectonic activity generated by the 1964 earthquake, up to the present helicopter- and ship-supported team efforts carried out mainly for assessments of mineral resources and geologic hazards.

Summary of Major Tectonostratigraphic Terranes

The concept that much of Alaska constitutes a gigantic structural collage of accreted tectonostratigraphic terranes has gained credibility as new stratigraphic and structural data have become available during the past decade.

This chapter will summarize the salient features of the cratonal parts of Alaska as well as those of the individual tectonostratigraphic terranes. Basic data to be discussed for each terrane, if available, include nature of crust and mantle, stratigraphy, paleomagnetic record, magmatic and metamorphic history, aeromagnetic and gravity data, geologic structure, and distribution and character of mineral resources.

In contrast to other volumes of this series that treat more southerly parts of the cordillera, a "time-slice" approach to data presentation will not be attempted for Alaska. The reason for this is that the accreted terranes are so heterogeneous and so varied in age that a coherent picture of the entire state can best be developed in terms of the terrane concept. For ease of presentation, the state will be divided into eight subareas (Fig. 1 and Outline), and the terranes of each will be discussed individually.

Crystalline Rocks

This chapter will focus on the distribution and significance of metamorphic and igneous rocks, stressing both the history of these rocks within the context of accreted terranes and also the post-accretionary histories that can be tied more closely to known plate-tectonics movements.

Major sections of this chapter will be the nature of metamorphic terranes; the age, lithology, and distribution of plutonic magmatic belts, pre-Cenozoic volcanic belts, Cenozoic volcanic rocks, and anatectically derived plutonic rocks; and the petrologic and tectonic evolution of the Aleutian Arc.

Major Structural and Terrane Boundaries

This chapter will provide data on the major structural features of Alaska and evidence for the displacement history of the terranes. It will include sections on the structural style and displacement of the major faults, the deformed,

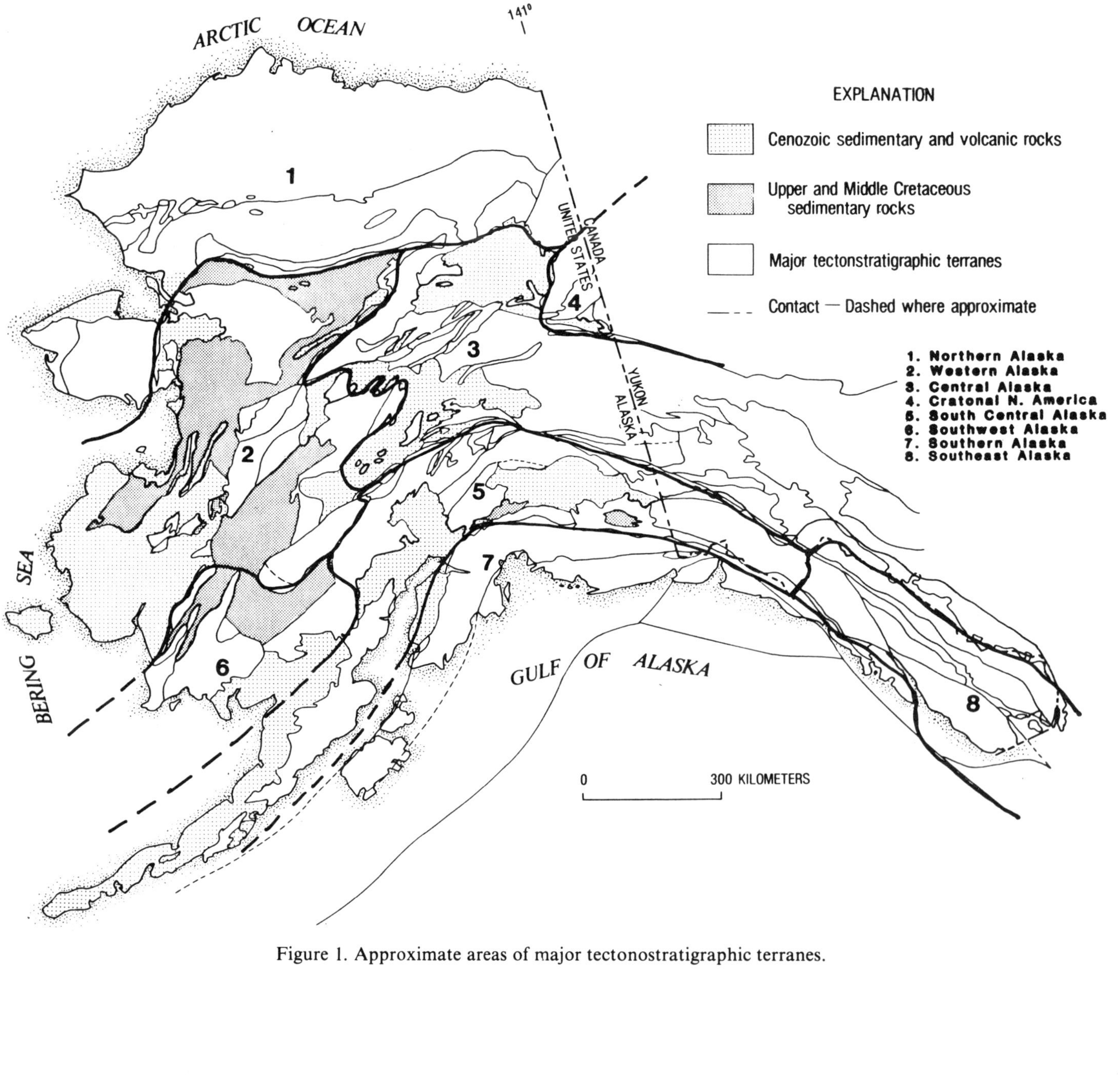

Figure 1. Approximate areas of major tectonostratigraphic terranes.

Outline

Chapter I. Introduction
Objectives

Chapter II. History of geologic and geophysical research

Chapter III. Summary of major tectonostratigraphic terranes
- A. Cratonal North America in Alaska
  1. Crust and mantle
  2. Stratigraphy
  3. Paleomagnetism
  4. Magmatism and metamorphism
  5. Geophysical data
  6. Structure
  7. Resources

  (Main topics to be considered for each terrane)
- B. Northern Alaska
- C. East-central Alaska
- D. Southwestern Alaska
- E. South-central Alaska
- F. Southeastern Alaska
- G. Gulf of Alaska margin
- H. Western Alaska

Chapter IV. Crystalline rocks
- A. Distribution and significance of metamorphic terranes
- B. Age, lithology, and distribution of pre-Cenozoic magmatic belts
- C. Age, lithology, and distribution of Cenozoic magmatic rocks (exclusive of the Aleutian Arc)
- D. Petrologic and tectonic evolution of the Aleutian Arc
- E. Anatectic plutonic rocks

Chapter V. Major structures and terrane boundaries
- A. Boundary faults
- B. Deformed flysch basins
- C. Ophiolite belts
- D. Structural style of terranes
- E. Paleomagnetic data

Chapter VI. Development of major post-accretionary basins
- A. Chukchi/Beaufort Sea continental margin
- B. Bering Sea continental margin
- C. Continental basins

Chapter VII. Quaternary geology
- A. Physiographic development
- B. Quaternary glacial history
- C. Permafrost
- D. Neotectonics

Chapter VIII. Mineral resources (time, space, and geologic relationships)
- A. Metallic minerals
- B. Petroleum
- C. Coal
- D. Geothermal resources

Chapter IX. Summary

Basic Data Maps to be Included in Alaska Volume
1. Terrane map/columns (1:2,500,000)
2. Neotectonic map
3. Glacial geology map
4. Geologic map
5. Metallogenic provinces map
6. Physiographic provinces map
7. Metamorphic map
8. Cenozoic magmatic belts map
9. Pre-Cenozoic magmatic belts
10. Correlation chart for post-accretionary rocks
11. Gravity map
12. Aeromagnetic–marine magnetic map
13. Seismicity map
14. Transects (reductions of continental margin and trans-Alaska transects)

collapsed flysch basins that lie between or below some terranes, the tectonic significance of ophiolite belts, the internal deformational styles and timing of deformation within the terranes, and a synthesis of the paleomagnetic data for relative movement of the terranes.

Development of Major Successor Basins

This chapter will focus on the post-accretionary history as recorded in major sedimentary basins containing overlap assemblages across two or more deformed accreted terranes. This history of overlap deposition begins in northern Alaska in mid-Cretaceous time and migrates to the south where the basal strata become progressively younger. Important successor basins to be discussed are both the inland and offshore basins of the Chukchi, Beaufort, and Bering Sea continental margins, as well as the continental basins of interior Alaska.

Quaternary Geology

Alaska's unique physiography, glacial history, periglacial phenomena, and neotectonic activity will be developed in this chapter. It will include sections on the physiographic development of the present landscape, the glacial history with its implications for migrations of early man across the Bering Strait, and the distribution and nature of permafrost as well as its importance to engineering works in the Arctic. The section on the neotectonics of Alaska will focus on the nature and distribution of active faulting, surface warping, and seismic activity relative to the present plate-tectonics configuration of Alaska.

Mineral Resources

This chapter will provide overviews and summaries of the distribution, significance, and genesis of mineral, fuel, and geothermal resources of the entire state. Much of the basic data will be presented earlier in individual sections dealing with tectonostratigraphic terranes, crystalline rocks, and successor basins. A section on metallic minerals will describe the location, geologic setting, and host rocks for the major mineral deposits, and will evaluate the relationship of the various mineral belts to enclosing tectonostratigraphic terranes. The distribution, evolution, and resource potential of Alaska's increasingly important petroleum basins will be summarized in a section on oil and gas. A section on coal will describe the occurrence of the enormous resources of this commodity throughout the state. Alaska's geothermal resource potential will be evaluated in the context of distribution of thermal springs, heat flow, and young volcanic belts.

Geologic Synthesis and Directions for Future Research

This chapter will evaluate the data on Alaskan geology and geophysics and will present a brief synthesis of the geologic and tectonic evolution of Alaska through time, based on these data. Emphasis will be placed on highlighting major problems and possible directions for future research in Alaskan geology.

The Cordilleran Orogen: Canadian Sector

H. Gabrielse
Geological Survey of Canada
100 West Pender
Vancouver, B. C.
Canada V6B 1R8

C. J. Yorath
Pacific Geoscience Centre
9860 West Saanich Road
Sidney, B. C.
Canada V8L 4B2

Chapter 1. Introduction

A. Preamble

The Canadian Cordilleran Orogen comprises the mountainous region of western Canada and includes the adjacent Pacific offshore area to the foot of the continental slope. Because the evolution of the orogen is linked directly to the dynamics of the nearby interacting lithospheric plates of the eastern Pacific Ocean, the Cordilleran volume deals with the oceanic region as far west as the spreading centres bounding the Juan de Fuca plate and the Queen Charlotte transform fault. (In this text, "Cordillera" implies "Canadian Cordillera" unless otherwise specified.)

In general terms the geology of this region can be related to rifting that formed the western margin of ancestral North America, subsidence and sedimentation along the resultant passive continental margin and, finally, its interaction with allochthonous terranes that produced the varied tectonic styles, regional metamorphism, the various plutonic and volcanic suites, and controlled the physiographic evolution leading to the present landscape. Current concepts on plate-tectonics play an important role in interpretations of the evolution of the Cordillera and, therefore, the integration of geophysical, structural, stratigraphic, and sedimentological data from studies underway in the Pacific offshore region is particularly cogent to an understanding of the ancient history of the orogen. These concepts are stressed by organizing the volume into chapters dealing with assemblages of rocks deposited or emplaced in environments that for the most part have modern-day plate-tectonics analogues.

This is an ideal time for an updated synthesis of Canadian Cordilleran geology. Since the last synthesis in 1970, a great deal of data have been collected and analyzed in the light of the revolutionary plate-tectonics concepts. These concepts have resulted in the following: (1) the identification of allochthonous terranes whose early history was unrelated to the region in which they are now found, and (2) the recognition of regional tectonic assemblages that reflect particular environments and that when arranged in temporal sequence record the tectonic evolution of the region.

In addition, much new and critical data have become available through isotopic age dating; paleomagnetic studies; geophysical studies along the continental margin; chemistry of volcanic and plutonic rocks, including strontium ratios; paleontological studies, particularly on conodonts, radiolaria, and foraminifera; structural syntheses of selected regions; and investigations of regional metamorphism and relationships to structure. All of these have led to new insights into the disposition of terranes and their degrees of allochthoneity; the significance of suites of plutonic and volcanic rocks in the tectonic evolution of the region; rates of uplift and denudation of plutonic and metamorphic belts; the dynamics of the interaction between western North America and the Pacific offshore region; the nature, timing, and significance of structural development; and the relationships of economic commodities to the geology.

These advances in the understanding of Cordilleran geology will be illustrated and emphasized by the accompanying new tectonic assemblage map, several cross-sections of the orogen, updated correlation charts, and a tectonic environment chart. The volume will better integrate mineral deposits with the geology with particular focus on their relationships to tectonic assemblages.

In conclusion, the volume will show the advances that have been made in Cordilleran geology over the past decade and a half during which much exciting research

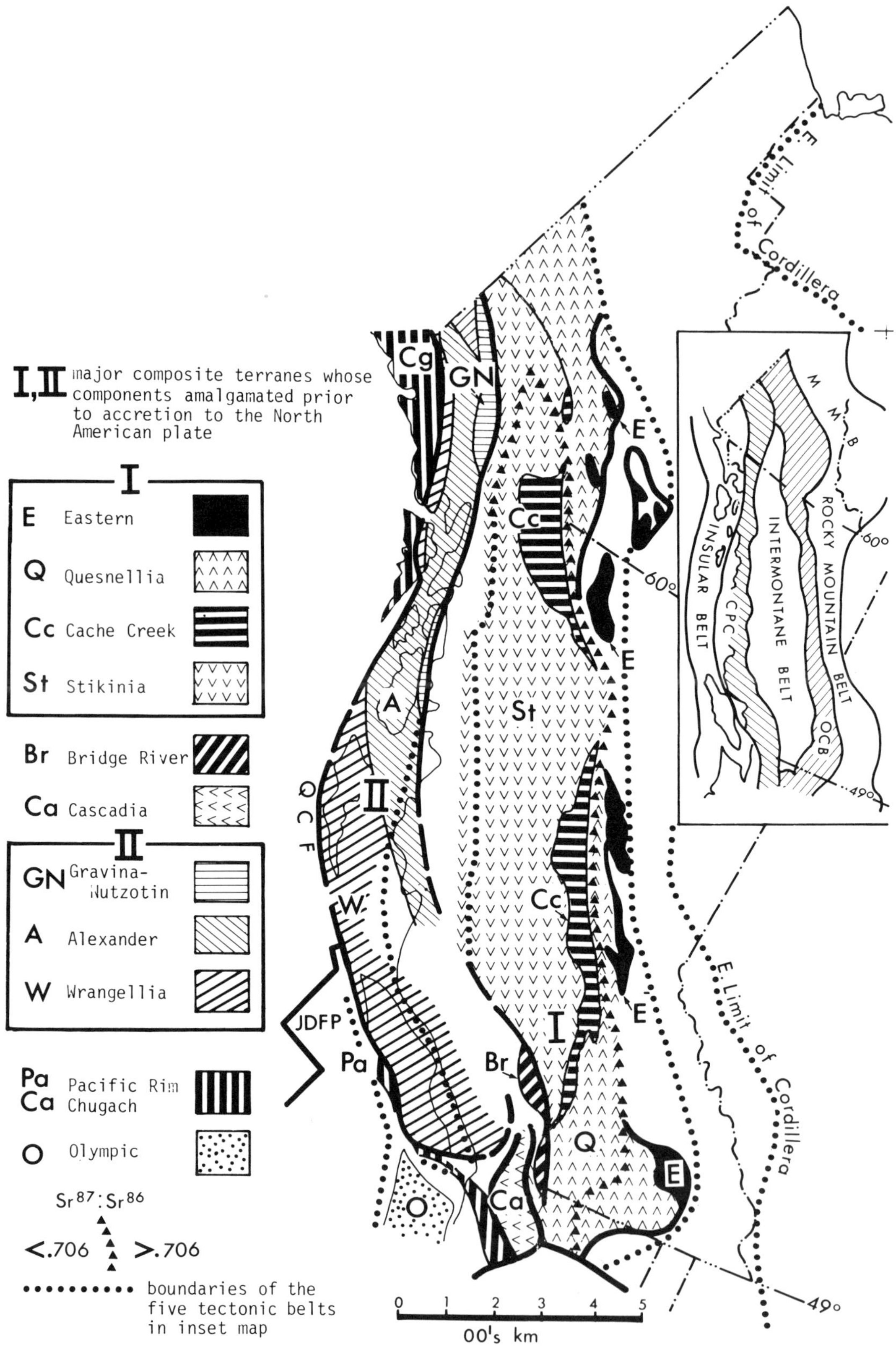

Figure 1. Distribution of terranes and tectonic belts in the Canadian Cordillera (Price et al., 1981). MMB, Mackenzie Mountain Belt; QCF, Queen Charlotte Fault; JDFP, Juan de Fuca Plate; CPC, Coast Plutonic Complex; OCB, Omineca Crystalline Belt.

took place employing many new approaches outlined above within the new conceptual framework of plate tectonics. This will result in a fresh, contemporary view of the Canadian Cordillera that contrasts greatly with the previous syntheses influenced largely by geosynclinal theory.

B. History of Exploration

Three stages of exploration may be stressed. The first involved the remarkable exploration trips during the late 1800s and early 1900s of G. M. Dawson, R. G. McConnell, Joseph Keele, and others. These expeditions and the 49th Boundary Commission Survey provided not only the first rudimentary information on the geology of the Cordillera but also, in some cases, on the physiography. Regional studies by the Geological Survey of Canada and generally more local studies by the British Columbia Department of Mines during the interval from the early 1900s to 1956 resulted in an understanding of the geology in various areas, but much of the region remained unexplored in a comprehensive way. The introduction of the helicopter in support of mapping at a scale of 1:250,000 greatly increased the amount and quality of data collected between 1956 and the present when most of the information on Cordilleran geology was obtained.

Chapter 2. Physiography, Physiographic Evolution and Glaciation

This chapter will be relatively short, probably less than 10 pages, and will include a summary of physiography and glaciation from the Quaternary volume. Emphasis, however, will be on physiographic evolution during the Cenozoic—integrating data on paleodrainage, uplift times and rates (fission-track studies), volcanism, erosion in the different environments, and sedimentation. This should provide the reader with some idea of how the landscape evolved rather than simply a description of the scenic wonders that surround us.

Chapter 3. Tectonic Framework

Three elements of the tectonic framework are the geological terranes, geological belts (Fig. 1), and tectonic assemblages. The terranes are fundamental in that they represent crustal blocks of markedly contrasting history. The boundaries of belts coincide in part with those of terranes but more importantly have been superimposed across them. Tectonic assemblages (used as the basis for an accompanying tectonic assemblage map) record the evolutionary history of the Cordillera.

A. Geological Terranes

In addition to ancestral North America the Cordilleran Orogen includes eleven allochthonous terranes (Figs. 1, 2). Faunal evidence from the Cache Creek terrane and paleomagnetic evidence for Stikinia and Wrangellia indicate that they originated far south of their present latitude. Although the terranes west of ancestral North America are

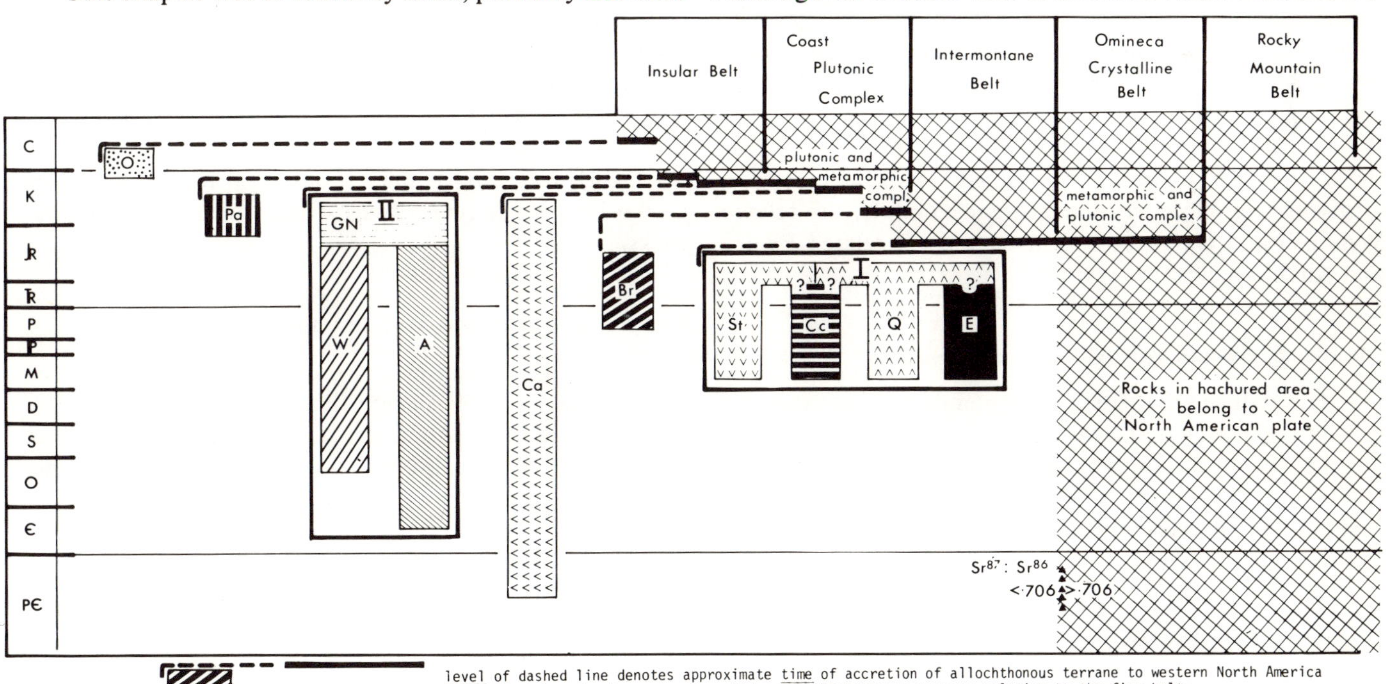

Figure 2. Times of accretion of allochthonous terranes to ancestral North America (Price, et al., 1981).

generally accepted as allochthonous, the relationships between terranes, the manner and direction in which they moved to positions of collision, and the age of accretion to one another and to ancestral North America are problems that remain, for the most part, to be resolved.

B. Geological Belts

Division of the Cordillera into five belts provides a convenient frame of reference relative to two zones characterized by plutonic and metamorphic rocks. They have been considered as core zones of orogens, but their relationships to bounding belts with regard to timing and deformation remain problematical.

C. Tectonic Assemblages

These are fundamental in describing the stratigraphy and in emphasizing the contrasts in stratigraphy between the various terranes. An accompanying coloured tectonic assemblage map on a scale of 1:2,000,000, (Tipper and others, 1981) is an essential part of the Cordilleran volume. Indeed, it should be possible to read much of the stratigraphy of the region from the map and the stratigraphic charts.

Chapter 4. Regional Geophysics and Geochemistry

A summary of geophysical and geochemical data will take a short chapter. Emphasis will be on information that contributes to an understanding of the regional framework, for example, crustal thickness, character of crustal layering, and trends and contrasts in magnetic anomalies. Geochemical parameters help to define the western margin of ancestral North America (initial $^{87}Sr/^{86}Sr$ 0.706 line), whereas the chemistry of plutons and volcanics can be interpreted in terms of tectonic environments, for example, arc-related Wrangell and Garibaldi volcanics and rift-related Edziza and Anahim volcanics.

Chapter 5. Basement Rocks

This will be another short chapter describing examples of well-documented ancient basement rocks: gneiss domes and wedges of the southeastern Cordillera; probably basement rocks in Omineca Mountains (Sifton and Butler ranges); and gneiss complexes of uncertain age in the Coast Plutonic Complex.

Chapter 6. Mid-Proterozoic Assemblages

Pre-Windermere rocks comprise two assemblages—one possibly pre-1335 Ma (Purcell, Wernecke) and another exposed in the northern Cordillera (post-Purcell-Wernecke) locally separated by a profound unconformity. Important mineral commodities in these rocks include lead, zinc, copper, and uranium.

The character and distribution of these rocks pose a number of problems in tectonic setting, age, correlation, and deformation. Do they represent remnants of a miogeoclinal prism along the western margin of ancestral North America? If they are aulacogen deposits, what lay to the west? If they are intracatonic basin deposits, what kind of dynamics determined the great crustal subsidence evidenced by the thickness of basin fill?

Chapter 7. Upper Proterozoic Assemblages

Strata of the Windermere Supergroup best fit a model of rifting, possibly between 770 and 900 Ma ago, rift sedimentation and volcanism, followed by subsidence and sedimentation along a passive margin. Major uncertainties exist, however, concerning correlations along the length of Cordillera and between typical Windermere rocks and the Rapitan Group in the Mackenzie Mountains. In the Mackenzie Mountains the Rapitan Group includes extensive cherty hematitic iron-formation underlying ancient glacial deposits. Stratiform copper deposits occur below the iron-formation. In the northern Rocky Mountains, diamictites in a typical Windermere succession are underlain by thin members rich in magnetite. Remarkably similar successions arranged in the same stratigraphic order are found in Australia. Rifting resulted in subsidence parallel with the new continental margin but also along trends almost orthogonal to the margin, resulting in abrupt thickness changes along the lengths of depositional basins. Lithologically, the typical Windermere rocks are identical with Upper Proterozoic rocks of the Appalachians, and similar environments of provenance and deposition are envisioned.

Lead, zinc, and copper in addition to the aforementioned iron occur typically as stratiform deposits in these rocks.

Chapter 8. Lower and Middle Cambrian Assemblage

Lower Cambrian clastic rocks lie unconformably on older strata along the eastern flank of the miogeocline, but little discordance is evident in offshelf environments. Commonly, the sediments thicken and become finer grained westward. Inner and outer detrital belts with an intervening carbonate belt are readily defined in most regions. In offshelf areas trace fossils are common in formations directly underlying strata mapped as Cambrian and are important for studies defining the base of the Cambrian system.

Although Lower Cambrian sediments suggest relatively even subsidence of the passive margin, Middle Cambrian sediments reveal marked facies differentiation characterized by rapid facies and thickness changes and the presence locally of thick debris flows. In places linear reefoid carbonates bound starved basins. These phenomena indicate rifting with the formation of grabens and horsts in the depositional basin.

Chapter 9. Upper Cambrian to Middle Devonian assemblage

Facies distributions and thicknesses of formations that comprise this assemblage reflect continued passive margin subsidence along the western margin of the continent. Thick Upper Cambrian and Lower Ordovician strata change facies gradually from east to west, suggesting relatively even subsidence. In western exposures, volcanic rocks, although common, are not voluminous. The remaining Ordovician to Devonian rocks show marked differentiation characteristic of platform, subsiding shelf and offshelf facies from east to west. The carbonate to shale transition is abrupt and is marked by debris flow deposits. In offshelf facies, volcanics are particularly abundant in Middle Ordovician strata but occur in every system. Facies and thickness distributions outline a number of basins and embayments, some of which have easterly trending margins (Peace River Arch, Liard Basin; Fig. 3).

The assemblage is particularly important as the host to large stratiform lead and zinc deposits, but there is no consensus about their origin, for example, sedimentary exhalative, volcanogenic, and so forth. Most hypotheses, however, involve a concept of rifting during sedimentation and mineral deposition.

Of fundamental importance is the behavior of the "passive margin" during the early Paleozoic. Rifting and volcanism are well documented. The offshelf character of western facies supports the concept of deposition on attenuated crust, and it has been proposed that rifting may have taken place in a back-arc basin rather than along the margin of an open Atlantic-type ocean. Lower Paleozoic strata on Cassiar Platform are lithologically similar to strata along the eastern margin of the miogeocline and perhaps have been displaced from originally contiguous rocks of the Cordilleran miogeocline by transform faulting. A similar argument can be made for rocks of the Alexander Terrane.

Separation of allochthonous and autochthonous rocks and correlation are problems of great significance in the Lardeau Trough region in the southern Canadian Cordillera. The importance of suggested Ordovician tectonic activity there remains obscure.

Chapter 10. Upper Devonian to Lower Mississippian Assemblage

A flood of thick, in part turbiditic clastic rocks closed

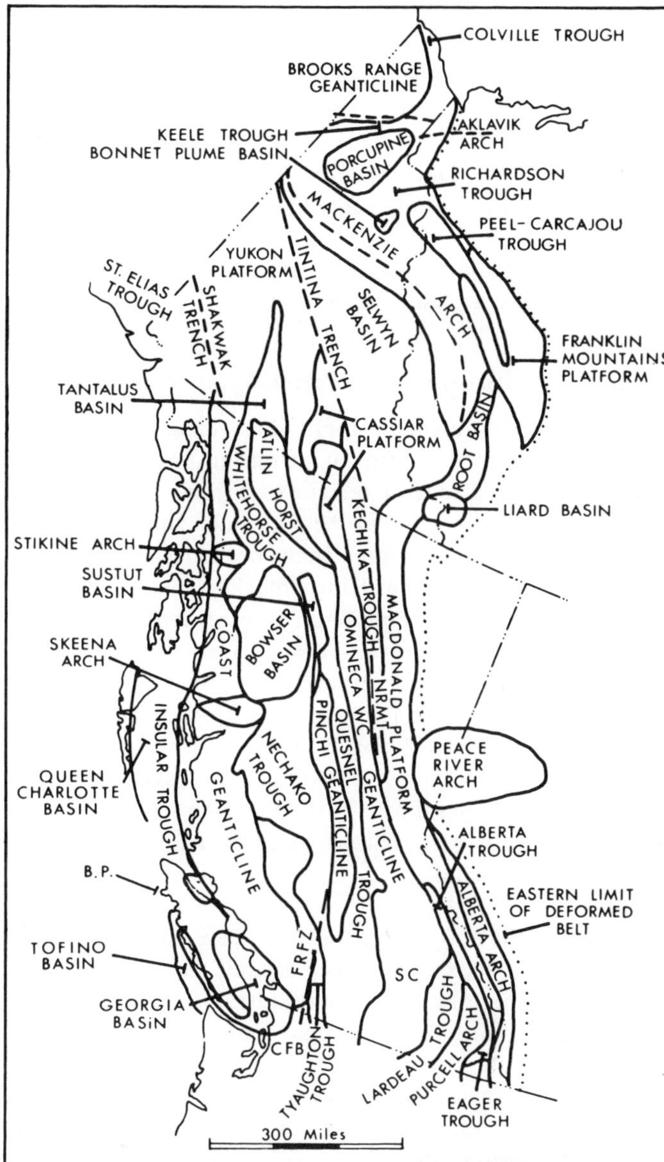

Figure 3. Tectonic elements of the Canadian Cordillera. B. P., Brooks Peninsula; NRMT, Northern Rocky Mountain Trench; WC, Wolverine Complex; SC, Shuswap Complex; FRFZ, Fraser River Fault Zone; CFB, Cascade Fold Belt (Wheeler et al., 1972).

the miogeoclinal depositional interval in the northern Cordillera. At least two great sequences of sediment are involved, each with somewhat different depositional regime and timing (Imperial and "Black clastic"). Carbonate deposition continued in the southeastern Cordillera interrupted by a black shale unit of Late Devonian to Early Mississippian age.

Correlation of the various clastic units and the nature of their environments of deposition are problematical. There is a striking similarity in the character of Late Devonian to Early Mississippian sedimentation in the northern Cordillera and that in the Cordillera of the western United States related to the Antler Orogeny. No contractional structures of comparable age have been demonstrated in the Canadian Cordillera, although Middle to Late Devonian zircon ages have been obtained from granitic plutons in the Omineca Crystalline Belt throughout the length of the Cordillera. Comparisons of the paleogeography and tectonics of Late Devonian to Mississippian time in the northern Cordillera, adjacent Alaska, and the Arctic are essential.

The assemblage hosts several large stratiform lead, zinc, and barite deposits in black shale.

Chapter 11. Mississippian to Permian Assemblages

The Upper Paleozoic rocks show facies changes not only from north to south and from east to west in the miogeocline but also from terrane to terrane. Mississippian strata of the miogeocline in the south are dominantly carbonates that thicken westward. In the northern Rocky Mountains they grade westward into black shale, and farther north clastic rocks are dominant. In northernmost Yukon, facies trends are east-west and basinal areas lie to the south. Pennsylvanian and Permian formations include much clastic material, but carbonate and persistent chert units are also important. Thick Pennsylvanian clastic rocks mark the collapse of the Peach River Arch. Coarse conglomerate characterizes the basal part of the Permian sequence in the northern Yukon and overlies a regional unconformity. The sub-Permian unconformity is present everywhere in the miogeocline.

West of the miogeocline lie diverse assemblages related to oceanic and arc terranes. The oceanic terranes are characterized by chert, basic volcanics, ultramafics, and discontinuous carbonate units. Some contain a distinctive Tethyan fauna but others, for example the Eastern Assemblage (Fig. 1), do not. Arc terranes include volcanic, clastic, and carbonate strata, the latter in some cases forming relatively widespread units.

The original relationships between the terranes and their relationship to ancestral North America is unknown. It is assumed that the Cache Creek oceanic rocks, on the basis of their Tethyan fauna, are far travelled. The lithology of these rocks is similar to that of the Eastern Assemblage, although generally the latter have been much less dismembered by tectonism.

Asbestos is by far the most important commodity in the eastern upper Paleozoic assemblages. Minor nickel and chromite occur locally in ultramafic bodies. Zinc, copper, lead, gold, and silver occur in massive sulfide deposits in oceanic rocks of the Sicker Group on Vancouver Island.

Chapter 12. Triassic to Middle Jurassic Assemblages

The miogeoclinal assemblage reflects a stable shelf environment with considerable variation in basin subsidence along the cratonic margin. Sediments were derived from the craton as shown by facies distribution and paleocurrent data. Remarkably fossiliferous strata spanning most of the Triassic are preserved in the northern Rocky Mountains. Thinner and less complete sequences occur in the southern Rocky Mountains and in the Yukon.

Island-arc sequences of volcanic, plutonic, and sedimentary rocks are dominant in the allochthonous terranes to the west. Particularly important are Karmutsen-Bonanza rocks of the Insular Belt and the Takla-Nicola rocks of the eastern Intermontane Belt. In the Intermontane Belt the volcanics range commonly from alkaline to calc alkaline. Andesite is most abundant, but some regions include considerable dacite and in a few places rhyolite. Lower Jurassic strata are dominantly detrital, derived largely from volcanic island arcs.

Plutons within the arc assemblages range from ultrabasic, characteristically in zoned or small bodies, to quartz monzonite and appear commonly to be at or near centres of volcanism.

Middle Jurassic sediments in the Stikine region record uplift of the Cache Creek Terrane and provide evidence for the development of major southwesterly directed fault and related structures reflecting the collision of Stikinia and ancestral North America during Middle Jurassic time.

The arc terranes are of great economic importance. The Takla-Nicola arc has been called the copper spine of the Cordillera because of the many large porphyry copper deposits related to the plutonic rocks. Molybdenum is important in some occurrences. The most productive lode-gold mine in the Cordillera is in Upper Triassic volcanics of southeastern Coast Mountains. Upper Triassic rocks host a large massive volcanogenic copper-zinc deposit in the Stikine region. Lower Jurassic arc volcanics in the area east of northern Bowser Basin contain gold- and silver-bearing quartz veins.

Outstanding problems concerning these assemblages are the relationships between the oldest plutons and the spatially related Late Triassic volcanics (they seem not to be comagmatic), the nature of the basement to the arcs ($^{87}Sr/^{86}Sr$ ratios are similar to those of oceanic island arcs), and the polarity of the arcs relative to Late Triassic and Early Jurassic basins (fore arc and back arc). Original spatial relationships between the various suites of volcanic rocks and their significance are controversial.

Chapter 13. Upper Jurassic to Mid-Cretaceous Volcanic Assemblages

These volcanics are rare in the eastern part of the Cordillera, but mid-Cretaceous andesitic volcanics, possibly related to mid-Cretaceous plutonic rocks, occur in western Selwyn Basin (Fig. 3). They are the only known volcanics in the Cordillera that appear to be coeval with the mid-

Cretaceous granite suite. Cretaceous volcanics occur in the western part of the Intermontane Belt south of latitude 54° N but generally are not voluminous. They are also present as pendants in the Coast Plutonic Complex. These marine Upper Jurassic to Lower Cretaceous volcanics are host to the Britannia copper-zinc and Northair gold deposits. The volcanics are difficult to correlate from one area to another.

Chapter 14. Upper Jurassic to Cenozoic Assemblages

These rocks contain the depositional and plutonic record of the processes that gave rise to the Cordilleran Orogen. Upper Jurassic foredeep clastic deposits in the Alberta Trough (foredeep) and similar deposits in the deeper Bowser Basin and Tyaughton Trough are evidence for uplift in the intervening region of the Omineca Crystalline Belt and of uplifts along Stikine and Skeena Arches (Fig. 3). This deposition with attendant deformation, plutonism, and metamorphism may be related to the continued collision of Stikinia with ancestral North America.

Similar deposition continued into the Early Cretaceous with clastics restricted to basins and troughs such as Alberta, Mackenzie, Colville, Tantalus, St. Elias, Tyaughton, and Queen Charlotte all related to nearby uplifts. Significant deformation of probable Early Cretaceous age took place in the western parts of the miogeoclinal regime.

Emplacement of mid-Cretaceous plutons along and near the Omineca Crystalline Belt was accompanied by widespread uplift of the orogen. A pre-Albian unconformity can be documented in most of the depositional basins.

Late Cretaceous to early Cenozoic deposition in the foredeep bounding the Mackenzie and Rocky Mountains was in response to eastward-migrating deformation and uplift. Local basins, for example, Sustut, Tyaughton, and Hecate, continued to receive clastic sediments from nearby sources.

From mid-Cretaceous through Cenozoic time, dextral transform faulting was an important factor in the evolution of the orogen. It had a marked effect on structural style, volcanism, and plutonism. Thus, there appears to be a direct relationship between dextral transcurrent faulting along the Tintina and Northern Rocky Mountain trenches and the Fraser River–Straight Creek fault zone, northeasterly directed thrust faults and northerly trending rifts. The latter were sites of important late Cretaceous to Eocene volcanism and plutonism.

The allochthonous, coalesced Wrangellia and Alexander Terrane, superterrane II of Figure 1, was accreted to Stikinia and related to superterrane I in Late Cretaceous time. The Coast Plutonic Complex appears to have been generated along the boundary zone between superterranes I and II, the focus of strong uplift in the early Cenozoic.

Cenozoic history is amenable to analysis with respect to the interaction of the western margin of North America and the offshore lithospheric plates. This is discussed in the following chapter.

A treatment of the Upper Jurassic to Cenozoic assemblages involves a host of problems related to timing and mechanism of colliding and accreting lithospheric plates, relationships of volcanism and plutonism to strain expressed in supracrustal rocks, and so forth. Is uplift and deformation pulsatory or continuous? Why are volcanism and plutonism restricted to discrete episodes if the generating mechanism (subduction?) is continuous? This chapter should perhaps deal with these problems only in a cursory manner leaving a broader synthesis to the chapter on Structural Style and Tectonic Evolution.

Mineral commodities of interest include molybdenum and tungsten with very large producing and potential mines. Important silver production has come from the Yukon, and tin occurrences are being explored in the southern part of the territory. Coal has been exploited locally in the western part of the Cordillera, and extensive deposits occur in the eastern Cordilleran clastic wedge.

Chapter 15. Upper Cretaceous and Cenozoic Volcanic and Plutonic Assemblages

These rocks range in composition from olivine alkali basalt typical of the Miocene-Pliocene flood basalts to andesite, rhyolite, and trachyte characteristic of Upper Cretaceous to lower Cenozoic volcanic centres. Their compositions, distribution, and mode of occurrence can be interpreted in terms of arc or rift environments related to the interaction of the North American and various oceanic plates. The effects of regional dextral transcurrent strain during this interval seem to be paramount. In addition, the possible role of "hot spots" in the mantle is being considered.

Many mineral commodities are associated with the assemblages, copper and molybedenum being particularly important. A resource that has received recent attention is geothermal energy. Young volcanic centres are now being explored for this purpose.

A great challenge for synthesis of Upper Cretaceous and Cenozoic volcanism and plutonism will be the integration of the distribution, character, and chemistry of these rocks in the Cordillera with data obtained from the ocean basins on sea-floor spreading, subduction, and transform faulting.

Chapter 16. Plutonic Rocks

At least seven distinct suites of granitic rocks have been recognized in the Cordillera. They present a contrast in age, chemistry, environment of emplacement, and relationships to volcanics. For example, alpine-type ultramafic rocks ranging from peridotite to gabbro are associated with oceanic assembleges; Alaskan-type ultrabasic bodies are related to arc volcanics; and Late Triassic to Middle Jurassic granodioritic to dioritic plutons are typically spatially related to arc volcanics. Devonian and mid-Cretaceous quartz monzonite to granite is characteristic of regions underlain by continental crust and may be the result of alpine-type subduction of the crust. Late Cretaceous and Cenozoic granitic rocks are in part tectonically related to regional dextral transcurrent strain. The origin of plutons in the Coast Plutonic Complex is enigmatic.

Much data on isotope ages and chemistry of these rocks can be synthesized.

Chapter 17, Regional Metamorphism

Most of the higher grade regionally metamorphosed rocks occur in the Coast Plutonic Complex and the Omineca Crystalline Belt. In recent years detailed studies have resulted in a much better understanding of the relationships between regional metamorphism, deformation, and plutonism. Nonetheless, major problems remain. Dating of metamorphism is particularly difficult, especially in the terranes where there has been more than one episode of metamorphism. Thus, the evolution of the gneiss domes, Shuswap and Wolverine Complexes, and so forth, still present great difficulties for analysis.

Belts of blueschist facies rocks and their ages are important in determing sites and times of subduction. They are known in Cache Creek melanges, in the western part of the Cascade fold belt, and along Tintina Trench.

Zeolite facies rocks are particularly well developed in Mesozoic volcanics of the Intermontane and Insular Belts. Their origin has been attributed to burial metamorphism and to circulation of thermal waters during volcanism.

Chapter 18. Structural Style and Tectonic Evolution

Structural style can be best described in relation to the five geological belts of the Cordillera (Fig. 1). Structural cross-sections of the orogen prepared for the Geodynamics Project will be invaluable for this chapter.

Treatment of the tectonic evolution must deal with the timing and consequences of Proterozoic rifting, the active depositional history of the so-called passive margin mid-Paleozoic rifting and/or orogeny and Mesozoic through Cenozoic plate collisions, accretion, and transform faulting. A synthesis will include a summary of sedimentation, volcanism, plutonism, metamorphism, and structural development in a sequential order. Contrasts between terranes will be highlighted. Generalizations concerning structural style are as follows:

1. *Foreland Thrust and Fold Belt* (Rocky Mountain Belt of Fig. 1). The structure is characterized by southwest-dipping, listric, imbricate thrust faults and related folds in sedimentary rocks that have been stripped off underlying crystalline basement. In eastern Mackenzie Mountains, several important contractional faults are east-dipping but probably are part of a conjugate system with the westward dipping faults. Structural types vary considerably depending upon the character of the stratigraphic assemblages involved. In the northern Yukon, upright open folds, some linked en echelon, and steep, in part transcurrent, faults are typical. Some of the faults appear to have been active in the Precambrian as have many faults (related to continental rifting?) jin the Mackenzie Mountain Belt.

Late-stage extensional faults are probably more abundant than formerly known, judging from local detailed studies. Perhaps the most important question concerning the structural style is the disposition of the crystalline basement—where and how does it become involved in the development of structures shown in the supracrustal rocks? This is important not only across the orogen but also along its length. Timing of deformation is still uncertain in many regions.

2. *Omineca Belt.* A great variety of structural styles result from differences in character of lithologic assemblages, superimposed deformation, and the metamorphic conditions at times of deformation; for example, infrastructural style of core complexes and metamorphic complexes versus supracrustal style of overlying strata. Also significant in the belt are the zones of intense cataclasis and mylonitization associated with regional transcurrent faults and eastward transported allochthonous sheets of arc and oceanic rocks. Westward-directed structures are common in the western part of the belt.

3. *Intermontane Belt.* The belt contains a great variety of structural styles influenced by the nature of the tectonic assemblages. Competent Mesozoic volcanics are little deformed although block faulted. The structures in successor basins vary from gently folded in younger basins to tightly folded strata in older basins, in part separated by decollement surfaces from underlying more competent rocks. Finally, incompetent melanges of the Cache Creek Assemblage are intensely deformed. The eastern rocks of the belt locally have been thrust southwestward, perhaps in response to collision of Stikinia with ancestral North America, and structures in westernmost rocks are commonly directed eastward in response to upward and outward movement of the Coast Plutonic Complex.

4. *Coast Plutonic Complex.* The plutonic style of the complex varies with depth. The deeper parts comprise allochthonous migmatite interleaved with metamorphic rocks passing upward into zones of para-autochthonous, steep-walled, tadpole-shaped plutons. These plutons have migmatitic tails or roots and intrusive heads that in the highest levels form intrusive tongue-shaped bodies. Many

structures are steep, indicating a predominant nearly vertical upward movement of plutons and blocks. Locally, north of lat 52° N, westward-directed recumbent folds are present. Strata in some of the roof pendants are gently folded but elsewhere may be tightly folded and highly schistose.

The Cascade Belt consists of a core of crystalline rocks flanked by outward-directed, in part gently dipping thrust faults. On the west side, metamorphic minerals indicate high-pressure, low-temperature metamorphism associated with subduction complexes. Nonetheless, in many places the stratigraphy and structure are coherent. Clastic wedges on the east and west and the presence of cross-cutting mid-Cretaceous to Late Cretaceous plutons suggest that the main thrusting took place during latest Early or mid-Cretaceous time.

5. *Insular Belt.* The thick competent Mesozoic volcanic rocks are broadly warped and cut by a great number of steep faults. Underlying incompetent strata are commonly tightly folded. Tight folding also is characteristic of rocks that include tectonic melanges in the Pacific Rim Complex. In general, younger strata in the belt are warped or broadly folded.

Chapter 19. Comparative Tectonics

It would be useful to compare the tectonics of the Canadian sector of the Cordillera in a general way with the tectonics of the western United States and Alaska. In addition, comparisons can be made with other orogens concerning the character of foreland deformation, nature, and involvement of basement in the evolution, the role of allochthonous terranes, and so forth.

Chapter 20. Regional Metallogeny

This chapter will include a summary of the kinds of mineral deposits related to the tectonic assemblages. Brief sections will discuss models of mineral formation in light of the Canadian examples, for example, porphyry copper and molybdenum, sedimentary iron formations, strata-bound shale-hosted lead-zinc deposits, and so forth.

Chapter 21. Summary of Outstanding Problems

These will be provided by the various authors and probably grouped under general headings. Some are discussed above, but many will appear during the course of this exercise.

References Cited

Price, R. A., Monger, J.W.H., and Muller, J. E., 1981, Cordilleran cross section, Calgary to Victoria, *in* R. I. Thompson and D. G. Cook, eds., Field Guides to Geological and Mineral Deposits: Calgary, Geological Association of Canada, Mineral Association of Canada, Canadian Geophysical Union.

Tipper, H. S., Woodsworth, G. J., and Gabrielse, Hugh, 1981, Tectonic assemblage map of Canadian Cordillera and adjacent parts of the United States of America: Geological Survey of Canada Map 1505A.

Wheeler, J. O., Aitken, J. D., Berry, M. J., Gabrielse, H., Hutchison, W. W., Jacoby, W. R., Monger, J.W.H., Niblett, E. R., Norris, D. K., Price, R. A., and Stacy, R. A., 1972, Cordilleran structural province, *in* R. A. Price and R.J.W. Douglas, eds., Tectonic styles in Canada: Geological Association of Canada Special Paper 11, p. 1–18.

The Cordilleran Orogen: Conterminous U.S. Sector

B. C. Burchfiel
Department of Earth and Planetary Sciences
Massachusetts Institute of Technology
Building 54–1010
Cambridge, Massachusetts 02139

Gordon P. Eaton
Department of Geology
Texas A & M University
College Station, Texas 77043

Peter W. Lipman
U.S. Geological Survey
Box 25046, MS 913
Federal Center
Denver, Colorado 80225

Robert B. Smith
Department of Geology and Geophysics
University of Utah
Salt Lake City, Utah 84112

Introduction

The Cordilleran sector within the conterminous United States extends from the offshore continental borderlands of the Pacific margin eastward as far as the Black Hills of South Dakota and the mountains of west Texas. It ranges from 800 to 1,600 km wide and is physiographically complex, consisting of high mountains, intervening lowlands, and plateaus that rise from the more gentle continental interior. The physiography of the Cordillera largely reflects the younger underlying structure, and the present orogenic belt has east and west limits coincident with its physiography.

As physiographically defined, however, (Fig. 1) the U.S. Cordillera thus includes diverse older sedimentary, igneous, and tectonic elements varying widely in age, geographic distribution, and geologic significance. Included are the margin of the Precambrian crystalline craton; an extensive cratonic sedimentary sequence developed along a passive margin in latest Precambrian and early Paleozoic time; a collage of terranes accreted to North America intermittently from middle Paleozoic time through the end of Mesozoic time; the igneous and tectonic record of successive plate-convergence events, probably beginning with Antler deformation in middle Paleozoic time and culminating in a fully developed Andean margin late in Mesozoic time; eastward Laramide migration of igneous and tectonic features related to plate convergence in early Cenozoic time; and gradual termination of subduction-related activity and inception of extensional and transform tectonic regimes later in Cenozoic time.

We have selected a time-slice organization for the synthesis of these diverse features. Ultimately, it is the grouping of events by their contemporaneity that permits an understanding of the kinematic and dynamic systems responsible for these events. This is best exemplified when plate-tectonics interpretations are applied to the geologic evolution of a region; a great variety of geologic events and features can be related across diverse geologic terranes and can be grouped into broad plate-tectonics kinematic and dynamic systems. The evolution of the region is then traced by the time-space relations of these dynamic systems.

The time-slice format permits examination of temporally related paleogeographic and geologic features. All areas and tectonostratigraphic units containing rocks and events of a particular time slice will be discussed. This approach focuses directly on broadly contemporaneous features that if properly interpreted can be grouped according to the dynamic systems responsible for shaping their evolution. Time-slice analysis forces us to examine the relationship between diverse geologic elements so that they can be grouped into meaningful genetic systems. Furthermore, it brings into focus those terranes that during that time slice may have been far removed from their present location.

Time-slice organization is dependent on areal or tectonostratigraphic subdivisions, which are second- or third-order elements in the organization. The boundaries of the time units selected must be elastic, because certain events and geologic features are time transgressive. Some events or tectonostratigraphic units may be included within a single time slice, even though their temporal limits fall outside the limits of that time slice.

One important concept developed in this synthesis is that the western Cordillera consists largely of oceanic and continental fragments that have been accreted, reworked, and reshuffled during its complex evolution. Some fragments were clearly foreign to North America before accretion, whereas others have been only displaced or rotated.

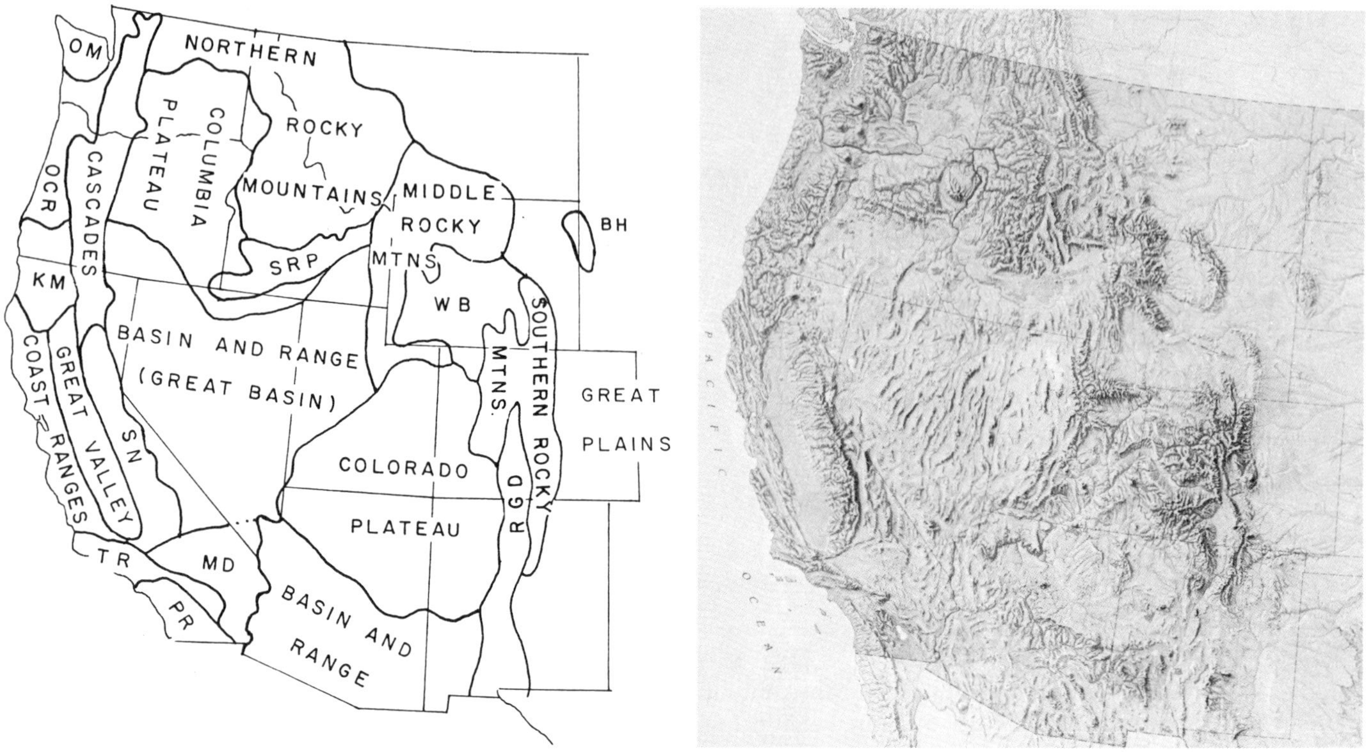

Figure 1. Physiographic regions of the western United States. OM, Olympic Mountains; OCR, Oregon Coast Ranges; KM, Klamath Mountains; TR, Transverse Ranges; SN, Sierra Nevada; MD, Mojave Desert; PR, Peninsular Ranges; SRP, Snake River Plain; WB, Wasatch Basin; RGD, Rio Grande Rift; BH, Black Hills.

The incorporation and modification of these fragments can be treated especially well within the time-slice concept.

Another important concept concerns the continual overprinting of plate-tectonics processes during the evolution of the Cordillera. Dynamic systems related to plate interaction can change rapidly, and the evolution of the Cordillera shows the reworking of features of one dynamic system by another. Time-slice organization brings this into focus and emphasizes the effects of the older systems on the newer ones. Since Cambrian time and perhaps earlier, western North America apparently has been adjacent to a large oceanic region, or paleo-Pacific. The geologic record suggests such continent-ocean interfaces are inherently unstable, developing into zones of plate interaction, and the complex overprinting of events in the Cordillera during much of Phanerozoic time results from the instability at continent-ocean interfaces.

Much of the basic data to be presented in this volume will appear on maps and charts. Principal maps will include a tectonic map, metallogenic map, paleogeographic maps, and geophysical maps (magnetic, gravity, heat flow, and seismicity). The charts will include correlation charts and geologic and seismic cross sections. We hope that these maps and charts will provide basic data for continuity throughout the volume.

The sections of the volume will summarize the major features of each time slice along with some of the major unresolved problems. The time-slice summaries presented here will undoubtedly be significantly modified when input becomes available from the authors responsible for each section.

A final section will contain summaries, discussions, and interpretations of time-independent features of the Cordillera, such as the evolution of magmatic and metamorphic belts; tectonic processes of convergent, transform, and extensional regimes; relations of sedimentation to tectonics; tectonics of accreted terranes; overall tectonic evolution; and metallogenesis.

Pre-Latest Precambrian Time: Formation of the Craton

Precambrian Crystalline Rocks

Precambrian crystalline rocks exposed within the Cordilleran orogenic belt are the westward extension of the Archean and early Proterozoic terranes of the North American craton. Although treated at length in a separate volume, Precambrian lithologies and structures have profoundly influenced the subsequent evolution of the Cordilleran region. Accordingly, pertinent aspects of the Precambrian are summarized here.

In the northern Cordillera, Archean gneisses, schists, granites, and some greenstones with ages greater than 2.5 b.y. crop out in Wyoming, Montana, and part of northwest Utah. They are in places overlain by early Proterozoic sedimentary rocks and have been reworked extensively so that they form a central core of older rocks surrounded by rocks deformed, metamorphosed, and intruded during early Proterozoic time. In Wyoming, rocks of this northern Precambrian terrane are bounded on the southeast by the northeast-trending Nash Fork–Mullen Creek fault zone, an early Proterozoic suture, which divides them from a broad polygenetic southern terrane consisting of several sequences of metamorphosed sedimentary, volcanic, and plutonic rocks that yield ages of about 1.8 to 1.6 m.y. The southern terrane contains no evidence of an older Archean basement and thus may represent a complex series of accretionary elements that added a vast amount of new material to the North American craton. Structural trends are generally northeast and are paralleled by prominent shear zones such as those of the Colorado mineral belt. A younger anorogenic plutonic sequence was intruded into this terrane about 1.4 to 1.5 b.y. ago, and postorogenic granite, syenite, anorthosite, and related rocks were intruded into similar rocks in southwestern California about 1.2 b.y. ago and in the southern Rocky Mountains about 1.0 b.y. ago.

The western extent of Precambrian crystalline rocks is based largely on indirect data. Crystalline rocks crop out in southwestern California only a few tens of kilometers from the present coast, and in east-central Oregon they end against faults. In both places oceanic rocks of Mesozoic age lie farther west, and the crystalline rocks represent the present westernmost extent of the North American craton. Evidence from depositional environments of younger sedimentary rocks and geochemical data from Mesozoic and Cenozoic igneous rocks suggest that the westernmost limits of the North American craton lies at depth in eastern Washington and east-central Oregon and trends southwest across central Nevada into the central Sierra Nevada. From there it trends south, bending southwest in the southern Sierra, and then trends southeast through the eastern part of the Peninsula Ranges of southern California and into western Sonora, Mexico. Although this is the present western limit of Precambrian crystalline rocks, some of the westernmost parts of the North American Precambrian craton have been juxtaposed by Phanerozoic structural events. Where Precambrian crystalline rocks underlie the Cordillera, their rocks and structures often influence younger features: the nature and extent of this influence remains a major focus of research.

Pre-Latest Precambrian Sedimentary Rocks

Unconformably overlying the Precambrian crystalline rocks is a thick sequence of largely terrigenous clastic sedimentary rocks whose distribution is limited to three areas: Montana and Idaho (Belt Series), central Utah (Uinta Mountain Group), and southeastern California (Pahrump Group). Thinner sequences in Arizona, such as the Unkar and Apache Groups represent presumably correlative rocks. Evidence suggests that at many places faulting was active during their deposition. The age and correlation of the rock units that make up these sequences are poorly constrained. The Belt Series was deposited between 1.5 and 0.8 b.y. ago, and it may be older than a deformational event dated at 1.35 b.y. Correlations between rock sequences in different areas, nature of their depositional environment, and plate-tectonics settings remain major problems. Location of the thick sedimentary basins for these rocks had marked effects on younger tectonic patterns and styles in the Cordillera.

Latest Precambrian to latest Devonian Time: Development of a Passive Margin

During the latest Precambrian to latest Devonian time a sequence of rocks was deposited whose distribution, variations in thickness, and facies suggest it was formed along a continental margin. The sequence consists of basal terrigenous clastic rocks of latest Precambrian and Early Cambrian age, overlain by carbonate rocks, with only thin intercalations of terrigenous, clastic rocks, of Middle Cambrian to Latest Devonian age.

The basal terrigenous clastic sequence unconformably overlies all older rock sequences. In most places it contains a diamictite unit near the base that, at least in part, is glacial in origin. This sequence appears to represent a continental shelf, passive margin-type sequence; however, slope, rise, and oceanic rocks of late Precambrian age are unknown. Important discontinuities in the distribution of these rocks occur in southern California where their regional trends strike southwest into the present Pacific Ocean, and in western Idaho where regional trends are truncated along the western end of the Salmon River Arch, a northwest-trending positive element developed in late Precambrian time.

It is still unclear whether rifting began during the deposition of the older sedimentary sequence of Belt and related rocks that underlies the latest Precambrian rocks. If so, did a continental margin develop that was the same as the margin for the latest Precambrian sequence? Evidence suggests that more than one rifting event took place and that the rifting event related to deposition of the latest Precambrian sequence represented the inception of the Cordilleran miogeocline.

During early Paleozoic time, a well-developed passive-margin sequence was deposited in the western Cordilleran region. An incomplete sedimentary section on the craton graded westward into a thicker and more complete continental shelf sequence. The shelf sequence continues unbroken from the terrigenous sequence of the upper Precambrian into a dominantly carbonate sequence. The areal distribution of thick sequences of lower Paleozoic shelf rocks is similar to that of the upper Precambrian rocks and has the same lateral discontinuities in southern California and in east-central Oregon where the Salmon River Arch remained a positive feature. The shelf character of these rocks suggests that they overlie an older Precambrian crystalline basement, and their distribution helps constrain the western limit of that basement.

Continental slope and rise and oceanic sedimentary rocks, thought to be the lateral equivalents of the shelf sequence, indicate a fully developed passive margin by Early Cambrian time. These rocks are now entirely allochthonous. Whereas strong evidence suggests that they were directly adjacent to the North American shelf, such problems as the origin of Middle Ordovician quartzites and Upper Cambrian arkosic rocks in the oceanic assemblage remain unexplained.

Beginning in Middle Ordovician time, accreted rock sequences are present that originally lay west of the North American passive margin. These sequences have no direct ties to North America, and their position relative to North America is unknown. They have reached their present position by accretion to North America at various times. The sequences contain rocks suggestive of oceanic volcanic arc and related environments. Ophiolitic rocks of probable Middle Ordovician age are present in the Klamath Mountains and North Cascades, and some ultramafic and mafic rocks in the Sierra Nevada may belong to these terranes. Shallow and deep marine sedimentary rocks and volcanic rocks are associated with the ophiolitic rocks, but their precise stratigraphic relations to the ophiolites and their correlation from place to place remain a problem. It is uncertain how many different terranes these areally separated sequences represent. The rock sequences suggest they were parts of island arcs active intermittently during Ordovician, Silurian, and Devonian times. Rocks in the Klamath Mountains were deformed in Middle Devonian time, and rocks in the North Cascade Range and the Sierra Nevada were deformed in pre-Middle Devonian time. In the Klamath Mountains Middle Devonian amphibolite-grade metamorphism accompanied eastward underthrusting of the central metamorphic belt beneath the Trinity ophiolite. Recently, Ordovician blueschists have been discovered in the Klamath Mountains, but their relation to the tectonic development of the arc or arcs is unclear.

The relation of these arc terranes to North America is problematic. Arc volcanic rocks are virtually unknown in the North American continental margin and adjacent rocks. Fossils from arc-associated rocks in the Klamath Mountains indicate a North American affinity, but there are no paleomagnetic or other data to suggest where these rocks originated. The geochemistry of the arc assemblages is poorly known, and the polarity of the arc or arcs is unknown.

Latest Devonian to Permian-Triassic Time: Initiation of an Active Margin

In early Mississippian time, the continental slope and rise rocks were thrust onto the North American continental shelf during the Antler orogeny. This orogeny is peculiar in that it contains no associated igneous or high-grade metamorphic rocks. It is represented only by the emplacement of a complex and poorly understood imbricated sheet of sedimentary rocks. It is not known if there were other parts to this orogen that have been lost or what dynamic system produced the orogeny. The Antler orogeny was the first major disturbance along the western margin of the North American continent since its origin in late Precambrian time and had major effects on sedimentary patterns in late Paleozoic time.

The Antler orogeny had created highlands on the western part of the continental margin, from which debris was shed eastward forming a clastic foredeep within the former shelf region. The Antler highlands were worn down and overlapped in Pennsylvanian and Permian times. West of the Antler highlands a sequence of continental rise, slope, and oceanic rocks (Havallah sequence) was deposited, but interpretation of its tectonic setting is dependent upon interpretation of the dynamic system responsible for the Antler orogeny.

During late Paleozoic time, sedimentation continued in the cratonal and eastern shelf regions as in early Paleozoic time but was disrupted through the central part of the craton by development of the Ancestral Rocky Mountains. Uplifts and basins stretched across the North American continent as far west as in the Oquirrh basin in the miogeocline, and local deformation may have extended farther west.

Several upper Paleozoic sequences that cannot be directly connected to each other occur west of rocks that probably were part of the North American continental margin. In the northern Sierra Nevada and Klamath Mountains, sedimentary and volcanic rocks that represent an intermittently active arc terrane overlie similar lower Paleozoic rocks that represent older parts of the same terrane. Similar rocks in the North Cascade Range are probably not correlative with the terrane exposed farther south, and other sequences within the western region also have unknown relations to adjacent terranes. Some are associated with ophiolites, oceanic sedimentary rocks, or melange containing some blocks that have a Tethyan fauna. Such oceanic melange is essentially continuous along the Cordillera, indicating that terranes outboard of this belt are probably exotic. The number of exotic terranes within the western region is still unknown, the polarity of arc sequences is unclear, and time of incorporation of these terranes into the orogen is poorly constrained.

Permian-Triassic into Late Jurassic Time: Development of an Andean Arc

During early Mesozoic time, an Andean-type margin was established along the western side of the North American continent and represented a major change in the tectonic development of the region. This Andean-type dynamic system dominates the evolution of the eastern part of the orogen throughout the Mesozoic, and dynamic systems related to the interaction of oceanic terranes and the North American continent dominate the western part of the orogen.

During the Permian-Triassic Sonoma orogeny, the continental slope-rise sequence and associated volcanic rocks were emplaced eastward onto the continental margin. An arc-continent collision appears to be the most reasonable dynamic system for this deformation, but how many of the upper Paleozoic rock sequences in the west are part of that arc is still uncertain. In southern California, deformation and Andean-type plutonism within the North American continental shelf and cratonal rocks can be broadly dated as Permian-Triassic, but its relation to Sonoma deformation farther north is unknown. It is possible that the Paleozoic trends in the California area were truncated at this time by left-slip transform faulting, scattering the fragments southward in California and Mexico.

Cratonal Mesozoic sedimentary rocks are largely nonmarine but with local intercalations of marine rocks. To the west, above the former Paleozoic miogeocline, sedimentary rocks of this age are uncommon, which makes reconstruction of early Mesozoic history in the central part of the Cordillera difficult. In the western part of what was probably continental North America, thick marine sedimentary and volcanic rock sequences appear to form an essentially continuous Andean-type arc complex. Within these sequences some rocks may represent reshuffled and disrupted parts of this Andean arc terrane, whereas others may be exotic, far-traveled fragments. Along the westernmost part of North America, an essentially continuous belt of very chaotic rocks (melange in part) contains ophiolites, deep marine sedimentary rocks, and volcanic and volcaniclastic rocks. Many of the ophiolite bodies present in the western part of the Cordillera were formed and accreted during this time period. The number of different terranes incorporated within this belt and their time of accretion are uncertain, as are the polarities of many of the arc sequences, their time and place of origin, their times of deformation, metamorphism and accretion, and their mechanisms of accretion.

Within the western and central part of the Cordillera, plutonic rocks form the core of one or more Andean arcs. Generally low-grade metamorphic rocks are present almost continuously along the western part of the Cordillera, and low- to high-grade metamorphic rocks are more irregularly exposed along its central and eastern part. Intraplate deformation and east-directed thrusting occurred within and east of the Andean magmatic belt, and similar deformation is present west of the belt where vergence is generally west-directed. Interspersed between these apparently compressional events, strike-slip and extensional faults developed both locally and regionally. The role of such faults in the overall development of the Cordillera is becoming more widely recognized.

Major uncertainties within this part of the Cordillera include the time-space relations of orogeny, arc magmatism and metamorphism, and the relation of these events to accretionary processes at the western margin of North America. We have yet to understand the dynamic systems that operated throughout this time period and must establish the contemporaneity of events more precisely before such reconstructions are feasible.

Late Jurassic into Late Cretaceous Time: Culmination of Igneous and Tectonic Activity along the Andean Margin

Throughout this time slice, the Andean dynamic system continued to dominate the tectonics of the central and eastern Cordillera and produced a foreland fold and thrust belt along its eastern side. Accretion of two major fragments, the southern part of the Wrangell terrane and the North Cascade Range, dominated tectonics along the northwestern part of the orogen. The nature and extent of accreted terranes farther south remain a problem.

The cratonal sedimentary sequence began with mainly nonmarine and locally marine rocks, but in Early Cretaceous time changed to a marine sequence with the development of a wide seaway in the continental interior. These marine rocks grade westward into thick foredeep clastics that were shed from and partly involved in the folded and thrusted areas in the eastern part of the orogen. In southeastern California these foredeep clastics are missing, and the extent, tectonic setting, and nature of clastic deposits farther south, as well as their relations to the Mexican geosyncline and plate boundary activity, remain a problem. In the central Cordillera the time-space relations of a broad and well-developed Andean arc or arcs are still poorly known, except in the south where there was a clear eastward migration of the arc extending into Late Cretaceous time. Plutons in the western part of the arc intrude the accreted terranes of earlier time periods, and a discontinuous forearc basin is present along the west side of the arc. The westernmost belt of rocks forms a greatly varied accreted sequence of oceanic rocks in the Franciscan complex, and correlative rocks exist to the north. Some of these accreted rocks represent trench melange, but how much of the complex consists of small fragments from a great variety of different terranes remains an unsolved problem. Larger fragments, such as the North Cascade Range and the Seven Devils terrane, may be North American, but the latter correlates best with the exotic Wrangall terrane of Canada and Alaska.

The older part of the east-directed foreland fold and thrust belt was developed along the eastern part of the Paleozoic passive-margin sequence and Belt Series rocks, except in southern California and Arizona where it developed within a cratonal sequence. To the west of this deformed belt discontinuous outcrops of metamorphic rocks exhibit contrasting ductile deformational style. Along the west side of the Cordilleran orogen, west-directed thrusts resulted from subduction and intraplate activity, but separation of these two thrust fault types remains a problem. High-pressure, low-temperature metamorphism has affected the accreted rocks and is an intimate but poorly understood part of the convergent process. The overall dynamic system or systems responsible for the development of the orogen is a major problem, and careful studies that establish contemporaneity of events are necessary. Strike-slip and extensional faults are present both locally and regionally, but their full significance is uncertain. The structural discontinuity in eastern Oregon which probably developed at this time may be related to strike-slip faulting.

Late Cretaceous and Eocene Time: Eastward Migration of Laramide Igneous and Tectonic Activity

During latest Mesozoic and earliest Cenozoic time, the pattern of igneous activity changed significantly, the foreland fold and thrust belt terminated in the central part of the Cordillera, and Laramide uplifts and basins developed within the North American craton.

In the craton, marine sedimentation continued to the end of Cretaceous time and locally into early Tertiary time. These marine sandstones and shales grade westward into thick orogenic sediments along the eastern margin of the foreland fold and thrust belt. From late Cretaceous until late Eocene time, large local basins developed adjacent to Laramide uplifts and were filled with subaerial and lacustrine sediments. Along a zone that trends north-south through the central Cordillera, volcanic and nonmarine sedimentary rocks are present only locally. Along the western side of the Cordillera are forearc-basin sequences as well as rocks that represent underthrust oceanic assemblages and contain probable far-traveled, exotic fragments. These sequences are well developed but discontinuous, as in Oregon and western Washington.

The Andean magmatic belt, which was continuous in the western region until about 80 m.y. ago, changed significantly during this period. At about the same time as Laramide uplift was initiated, the belt began to sweep eastward, became discontinuous, and finally became inactive in mid-Tertiary time. Eastward sweep and extinction of magmatic activity was diachronous along the U.S. Cordillera, culminating in the northern Rocky Mountains about 70 m.y. ago, as recorded by the Boulder batholith and associated volcanics, but continuing in the central and southern Rockies in the Arizona–New Mexico upper province and along the localized northeast-trending Colorado mineral belt until about 55 m.y. ago. These Laramide events are the oldest that span the full width of the U.S. Cordillera as discussed in this volume.

Orogenic development continued along the west side of the belt, including accretion of oceanic rocks and fragments. Right-slip displacement is evident in the North Cascade Range, and paleomagnetic studies indicate large rotations of the Siletz River and southern Sierra Nevada areas. Oceanic metasedimentary rocks were emplaced along a low-angle surface beneath the magmatic belt and Precambrian rocks of southern California at roughly the same time as the Laramide uplifts and development of mylonitic fabrics and penetrative deformation in southeastern California and Arizona. The foreland fold and thrust belt continued to develop until the late Eocene in the northern and southern Cordillera but was inactive in a central segment from northern Utah to southern California. East of the inactive segment of the thrust belt and partly overlapping it to the north and south, the basement was compressed and elevated and adjacent areas depressed as Laramide deformation broke the continental crust of the craton. This partly rejuvenated the Paleozoic Ancestral Rocky Mountains structures and extended as far east as the Black Hills in more subdued style. The mechanics of Laramide deformation are controversial, and its relation to changes in the geometry of subduction and the magmatic belt remain important research topics. Reconstruction of dynamic systems operative during this time period depends strongly upon establishing the contemporaneity of events across more than 1,500 km of the orogen.

Eocene to Recent: Transition to Extensional and Transform Tectonics

Sedimentation, igneous activity, and tectonics of this period record a gradual transition from the plate-convergent regime that dominated Cordilleran tectonics for several hundred million years to a gradually enlarging extensional regime that reflected transform motion between Pacific and American plates.

In the early part of this time period, Cordilleran sedimentation may be considered in three parts: a narrow belt of largely marine deposits along the Pacific margin, local continental basin-fill deposits in the central and eastern mountainous regions, and a formerly widespread sheet of continental deposits east of the Rocky Mountains front that are mostly preserved farther to the east. Continental rise and shelf sediments of the Pacific margin include subduction-related melange deposits; facies are variable and record complex local offshore topography related to rapidly changing plate boundaries. The local basin-fill deposits within the Cordillera reflect a continuation of post-Laramide orogenic sedimentation and record the initiation of extensional deformation and associated graben formation, especially in the northern Rocky Mountains and along the Rio Grande rift zone. Many of these deposits are poorly understood because of intricate local facies changes and difficulty in dating subaerial sedimentary sequences. East of the mountainous region, remnants of thin, widespread post-orogenic sediments indicate continued erosion of the Laramide uplifts, although more slowly than in Eocene time.

In the latter part of this time period, deposition of marine sediments continued along the Pacific margin, accumulating to great thicknesses in local basins adjacent to the San Andreas and related transform faults, and filling the now inactive submarine trench west of the Cascade volcanic arc. In the central and eastern Cordillera, thick continental sediments also accumulated in local grabens related to regional extensional deformation. These basin-fill deposits are lithologically variable, reflecting proximity to rugged local topography and varying tectonic histories of the different basins.

Igneous activity during the early part of this period consists mainly of calc-alkaline volcanics. These rocks show complex space-time-compositional patterns that have been interpreted as continuing retrograde activity of the Andean arc as the descending slab slowed, steepened, and deformed internally, although this interpretation is still controversial. As this seemingly arc-related volcanism shifted to the west and south, a contrasting assemblage of dominantly basaltic or bimodal basalt-rhyolite activity began diachronously about middle Miocene time in areas of extensional deformation such as in the Rio Grande rift zone and in the northern Rocky Mountains. In addition to their significance for plate-tectonics interpretations, these rocks are associated with many important base- and precious-metal ore deposits.

During the later part of this time period, arc volcanism continued in gradually diminishing intensity in areas north and south of the growing transform boundary. At present, arc volcanism is represented within the United States only by the relatively inactive Cascade Range that lacks a well-defined seismic Benioff zone, presumably because of the small size, youth, and relatively slow convergence rate of the oceanward plate. East of the Miocene Cascade arc a great outpouring of basaltic lava on the Columbia Plateau may reflect weak back-arc spreading. Subsequently, the Snake Rive Plain became a complexly evolving locus of volcanism in which linear sequences of silicic centers, followed by basaltic activity, migrated both to the northeast toward the Yellowstone region and on a reduced scale to the northwest toward Newberry volcano. These volcanic features, which have been interpreted as the result of a major asthenospheric hotspot or as migration of magma along ancient lithospheric flaws, clearly represent unusual events not yet incorporated into the overall tectonic evolution of the Cordilleran region. East of the transform boundary, the dominant late Cenozoic volcanism consisted of relatively small discontinuous basaltic volcanic fields, although local larger centers were more compositionally diverse and major silicic ash-flow calderas formed at the Jemez Mountains and in Long Valley. During the later part of this time period, extension continued blocking out most of the Basin and Range province.

Tectonically, much of the Cordilleran region was relatively quiescent during the early part of this time slice. Notable exceptions are the complex deformation that accompanied the beginning of Pacific–North American plate interaction along the Pacific margin, the development of metamorphic core complexes associated with low-angle faulting and mylonitization in the central Cordillera, and the initiation of the Rio Grande and northern rifts, which seem to represent intracontinental back-arc extension east of continuing Andean arc volcanism farther west in the Cordillera. Timing and correlation of many of these events are poorly constrained and worthy topics for more intensive study.

Significant uplift of the entire Cordilleran region, including the Colorado Plateau and adjacent parts of the High Plains, is generally associated in space and time with the increased volcanism; both may reflect a deep thermal event. Maximum regional uplift occurred along the southern Rocky Mountains part of the Rio Grande rift and in the Yellowstone area, regions that had especially intense volcanism. Tectonics of the continental margin of the southern Cordillera continued to be dominated by transform faults and associated structures. The last major change occurred about 5 m.y. ago when the plate boundary shifted onto the North American continent as marked by initial opening of the Gulf of California and the present trace of the San Andreas fault. The continental borderlands of southern California developed extensional basins related to the transform system, and many areas along the margin have been rotated dextrally.

Contemporary Tectonics and Lithospheric Structure

Preexisting zones of weakness inherited from previous episodes of tectonism influence the contemporary deformation of the Cordillera. Along the west coast, the San Andreas fault now forms an active boundary between the obliquely converging North American and Pacific plates. Average rates of convergence between these plates exceeds 6 cm/yr. Individual segments of the San Andreas fault, however, demonstrate both aseismic slip and variable slip rates, some less than 5 cm/yr, suggesting that part of the interplate strain may be accommodated internal to the North American plate. Intraplate strain release mechanisms of the Cordillera have been related to several hypothesized mechanisms: (1) intraplate lateral shear between the stable part of the North American plate and the Pacific plate; (2) influence of Farallon plate subduction and related lithospheric diapirism; and (3) crustal intrusion and concomitant extension.

In the recent decade regional geophysical data have been used to elucidate the lithospheric structure of the Cordillera and to evaluate various kinematic and dynamic models of intraplate deformation. Modern earthquake data show broad zones of epicenters around the southern, eastern, and western parts of the Great Basin. Secondary seismicity occurs in western Montana, central Idaho, and along the Rio Grande Rift. Subduction-related seismicity has been demonstrated for the Oregon-Washington coast and reflects low rates of convergence. These broad zones of earthquake activity can be interpreted to infer elastic strain release around the margins of subplate blocks influenced by intraplate oblique convergence. Earthquakes within the western Cordillera are shallow and seldom exceed 15 km in depth. Focal mechanisms confirm the lateral-plate convergence along the San Andreas fault, whereas within the interior of the Cordillera variable solutions primarily of normal and strike-slip faulting mechanisms characterize the intraplate seismicity. A common characteristic of the intraplate mechanisms is the predominance of west- to northwest-trending minimum compression axes (T-axes) suggesting that the interior of the western United States is responding to west to northwest extension.

Regional gravity anomalies suggest relative isostatic compensation for structural features with dimensions greater than ~25 km. The magnetic field responds primarily to the presence of crystalline basement and has been used to infer the presence or absence of Precambrian basement. Regional heat-flow patterns show normal to low values along the San Andreas fault and low heat flow in the Sierra Nevada. Heat flow increases to above normal values across the Great Basin and along the Cascade Range, whereas normal to low heat flow characterizes the Colorado Plateau and Rocky Mountains provinces. The abrupt heat-flow transition at the boundaries of the Great Basin suggests shallow and recently active heat sources interpreted to reflect intraplate extension such as crustal underplating, dike and sill intrusion, and lithospheric upwarping by mantle mechanisms.

Trilateration data show north-south compression along the San Andreas fault. However, similar measurements across the Great Basin suggest low strain rates and a complex strain pattern of horizontal compression to extension. Trilateration data suggest east-west compression for the coastal area of Washington, whereas the Columbia Plateau appears to be in relative strain quiescence. Level-line reoccupations demonstrate that uplift of several millimeters per year associated with the San Andreas fault in southern California and the Great Basin shows areas of relative uplift or eastward regional tilt. Local uplift anomalies as large as 15 mm/yr have been documented across Cenozoic hot spots such as Yellowstone and the Rio Grand rift.

While the regional geophysical data are relatively dense, their interpretation remains in many cases equivocal. Important problems remain: (1) There is little correlation on a one-to-one basis of earthquakes and late Cenozoic faulting within the interior of the Cordillera. New seismic reflection data suggest that shallow-dipping listric faults may accommodate extensional slip. (2) Lithospheric structure is anomalous within the Cordillera with crustal thicknesses as small as 25 km beneath the Great Basin but thickening to 50 km beneath the Sierra Nevada and 45 km beneath some parts of the overthrust belts of the Cordillera. Upper mantle velocities are low beneath the Great Basin, suggesting the possibility of partial melts. These properties demonstrate an important problem, that is, how is the time-spatial transgression of Mesozoic compressional tectonics related to variations in crustal thickness, heat flow, and the current extensional regime of the western interior? (3) Intraplate deformation of the Cordillera is related both to brittle and nonbrittle processes. What is the relative balance of these mechanisms and how have they affected the geometry of overthrust faulting, folding, and later extensional, normal faulting?

Tectonic Synthesis

A tectonic synthesis will include summaries and discussions leading to interpretations of the tectonic evolution of the Cordilleran orogen. Such a synthesis will include many of the time-independent features of the Cordillera and overall development and processes represented in Cordilleran evolution. The exact nature and content of this chapter will be determined largely by what will be covered in the time-slice sections of the volume, but we would anticipate such subjects as (1) tectonic processes of accreted terranes, (2) development of the foreland fold and thrust belt and Laramide structures, (3) evolution of metamorphic belts and their relation to tectonism, (4) evolution and genesis of magmatic belts, (5) extensional tectonics and core complexes, (6) strike-slip faults and their importance in Cordilleran evolution, (7) patterns of sedimentation and basin analysis, and (8) Cretaceous to Holocene plate kinematics and tectonic models.

The Geologic Regions of Mexico

F. Ortega-Gutiérrez
J. C. Guerrero-García
Instituto de Geología, UNAM
Ciudad Universitaria—Delegación Coyoacán
04510 México, D. F.

Introduction

The main objective of this volume will be to present the state of modern knowledge about the geology of the most important geologic regions of Mexico (Table 1). These regions were carefully selected on the basis of their geographic, chronological, and tectonic characteristics in order to enhance regional discussion and correlation with adjacent areas.

Mexico is a country endowed with an enormous variety of geological systems and mineral wealth. Doubtless, the geologic record of Mexico and its present tectonic activity will reveal many elements of basic importance to the resolution of some of the fundamental problems faced today by the science of geology.

In the recent geologic past, hundreds of thousands of cubic kilometers of volcanic material of felsic composition were "added" in a few million years to the Mexican continental crust. Where did this material come from? Was it extracted from the mantle, or from the fusion of older continental crust?

The answers to these sorts of questions could influence in a most definite way the presently debated theories explaining the petrologic evolution of the crust and mantle of the earth since the beginning of geologic history.

The study of recent tectonic activity in the Gulf of California and in the Acapulco Trench can be of primary importance to the understanding of those geologic processes that have shaped the lithosphere at least in Phanerozoic time, that is, ocean expansion and contraction.

This volume, as conceived, will deal with the geology of all Mexico: two billion years of geologic history written across two million square kilometers of exposed land. However, even though the Cordilleran orogenic system encompasses most of this territory (Figure 1), the eastern part of Mexico is certainly more akin to the evolution of the eastern margin of North America, including elements of the Grenville, Appalachian-Ouachita, and Gulf Coast tectonic regimes. The distribution of the selected regions of

TABLE 1. GEOLOGICAL REGIONS OF MEXICO, A PROPOSAL OF SUBDIVISION

I. Introduction
II. History of research
III. Physiography
IV. Geophysics
V. *Precambrian regions*
   1. Sonoran
   2. Oaxacan (Grenville)
   3. Others
   4. Precambrian evolution
VI. *Paleozoic regions*
   1. Cordilleran
   2. Huastecan (Appalachian-Ouachitan)
   3. Paleozoic evolution
VII. *Mesozic regions*
   A. Earlier Mesozoic
      1. Eogulfian
      2. Cordilleran
   B. Later Mesozoic
      1. Atlantic-Tethyian
      2. Cordilleran
   C. Mesozoic evolution
VIII. *Cenozoic regions*
   1. Gulf Coastal Plain and Yucatan Peninsula
   2. Continental sedimentary and volcanic basins
   3. Sierra Madre Occidental
   4. Transmexican Volcanic Belt and late Cenozoic volcanism
   5. Baja California and adjacent Pacific islands
   6. Gulf of California
   7. Cenozoic evolution
IX. Summary and conclusions

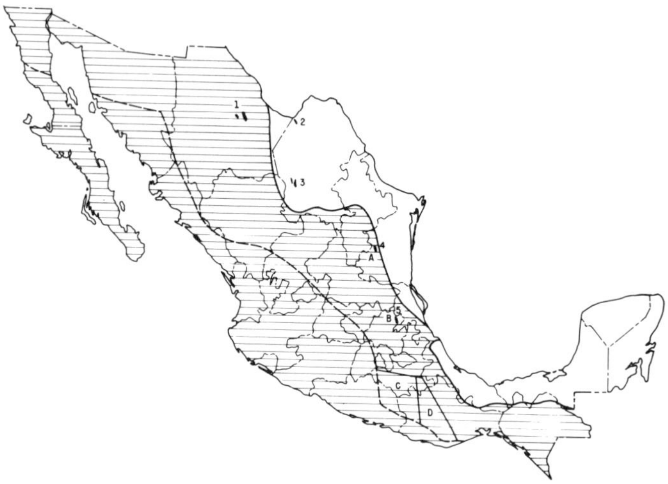

Figure 1. Major geological provinces of Mexico. Ruled pattern, regions affected by or generated during the Cordilleran tectonic regime. Unruled pattern, Gulf Coast province. Dashed line indicates, to the east of it, the real or inferred presence of Precambrian continental crust. 1, Sierra del Cuervo and Placer de Guadalupe Paleozoic areas. 2, Sierra del Carmen. 3, Las Delicias. 4 and A, Ciudad Victoria. 5 and B, Molango. C, outcrop areas of the Paleozoic Acatlán metamorphic Complex. D, outcrop area of the granulitic Oaxacan Complex.

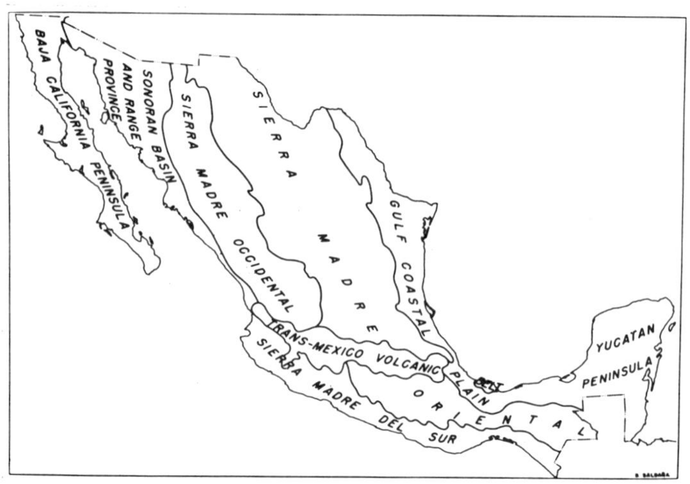

Figure 2. Physiographic provinces of Mexico.

Mexico, unfortunately, cannot be shown clearly on a small-scale map because of the poor knowledge, complexities, and composite nature of the elements used in their definition.

Accordingly, in Figure 1 only the two major tectonic systems of Mexico are shown, together with those areas in which overprinting by these systems is minor. The latter are small geologic "windows" that preserved the record of previous or independent orogenic history.

The classical tectonic subdivision of the North American plate into its well-known western and eastern continental margins enclosing a Precambrian craton may be applied easily to the northern part of Mexico where its maximum width exceeds 2,000 km. However, the country quickly thins out to the south and at the Isthmus of Tehuantepec is only 200 km wide and yet the same major elements of the North American tectonic framework still can be discerned, albeit highly disturbed. Consequently, the geologic constitution of southern Mexico is overwhelmingly complex, making subdivision into tectonic provinces difficult.

If we take, for example, the Sierra Madre del Sur province extending from the State of Jalisco to the Isthmus of Tehuantepec, it is possible to find in its geologic record at least four major sutures that separate small terranes or tectonic microprovinces with contrasting geologic evolution. Nevertheless, each microprovince is extremely important in the study and understanding of the geologic history of the entire region (Figure 2).

From the tectonic point of view, Mexico is bordered by three contrasting oceans: the Pacific, which is being consumed underneath Mexico in the Middle America Trench; the Gulf of California, a small ocean recently born from the mantle and in the process of expansion; and the Gulf of Mexico, an interior ocean that now apparently remains static. The major lithospheric evolutionary processes responsible for those oceans have shaped the internal architecture of most geological provinces of Mexico during late Phanerozoic time. But looking still deeper into geologic time and crustal depth, it is possible to find scattered and extensive outcrops of rocks that date back to Precambrian and early Phanerozoic times. These are the exciting pages of the geologic history of Mexico that will be explained in this volume.

The outlined subdivision of the geological regions of Mexico (Table 1) is based primarily on time, secondly on inferred tectonic history, and thirdly, where the region is large and physiographically well defined, on geographic distribution. However, because the geologic data concerning each of the regions proposed are extremely variable in quantity and quality, the integration of a volume like this is a difficult task. It is necessary at this stage to mention some severe limitations that the work will have to inevitably face:

1. There are very few factual data published on the geology of Mexico in comparison with its territorial extension and the great variety of geologic problems encountered.

2. Sedimentary marine basins, because of their present wealth and oil potentiality, have been relatively much more studied, in detriment to the extensive crystalline and volcanic belts that underlie most of the Mexican territory and adjacent oceans.

3. It is bad fortune that in Mexico geologists and geophysicists know each other's work too little. The continuously increasing geophysical information obtained by

several national and foreign institutions in numerous regions of Mexico has been rarely used to improve or constrain proposed geologic models' and vice versa, the available geologic information is not infrequently disregarded when geophysical data are converted into a geologic model of any certain region.

Thus, given the poor knowledge of Mexican geology, it is easy to ask still a great number of fundamental questions referring to the evolution of the proposed geological regions of Table 1.

These questions, however, are only those that come immediately to our minds from our personal interests and experience in geologic work; many others, probably more fundamental questions, will certainly be raised by contributors to this volume within their fields of geologic interests. But it is hoped that besides the vital importance of detecting several of the main problems of Mexican geology, those eventually engaged in the writing of this volume will make their best effort to come out with the best possible partial or complete answers to the questions posed.

Precambrian Evolution

The Precambrian rocks of Mexico are in absolute minority, somewhat scattered, and probably the least studied from the structural and petrologic points of view. Consequently, the reconstruction of the Precambrian evolution of Mexico has to be rather speculative. The two major Precambrian provinces belong to the Proterozoic. The Sonoran province of northwestern Mexico includes rocks of the lower, middle, and upper Proterozoic, while the Oaxacan province of southern Mexico belongs entirely to the middle Proterozoic.

Despite the small extent of Precambrian outcrops in Mexico, the understanding of its geology is of great importance to the reconstruction of the paleogeography of old continents. The "Grenvillian" age of the Precambrian rocks of southern Mexico, for instance, together with the identification of similar rocks in small windows and wells along most of the eastern margin of Mexico, and possibly as far south as the border with Guatemala in Central America, would imply that this part of Mexico maintained its geological integrity throughout Phanerozoic time, or that if displacements can be documented, these were of little importance.

Also of major significance is the possibility that the Grenville Belt is continuous from Labrador to Central America. The location of the Grenville Front in the subsurface basement terranes of Mexico is of great relevance to proposed displacements of the southern region of the North American craton in relation to the Mesozoic origin of the Gulf of Mexico. The description of little-known or unknown Precambrian terranes throughout the world and the interpretation of their petrologic and tectonic history are of great value if uniformitarian or nonuniformitarian plate-tectonics concepts are going to be applied to Precambrian lithospheric evolution.

Some of the basic questions to be addressed for this early part of the geologic history of Mexico are the following:

1. Besides the Sonoran terrane of northwestern Mexico and Oaxacan in the south, are there any other Precambrian belts exposed elsewhere? Granulite facies banded gneisses reported from the State of Sinaloa and the recently dated late Proterozoic granitic rocks in the State of Chiapas seem to be new candidates that may extend the record of Precambrian history of Mexico.

2. What is the subsurface distribution of Precambrian terranes in Mexico? Have all these terranes formed part of a single craton during Phanerozoic tectonism?

3. Is the Oaxacan province, as traditionally accepted, an extension of the Grenville province of the eastern Canadian Shield? Striking similarities in age (950–1,100 m.y.) and the widespread occurrence of granulites, anorthosite massifs, and platform-type sediments in the paragneissic protolith in both provinces strongly support the correlation. The continuation of the Oaxacan granulites toward Central and South America remains an outstanding problem.

4. Can conventional plate-tectonics models be applied to explain the origin and evolution of the Sonoran and Oaxacan terranes? The pervasive granulite facies metamorphism that affected the entire Oaxacan province and the abundant calc-alkalic magmatism present in the 1.6 to 1.8 b.y. crystalline terranes of the Sonoran province seem to require the operation of normal (i.e. present) plate-tectonics mechanism in their evolution.

The union of the exposed Precambrian terranes of western and eastern Mexico at depth, beneath the Phanerozoic cover of central and northern Mexico, poses fundamental problems such as the possible continuation of the Grenville Front in this region and its displacement by alleged sinistral transverse faults, such as the Torreón-Monterrey lineament and the Mohave-Sonora megashear.

Finally, the western continental margin of Mexico includes several accreted terranes, most of which are poorly dated or completely unknown. It is certain, however, that Cordilleran tectonism strongly affected this margin to the point of almost complete elimination and reworking of older sialic crust. It is possible that Precambrian relicts will be eventually identified in these regions and may shed some light on the nature and provenance of these terranes.

Paleozoic Evolution

The understanding of the Paleozoic orogenic history of Mexico is particularly difficult because of the scattered distribution and intense metamorphism and deformation of the majority of the exposed areas. However, the questions raised by the nature of these terranes are of central importance to the comprehension of the early Phanerozoic evolution of the continental margins of North America, as will become clear from the following questions:

1. Although the existence of an "Appalachian-Caledonian" metamorphic belt in southern Mexico (Acatlán Complex) has been well established, are the metamorphic rocks of northeastern Mexico (Granjeno Schist) also related to this orogenic system? The presence in both complexes of ophiolite-like assemblages affected by high pressure/low temperature metamorphism, dated as early Paleozoic in the Acatlán Complex, indicates that oceanic lithosphere was consumed in a subduction zone and an ocean disappeared in this region. Was this Paleozoic ocean part of Iapetus?

2. Isotopic Ordovician, Devonian, and late Paleozoic ages have been obtained from the metamorphic rocks of these two complexes, the tectonothermal evolution of which is extremely complex. How may the multiple deformation and metamorphism of the Acatlán Complex and Granjeno Schist be related to well-established orogenic phases of the Appalachian, Caledonian, and Ouachitan tectonic systems?

3. The metamorphic belts of Central America, such as Chuacus and Omoa, include rocks with a tectonic history in many ways comparable to that of the Paleozoic crystalline belts of southern Mexico. Is it possible then to reconstruct, across major Mesozoic and Cenozoic faults, suspected former continuity of all these terranes?

4. Early and late Paleozoic sedimentary rocks occur adjacent to the Granjeno Schist and Acatlán Complex, which are partly of equivalent age. To explain this relation, major tectonic discontinuities or unconformities have been proposed without general agreement. This problem is perhaps the most important that may be raised concerning the Paleozoic evolution of eastern Mexico.

5. Given the answers to these essential geologic problems posed by the Paleozoic terranes of eastern Mexico, how can they be fitted into the reconstruction of Mesozoic Pangea? And how were these terranes displaced and deformed during the evolution of the present continental margins?

The Paleozoic record of western Mexico as known today seems represented almost exclusively by sedimentary rocks related to the evolution of a passive continental margin. There are terranes, however, such as those of northern Sinaloa and in the southern tip of Baja California, that are composed of low- and high-grade schists, gneisses, and migmatite that could include elements of Paleozoic or even Precambrian orogenic history.

Mesozoic Evolution

The breakup of Pangea at the end of Paleozoic time marked a fundamental, still poorly understood geologic revolution that had the most visible effects in the constitution and modification of the lithosphere underlying Mexico. The formation of the Gulf of Mexico and the convergence of the western margin of North America constitute the geodynamic framework in which the Mesozoic systems of Mexico evolved.

These systems in general are better known, more widespread, and the most important of Mexico due to their actual and potential content in metals and hydrocarbons. However, even the well-known Sierra Madre Oriental, a fold-and-thrust belt formed by Jurassic-Cretaceous marine sediments, still poses fundamental questions related to its evolution:

1. From the nature and present distribution of sedimentary rocks in the Sierra Madre Oriental, what can we infer about the paleogeography and paleotectonics of the corresponding basins and environments of deposition?

2. What is the precise nature of its structural geometry and tectonic style? Can the latter be geographically zoned?

3. How much is it known about the vertical dimension of this major orogenic belt?

4. Where are the "roots" of this system?

5. What is the meaning of the major bends that the system undergoes in northern Mexico between Monterrey and Torreón, and apparently just south of the Transmexican Volcanic Belt?

The northwestern margin of Mexico, including the Peninsula of Baja California, may be naturally related to the Cordilleran evolution of Western North America during Mesozoic time and, therefore, major questions (and solutions) could be quite similar in both areas. Nevertheless, perhaps a unique and most significant question that has been asked in relation to the Mesozoic evolution of this part of the Mexican Cordillera pertains to the role of the Mohave-Sonora megashear (Silver and Anderson, 1974) as a transform boundary connecting the Pacific and Atlantic megatectonic lithospheric systems in middle Mesozoic time. Geologic enigmas, such as the origin of the Gulf of Mexico and the present distribution of pre-Mesozoic basement terranes in northern Mexico, have found in this concept a sound working hypothesis for attempting their final solutions.

By contrast, the southwestern margin of Mexico from the State of Jalisco to the very border with Guatemala is an array of complex, crystalline, and volcano-sedimentary terranes, the nature and interrelationships of which cannot so easily be explained by "Cordilleran" models. The Precambrian crust (Oaxacan Complex) approaches to within 30 km from the Pacific coast, and the Xolapa Complex, an enigmatic high-grade plutono-metamorphic belt extending for more than 600 km along the Pacific coast, still remains essentially unknown.

Some of the very important questions that may be raised concerning the Mesozoic evolution of south and southwestern Mexico are the following:

1. To what extent can the "accreted terranes" concept be applied to include *allochthony* as the most important factor in the understanding of present relationships among (tectono-stratigraphic) terranes and their evolution?

2. What is the age and the tectonic nature of the Xolapa Complex?

3. What is the age and nature of the northern limit of the Xolapa Complex against the "Appalachian-Grenvillian" Paleozoic continental margin?

4. Why, in the first place, are we including the Xolapa Complex in this Mesozoic context if, in fact, it has been traditionally considered a Paleozoic or even Precambrian terrane?

5. The Xolapa Complex is the largest metamorphic belt of Mexico (20,000 to 30,000 km^2) and also the least known. Radiometric dates from early Paleozoic to middle Tertiary time have been obtained by several methods on its minerals and rocks. The oldest (indisputable) stratigraphic sequence unconformably overlying the Xolapa Complex is of late Tertiary age!, even though sedimentary, metamorphic, or volcanic sequences assigned to the Paleozoic or Mesozoic occur in close contact relations with the Xolapa Complex.

The most significant geologic feature of this complex is related to its metamorphism. The widespread occurrence of migmatite and granitic orthogneisses in which minerals such as cummingtonite, cordierite, and sillimanite (wollastonite in associated calcareous units) indicates low-pressure and high-temperature metamorphic conditions. Thus, is the Xolapa Complex the low P/T component of a dismembered paired metamorphic belt, the high P/T complement of which was laterally displaced? If so, when did this occur?

6. Measured strontium isotopic ratios in certain rocks of the Xolapa Complex indicate a rather primitive protolith, raising this question: Is the Xolapa Complex a good example of the crustal roots of a magmatic arc, and does it include reworked elements of older arcs? Is there any pre-Mesozoic continental crust involved?

The nature of the continental lithosphere in southern Mexico, beyond the limits of the Xolapa, Acatlán, and Oaxacan complexes, appears to undergo a major change in structure and composition. Relief suddenly rises to more than 3 km above sea level along a front of strongly mylonitized rocks of obscure stratigraphic parentage, and the sector of the Sierra Madre Oriental exposed to the east of this front is composed of intensively deformed sedimentary and epimetamorphic rocks of Mesozoic age. Some of these rocks were deposited over oceanic crust of equivalent age. The tectonic significance of this thick mylonitic belt separating "basin" from "platform" Mesozoic sedimentary facies remains an essential problem: Do the major mafic-ultramafic complexes exposed along this contact mark a collisional suture where former oceanic lithosphere was consumed? Are there any magmatic arc or subduction assemblages represented within this possible Mesozoic suture? What is the timing of the multiple orogenic phases represented here, and how do they correlate with those of the less deformed northern sector of the Sierra Madre Oriental?

Cenozoic Evolution

During this time most of the Mexican territory emerged from the seas, and a complex period of geologic evolution was initiated. Uplift, erosion, and calc-alkalic volcanism were the dominant processes of this final part of the geologic evolution of Mexico.

In early Tertiary time, the eastern part of Mexico was the site of compressive orogeny (Hidalgoan or Laramide) and the Sierra Madre Oriental fold-and-thrust belt was formed. Western Mexico was probably facing a continuous subduction zone, and above it, a parallel belt of andesitic volcanism was emplaced on emerged continental crust. On the other hand, vast areas of present southeastern and eastern Mexico were undisturbed and still under the sea.

Subducting lithosphere in the west underwent a main period of intensified activity during Oligocene-Miocene time, and as a result, the largest ignimbritic province on earth, the Sierra Madre Occidental, was emplaced.

Major reorganizations of plate geometry at the end of early Tertiary time seemingly sliced a large piece of continental crust off the southern part of Mexico. The present truncated nature of this margin has been explained in this way. The uplift and dominant lateral shear stresses resulting from this process probably were the dynamic systems that generated the folded and faulted province of the Sierra Madre de Chiapas. Abundant calc-alkalic magmatism also invaded this part of the crust during this time.

As a consequence of continued plate convergence in late Cenozoic time along southern Mexico, intense calc-alkalic volcanism has been building the east-west–trending chain of major strato-volcanoes that characterizes the Transmexican Volcanic Belt.

Meanwhile, plate divergence by sea-floor spreading in northwestern Mexico has formed the Gulf of California in the last few million years. The attending tholeiitic and alkalic volcanism extended not only along the gulf and adjacent areas, but it seems that related extensional regimes are propagating across northern, eastern, and central Mexico. Volcanic alkalic centers such as La Primavera in the State of Jalisco, Tuxtla in Veracruz, and the common occurrence of alkalic basaltic fields of late Cenozoic age in northern and eastern Mexico attest to the present tensional state of stress of the major part of the Mexican lithosphere.

Main problems related to the Cenozoic evolution of Mexico concern the reconstruction of the geometry and geodynamic framewrk responsible not only for the magmatic and structural evolution of Mexico during this time but also for the formation of the majority of its mineral deposits. Some of the most important questions in this context may be the following: Was there really a continuous subduction system along the entire western margin of Mexico during most of Tertiary time, and if so, is there a record of concomitant calc-alkalic magmatism in the Tertiary volcanic provinces of the Mexican continental interior? What is the plate-tectonics significance of the Tertiary calc-alkalic and alkalic volcanisms of western and eastern Mexico, respectively? Where is, if any, the missing piece of continental margin whose displacement would be responsible for the apparent truncation of southern Mexico against the Pacific basin? Finally, why is the Transmexican Volcanic Belt oblique to the Middle America Trench?

Conclusions

These are many of the basic questions that will be addressed in this volume in search for the most consistent answers deductible from available data. Unfortunately, lack or scarcity of isotopic dates for most of the geological regions of Mexico (there is about only one isotopic age per each 5,000 km^2) will prove to be one of the most important limiting factors for the proper understanding of the geologic evolution of the chosen regions. The other two major limiting factors are the existence of large areas without geologic or even topographic maps and the paucity of geophysical information about the deep structure of the crust and upper mantle beneath Mexico.

Questions such as to what precise distance can the Precambrian crust outlined in Figure 1 be extended westward underneath the Cordilleran System will be particularly difficult to answer without the data of gravimetric, magnetometric and seismic studies.

The recent account of widespread volcanic fields in northern and central Mexico, carrying samples as xenoliths from the underlying deep crust and upper mantle, will prove very useful to mitigate the present lack of geophysical information and to delineate the possible amount of tectonic overlap that the Pacific continental margin of Mexico underwent during the Mesozoic tectonic regimes.

The geologic history of mineral deposits cannot be divorced from other aspects (*i.e.*, tectonic, petrologic, etc.) of geologic evolution, because these unique materials, as a rule, are intimately associated in space and time with the petrotectonic units in which they occur. However, inasmuch as a special volume on the economic geology of Mexico is planned within these regional syntheses on the geology of North America, the nature and origin of the associated mineral deposits in the geological regions of Mexico will not be discussed in any detail in this volume. It is highly desirable, however, that brief mention of this topic is made, since the nature of the mineral deposits is often of great help to the better comprehension of the tectonic character of any geological region under investigation.

The actual task in front of those experts that will be contributing to the integration of the present volume is not only one of compilation and condensation of the existing data into a coherent article. Substantial new ideas will certainly result as the scattered and often apparently unrelated geological and geophysical information is critically analyzed and logically assembled.

The integration of a volume of this nature should be considered most timely, because the output of data from national and foreign earth sciences institutions has been enormous in the last decade, relative to the extremely few scientists currently engaged in the study of the foundations of the geology of Mexico.

It is anticipated that this work on Mexican geology will be a most stimulating and influential book and a breakthrough for years or decades to come, for it will constitute the first integral attempt, armed with the most modern concepts of geologic evolution, directed to the understanding of the geologic history of Mexico.

Cited Reference

Silver, L. T., and Anderson, T. H., 1974, Possible left-lateral early to middle Mesozoic disruption of the southwestern North American craton margin: Geological Society of America Abstracts with Programs, v. 6, p. 955.

The Arctic Region

Arthur Grantz
U.S. Geological Survey
345 Middlefield Road
Menlo Park, California 94025

Leonard Johnson
Office of Naval Research
Arlington, Virginia 22217

J. F. Sweeney
Pacific Geoscience Centre
9860 West Saanich Road
Sidney, B. C.
Canada V8L 4B2

With contributions from: D. L. Clark, Olav Edholm, R.K.H. Falconer, D. M. Hopkins, Yngve Kristoffersen, H. C. Larsen, and Jorn Thiede

Introduction

The synthesis of the geology of the North American plate in the Arctic Ocean (Fig. 1) will cover the area between the Arctic (Nansen) Mid-Ocean Ridge and its extension into the Siberian continental shelf on the Eurasian side, and the Arctic continental margins of Greenland, Canada, Alaska, and Chukotka on the Amerasian side (Fig. 2). This predominantly oceanic region consists of a complex of ridges and plateaus of diverse and, in part, enigmatic origin, several oceanic deeps, and thick continental-rise sedimentary prisms. The margins of the region consist of an active spreading center along the Nansen Ridge, an active fracture zone off northern Greenland, and older passive margins off Canada, Alaska, and Chukotka.

The bathymetry and geography of all the world's ocean basins and their borderlands are now known—except for the Arctic. Although apparently detailed maps of this region have been published, the soundings supporting the bathymetry shown are sparse, resulting in contours that only grossly reflect the true bottom physiography. In the Arctic, geologic knowledge is so limited that suggestions about the nature of the Alpha-Mendeleev Ridge, for example, range from continental crust to fossil subduction zone or extinct spreading ridge. The hostile environment over most of the Arctic Ocean has thus far prevented a systematic study of this important subbasin of the world ocean. Despite numerous expeditions into its marginal regions and repeated attempts to visit and probe its interior, the deep Arctic Ocean remains a frontier for geologic exploration.

Knowledge of the Arctic is important for more than just solving Arctic problems, and its plate tectonic evolution is closely linked to the evolution of the Atlantic and Pacific basins as well as to the continents that almost surround it. For example, the separation of Svalbard and northeast Greenland about 36 m.y. B.P. (La Brecque et al., 1977) may have been related to global changes in the direction of plate movement and led to major changes in North Atlantic water circulation, sedimentation and climate. An understanding of past and present plate movements and accretionary tectonics in the Arctic will be required before a complete model of even late Mesozoic and Cenozoic worldwide plate motions can be achieved. These motions and the structure, paleontology, and paleoenvironment of the Phanerozoic sedimentary rocks of the circum-Arctic and its continental shelves are highly relevant to the exploration for hydrocarbons. Large hydrocarbon discoveries in northern North America and the presence in the Arctic of large sedimentary basins of diverse ages and of wide continental shelves suggest that the Arctic may contain major resources of oil, gas, and coal.

The following pages provide a brief summary of what is presently known about various areas or topics related to the geology of the Arctic and what the significant problems are perceived to be. Many of the statements are derived

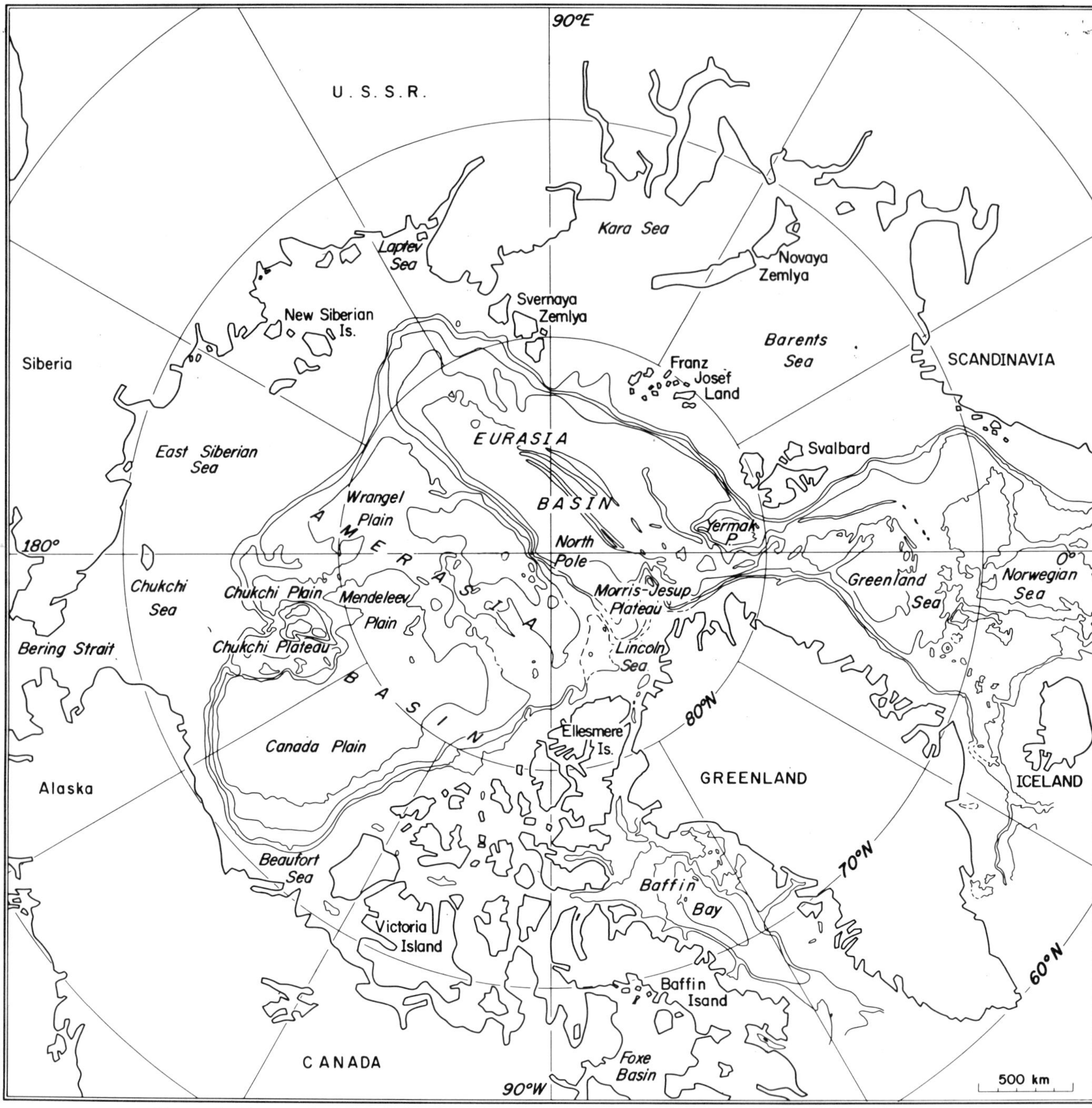

Figure 1. Geography of the Arctic region. Bathymetry in kilometers; 500-m contour also shown.

from the results of a workshop to define a 5-year plan for eastern Arctic research (Falconer and Thiede, 1980).

Regional and Structural Problems

Norwegian-Greenland Sea

The breakup between Norway and Greenland took place about 57 m.y. B.P. and was associated with what are now marginal escarpments (Vöring and Faeroe-Shetland escarpments in the east and Greenland escarpment in the west, Fig. 3). Initiation of sea-floor spreading on the Mohns Ridge was concurrent with breakup and proceeded in a geometrically straightforward manner, whereas spreading in the areas north of the Greenland Fracture Zone and south of the Jan Mayen Fracture Zone was more complex.

The area south of the Jan Mayen Fracture Zone is generally acknowledged to have contained a spreading cen-

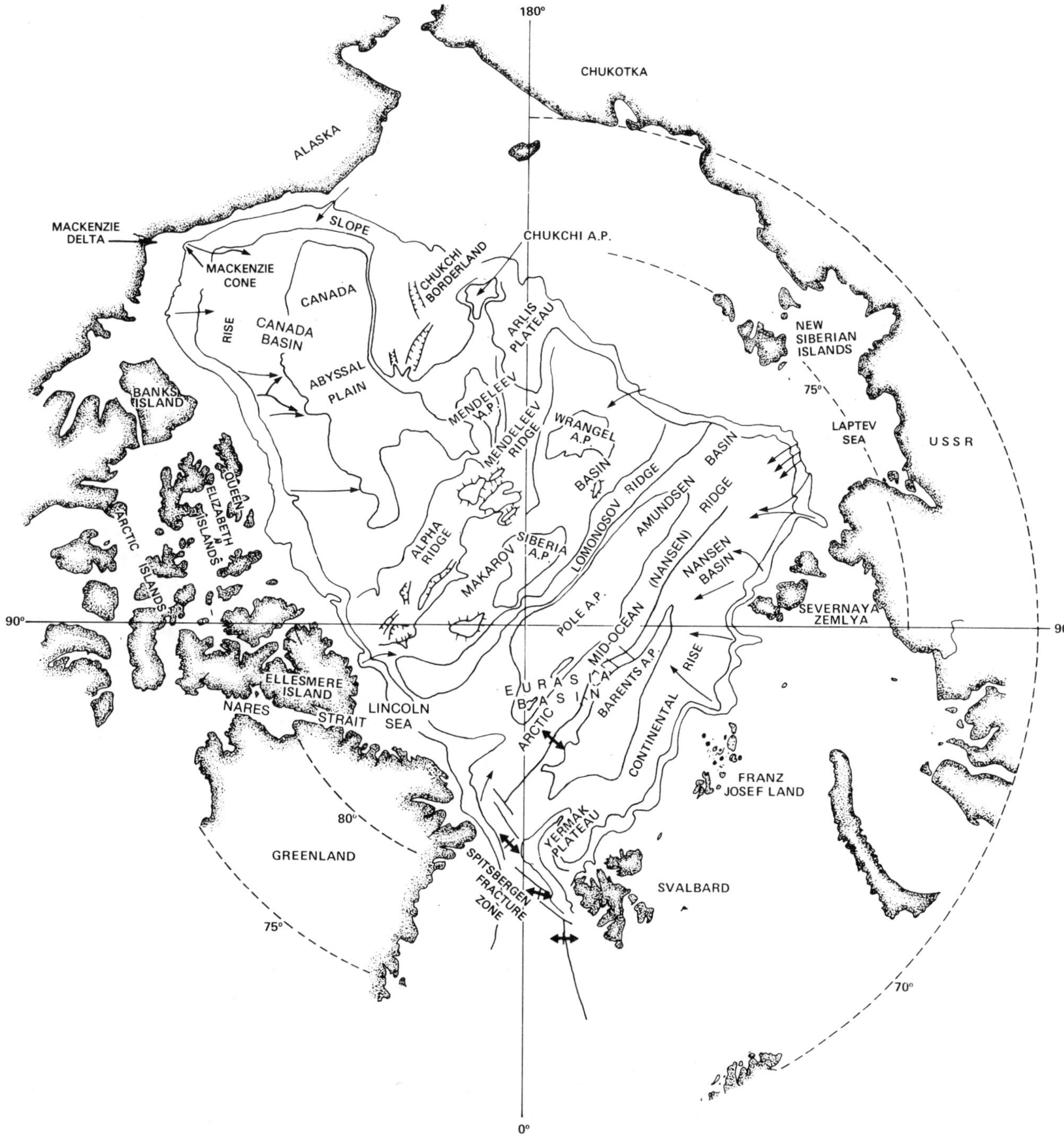

Figure 2. Physiographic diagram of the Arctic Basin. Thin arrows indicate submarine canyon systems; thick double arrows indicate axis of spreading center.

ter before the establishment of the presently spreading Kolbeinsey Ridge, which may have been active since magnetic anomaly 6C time (24 m.y.B.P.). The earlier spreading axis lay to the east of Kolbeinsey Ridge in the Norwegian Basin. Spreading there began about 57 m.y.B.P. (about anomaly 24) and ended between 36 m.y.B.P. (anomaly 13) and 24 m.y.B.P. During the shift of the spreading axis from the Norwegian Basin to the Kolbeinsey Ridge, a fragment of the Greenland continental shelf was broken off and is now the Jan Mayen Ridge (Fig. 3).

North of the Greenland Fracture Zone, sea-floor spreading in the Norwegian-Greenland Sea has apparently taken place only since about anomaly 13 time (36 m.y. B.P.). The present spreading axis, the Knipovitch Ridge, is probably not older than 5 to 6 m.y. No major opening existed prior to 36 m.y. B.P., and the motion

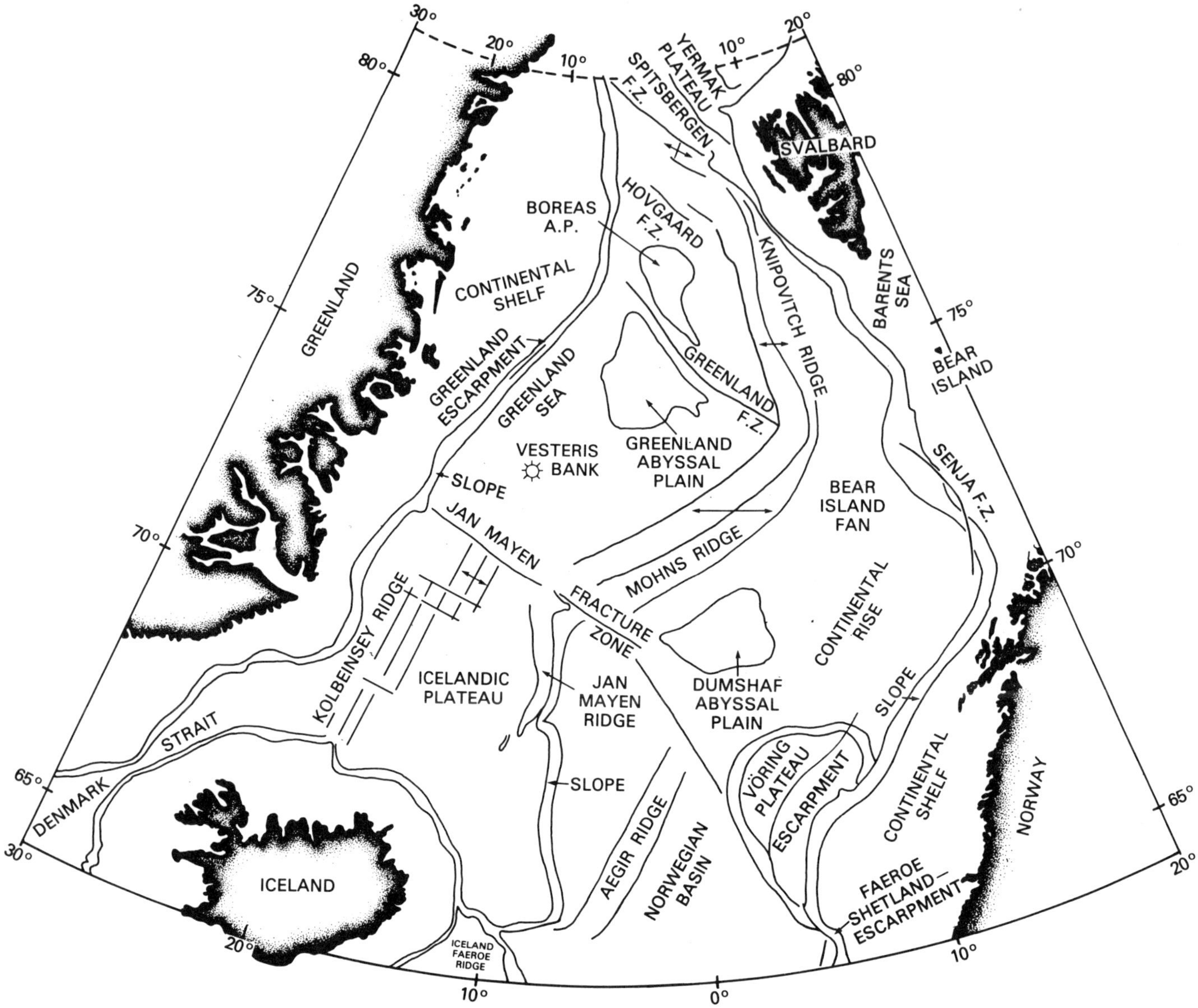

Figure 3. Physiographic diagram of the Norwegian–Greenland Seas. Axis of spreading center is denoted by double-ended arrows.

between Greenland and Svalbard was primarily one of shear along the Greenland and Senja Fracture Zones. Significant problems in need of resolution include: (1) the spreading history of the region between the Greenland and Spitzbergen Fracture Zones, where the spreading ridge axis appears to have changed position at least once since 36 m.y. B.P.; (2) the structural history and vertical motion of the surrounding continental margin; and (3) the structure and evolution of the Spitzbergen Fracture Zone in space and time. For example, does the fracture zone lie entirely in oceanic crust, and is it subdivided into short ridge segments or is it one large and continuous transform fault? If it is segmented, do any of the segments bound remnants of continental crust? What is the relationship of the fracture zone to the Tertiary compressional deformation of Svalbard?

Eurasia Basin

The plate-tectonics evolution of the Eurasia Basin is now reasonably well known from detailed aeromagnetic surveys, scattered bathymetric data, and regional plate kinematic considerations. The basin is mainly the product of sea-floor spreading along the Nansen Ridge over the 50 to 60 m.y. since anomaly 24 (late Paleocene) time. A narrow continental fragment that was rifted off Eurasia, the Lomonosov Ridge, forms one of the borders. Because the part of the Nansen Ridge nearest the New Siberian Islands closely approaches the pole of rotation between the Eurasian and North American plates, which now lies in the Siberian Platform, its spreading rates are exceptionally slow: half rates of 0.2 cm/yr (Siberian Platform) -0.6 cm/yr (near Greenland).

During the short period of seismic monitoring over the crestal mountains of the western flank of the Atlantic end of the Nansen Ridge, microearthquakes were recorded at a rate of 1 to 2 events per hour. Epicentral separations were 15 to 25 km. Two parallel epicentral lineations were found, a main one that coincides with the median valley and another, less active one that follows the northwest wall of the rift valley. This activity could have been related to faulting along the rift valley wall. A crustal model for the western crestal region, based on refraction profiles, gave compressional velocities of 4.6 km/s at the sea floor, 6.1 km/s at 1.5 km depth, and 7.9 km/s at 6 km depth. In the median valley a 7.2 km/s refractor was found at a depth of 3 km.

Significant problems in need of resolution include: (1) the nature and origin of magnetically smooth crust between anomaly 24 and the basin margins; (2) the origin of dramatic along-strike changes in magnetic amplitude and, at least along the rift axis, correlative changes in water depth (Do the areas of low-magnetic amplitude correspond to basement deeps floored by crust with an exceptionally thin basaltic layer? Is such along-strike variability characteristic of extremely slow spreading axes?); (3) the composition of basaltic material along the Nansen Ridge axis and comparison of its along-strike variability, if any, with variations in amplitude of magnetic anomaly, seismic structure, and bathymetry; (4) the presence or absence of hydrothermal activity along the Nansen Ridge axis; and (5) the question of whether the ends and the Amerasian side of Lomonosov Ridge meet adjacent terrains at sheared, rifted, or composite structures.

Needed research in the Eurasia Basin includes seismicity studies, and particularly fault plane solutions, for which a local seismograph network and ocean-bottom seismometers would be especially useful. Routine bathymetric, gravity and magnetic surveys, and bedrock sampling are also needed.

Lomonosov Ridge

Preliminary analyses of geophysical measurements and sediment samples collected during the Lomonosov Ridge Experiment (LOREX) in 1979 indicate that in the vicinity of the North Pole the Lomonosov Ridge consists of tilted en echelon fault blocks. Bottom sediments in the adjacent abyssal plains lap onto and abut unconformably against the ridge flanks. These sediments may consist in part of material derived from the ridge by current scour, but which part is unclear. The crust beneath the ridge is divided into two P-wave velocity zones, a 4.7 km/s layer about 6 km thick over a 6.6 km/s base. The crust extends to a depth of about 27 km, and the topographic portion of the feature has an average density modeled to be not greater than about 2.5 g/cm^3. The combination of ridge morphology, structure, average density, P-wave velocity, and sediment character suggests that the ridge is composed of continental material. The nature of the intersection of the Lomonosov Ridge with the Siberian and Canadian continental margins is a significant problem in need of resolution.

Alpha-Mendeleev Ridge

The enigmatic Alpha-Mendeleev Ridge is a large, bathymetrically irregular feature that roughly bisects the Arctic Basin (Fig. 2). Seismic reflection data indicate that between 100 and 1,200 m of sediment overlie structurally irregular basement on the ridge. Unreversed refraction data have suggested a 2.0 km/s surface layer 300 to 500 m thick that overlies a 4.7 km/s layer 2.8 km thick and a 6.4 km/s layer of undetermined thickness. The ridge is characterized by a free-air gravity high and high amplitude, short wavelength magnetic anomalies. The latter are linear for short distances and appear to be related to topography, but they do not resemble sea-floor spreading anomalies. Heat flow on the ridge, as on Yermak Plateau, is higher than in the adjacent basins. Marine fossils of Maastrichtian age (70–65 m.y. B.P.) and of Paleocene or Eocene age were found in what appear to be slumps of unconsolidated siliceous sediment in Pliocene-Pleistocene cores taken on Alpha Ridge.

Before a detailed kinematic model of the Arctic Ocean can be made, it will be necessary to determine the crustal structure and origin of the Alpha-Mendeleev Ridge. At present there are four leading hypotheses concerning its origin—subsided or allochthonous continental crust; former spreading center; intraplate aseismic ridge; or former subduction zone. A composite origin cannot be ruled out. Necessary data to determine its origin include: (1) determination of the crustal velocity structure, depth to Moho, and seismic velocity of the upper mantle; (2) shallow and intermediate reflection measurements to identify structures within the ridge and gravity measurements to determine ridge density; and (3) dredging and coring to recover samples of ridge basement. Given the thickness of the present sediment cover, this may not be possible, but steep escarpments up to 700 m high in the crestal region may expose pre-Pleistocene sediment or basement rocks. Interpretations of the age and origin of the ridge will be strongly constrained by such samples. An integrated geophysical-geological study, CESAR, (Canadian Expedition to Study the Alpha Ridge) is planned by the Canadian Government for 1983 to attempt to gather part of this critical information.

Morris Jesup Plateau

The area of the Yermak Plateau and Morris Jesup Plateau may have been influenced by an Iceland-like "hot spot." The once intact Yermak–Morris Jesup aseismic ridge reached its maximum development in late Eocene–early Oligocene time, when it probably formed an emergent, volcanic landmass. This aseismic ridge system may have formed a continuous land bridge from Svalbard to northeast Greenland until early Oligocene time, which contributed to blocking of the exchange of water between the Arctic Basin and the incipient Greenland Sea. In early Oligocene time (anomaly 13) a deep rift separated the Yermak Plateau from the Morris Jesup Plateau. Significant problems in need of resolution include the framework geology of the Yermak Plateau and Morris Jesup Plateau. Do these areas resemble the Iceland-Faeroe Ridge or the Vöring Plateau in crustal structure and origin? Are they features of "hot spot" origin and characterized by iron-titanium–rich basalts?

Chukchi Continental Borderland

The Chukchi borderland, which is 400 km wide and 600 km long and includes the Chukchi Cap, extends northward into the Arctic Basin from the Chukchi Shelf. The borderland consists of a cluster of shallow, flat-topped plateaus and ridges with an intervening deep basin and sea valleys. The summit of the Chukchi Cap, the largest plateau in the borderland, has a diameter of about 100 km and a minimum depth of less than 300 m. The surface of the cap was apparently gouged by ice during times of low Pleistocene sea level. Soviet workers infer from dredging and from magnetic and seismic data that the Chukchi Cap is a relic of the Precambrian platform (Yu. E. Pogrebitsky, 1981, written commun.), and extrapolation of geologic data from the Chukchi Shelf indicates that the plateaus of the borderland may be underlain by early Paleozoic and Precambrian bedded rocks. A Cretaceous-Tertiary basin observed on seismic reflection data beneath the northern Chukchi Shelf, opposite the borderland, suggests that the plateaus of the borderland may have been rifted from the shelf in Early Cretaceous time. Additional seismic refraction and potential field data, dredging and coring, and seismic reflection profiles are needed to determine the structure and composition of the Chukchi borderland and its relation to the Chukchi Shelf.

Amundsen and Nansen Basins

The Amundsen Basin overlies the northern flank of the Nansen Ridge and the Pole Abyssal Plain (Fig. 2). Crustal refraction experiments indicate that the abyssal plain is underlain by oceanic crust that is up to 10 km thick close to the Lomonosov Ridge. Structural models suggest that the crustal section close to the Nansen Ridge consists of approximately 2 km of high-velocity sediment, a very thin layer 2, and a thickness of 5 km from the top of the basement to the Moho. Very high intensity shear-wave energy has been identified with phase velocities appropriate for layer 3 propagation. Nansen Basin, which overlies the southern flank of the Nansen Ridge, exhibits shallower water depths than the Amundsen Basin. The underlying crust has greater sediment thickness and depth to basement, lower amplitude magnetic anomalies, and 5 to 10 percent higher spreading rates than the crust beneath the Amundsen Basin. Needed research in the Amundsen and Nansen Basins includes attempting to determine (1) the cause for the asymmetry of both the spreading rate and basement depth across the Nansen Ridge, and (2) whether the thin crust of the Amundsen Basin is also found in the Nansen Basin and if this phenomenon is related to the slow spreading rate of the Nansen Ridge.

Makarov Basin

Plate-tectonics interpretation suggests a complex history for the area between the Lomonosov and Alpha Ridges. Recent aeromagnetic data indicate linear anomalies parallel to the Lomonosov Ridge implying an accretionary origin for the basin, perhaps during the interval 80 to 47 m.y. B.P. (anomalies 19–34). The association of basement relief with magnetic anomalies close to Lomonosov Ridge suggests, however, that at least part of the basin may have been formed by other processes. Above the crust is a sedimentary layer approximately 1 to 2 km thick. Total depth to the Moho is 8 to 9 km. More aeromagnetic data, especially toward the Canadian and Siberian ends of the basin, are required to confirm the reported magnetic anomaly correlations. The source of the magnetic anomalies and the reason for the shallow depth of the basin need to be discovered. The extent and distribution of basement structure within the basin would be revealed by shallow seismic and gravity profiling.

Canada Basin

Regional geological and aeromagnetic studies and the structure of its continental margins all suggest that the Canada Basin opened by sea-floor spreading during late Mesozoic time, but the specific timing within the late Mesozoic is debated. A tentative sea-floor spreading model for the Canada Basin constructed from the aeromagnetic anomaly data suggests that accretion began about 153 m.y. B.P. (Oxfordian; anomaly M-25) and continued until 127 m.y. B.P. (Valanginian; anomaly M-12). The maximum distance between the oldest and youngest anomalies is 292 km, which implies a half-spreading rate of 1.3 cm/yr or a total opening rate of 2.6 cm/yr. An alternative identification assigns the youngest anomaly to M-3 (116 m.y. B.P., late Barremian). Stratigraphic and structural studies on the North American polar continental margin suggest that sea-floor spreading began in the interval 135 to 120 m.y. B.P.

with initial but limited rifting taking place during Early Triassic time along the Canadian segment and at the outset of Jurassic time along the Alaskan segment.

The deep part of the Canada Basin, 150 km seaward of the western Alaska continental slope, is underlain by 4 km of sediment with an estimated mean velocity of 2.7 km/s. Below the sediment is a basement layer 4 to 6 km thick with a velocity of approximately 4.9 km/s. If velocities of 7.7 to 7.9 km/s are taken to be upper mantle, the depth to the Moho is 13 to 16 km. Farther to the east, close to the MacKenzie Delta, the Canada Basin is underlain by a large sediment cone at least 6 km thick with seismic velocities reaching at least 3.8 km/s. Cores indicate that the uppermost sediments in the Canada Basin are turbidites and CDP seismic reflection profiles of the basin off Alaska suggest that turbidite-bearing sediment comprises most of the core. Beneath the southeastern part of the Canada Basin, and perhaps elsewhere, the sediment pile is disrupted by numerous large diapiric folds, and could contain hydrocarbon deposits. Additional seismic refraction and reflection profiles are needed to better define the crustal structure of the Canada Basin and the thickness and configuration of its sedimentary fill and diapiric folds.

Continent-Ocean Margins

The continental margins surrounding the Arctic Ocean and Greenland Sea are of passive type and include both rifted and sheared segments. Except for the Alaskan Beaufort Shelf, these are the least known continent-ocean margins in the world. Large segments are devoid of published geophysical and geological data, and some are even poorly known bathymetrically. Understanding of these margins is crucial for establishing the geological history of the Arctic Basin and adjacent areas and for gaining important information about the hydrocarbon potential of these regions.

The East Greenland continental margin, along the west side of the Greenland Sea, can be divided into a northern area showing evidence for suturing of Hudsonian, Grenvillian, and Caledonian age followed by post-late Caledonian molasse sedimentation and Mesozoic rifting, and a southern area that apparently acted as a cratonic block from early Proterozoic to middle Cretaceous time. East Greenland was finally separated from northwestern Europe by sea-floor spreading in latest Paleocene to earliest Eocene time and is now bounded by a passive margin with rifted and sheared segments. On land, between 68° N. lat and Scoresby Sound, Tertiary rocks along the East Greenland margin consist of thin predrift sediments overlain by 1 to 7 km of late Paleocene basaltic lavas that were extruded immediately prior to breakup. The resulting subsidence of the shelf led to the accumulation of 2 to 8 km of postbasaltic sediments offshore, while the adjacent land areas were uplifted 1 to 2 km. Initiation of spreading along the Kolbeinsey Ridge during late Oligocene time was accompanied by renewed tectonism along the East Greenland margin between 68° and 72° N. Finally, the shelf experienced strong Miocene progradation. Significant problems remaining to be resolved include: (1) the history and plate-tectonics significance of the Tertiary orogeny in northeast Greenland; (2) the apparent westward migration of the Knipovitch Ridge and its relationships to the adjacent Greenland margin; and (3) the nature of the northeastern Greenland margin, which is almost totally unexplored. It presumably consists of a rifted margin on the south and a sheared margin on the north. The northeast Greenland margin may have high hydrocarbon potential.

As previously mentioned, stratigraphic relations and seismic reflection records indicate that the present North American margin of the Canada Basin was probably formed during Early Cretaceous time following the initiation of a rift valley system in earlier Mesozoic time. Breakup and rifting appear to have begun not later than about 120 m.y. B.P. Similarities in structure and stratigraphy between the Alaskan and Canadian margin segments suggest that northern Alaska was once connected to the Canadian Arctic Islands and that the two areas separated by rotational rifting about a structurally complex pivot zone in the MacKenzie Delta region. The rifting process produced the Canada Basin. Tectonic effects related to the rifting appear to increase in complexity along the polar margin as the presumed pivot zone is approached. In Early Paleozoic time, during an earlier tectonic regime, a continental block approached what is now the North American polar margin from the general direction of the present Arctic Basin. Oceanic crust was progressively eliminated between this block and Arctic Canada until the landmasses collided in mid-Paleozoic time on the Arctic Basin side of the Queen Elizabeth Islands. It has also been suggested that accretionary tectonics may have been an important post mid-Paleozoic process in the Arctic. In the present tectonic regime, thick young sediment wedges are building out onto the floor of the almost landlocked Arctic Basin. These wedges appear to be the main source of magnetic anomaly lows and free-air gravity anomaly highs along the margins of the Canada Basin off Canada and Alaska. Associated seismicity along the margin at the head of the MacKenzie Cone may be produced by the weight of the thick prograding sediment prism in combination with a reduced threshold for seismic activity at the continental boundary.

Adequate understanding of the structure and tectonic history of the margins of the Arctic Basin and Greenland Sea will require substantial amounts of integrated geophysical and geological data. Important questions are the character of the main segments of the continent/ocean boundary in the Arctic, the thickness and age of the continental terrace sedimentary prisms, the history of postrift continental shelf subsidence, and the role of volcanism in the development of the Arctic margins. Basic tectonic questions include: (1) character and history of the interaction between the Nansen spreading ridge and the continental margin where these intersect in the Laptev Sea; (2) nature of the junctions of the Chukchi Continental Borderland, Alpha-Mendeleev Ridge, Lomonosov Ridge, and Morris Jesup Plateau with the Siberian, Canadian, and Greenland margins; (3) structure of the continental shelf and slope north of Ellesmere Island and Greenland; and (4) information about the prolongation of the Nares Strait lineament across the continental shelf in the Lincoln Sea.

Evolution of the Arctic Ocean Paleoenvironment

Late Mesozoic–Paleogene History of the Arctic Ocean

Only two deep-sea cores provide any pre-Neogene history for the central Arctic. FL-437, from the Alpha Ridge about 500 km WNW of northern Ellesmere Island, contains Maastrichtian silicoflagellates and diatoms in orange-yellow siliceous sediment within a section of brownish Pliocene to Pleistocene mud. Although the siliceous sediment apparently constitutes a slump, the fossils indicate that the region was formerly the site of a warmer, ice-free ocean that contained a normal marine biota, and was neither brackish nor fresh, nor isolated from the larger ocean basins, as has recently been proposed. A slump block of orange-yellow siliceous sediment with Paleocene to Eocene silicoflagellates and diatoms has also been found within a core (FL-422) of brownish Pliocene(?) and Pleistocene mud that was taken from Alpha Ridge about 115 km southeast of core FL-437. The fossils indicate that the relatively warm water conditions of the latest Cretaceous in this region continued to the early Tertiary.

Midway between the Chukchi borderland and the west end of the Alpha Ridge, core FL-224 from the abyssal plain contains abundant dinoflagellates of mainly late Albian–earliest Cenomanian age. Latest Cretaceous and Holocene forms are also present. Core B-25 from the crest of the Lomonosov Ridge also contains mid-Cretaceous dinoflagellates and late Cenozoic pollen in addition to Upper Devonian spore fragments. The pre-Cenozoic microfossil populations in B-25 are reworked and thermally altered. An ice-rafted origin for these deposits is likely.

Late Mesozoic and Paleogene sediment from the Arctic Basin may contain the biologic record of a near-polar ocean with both bottom and surface water masses that were relatively warm. This is a unique environment without analogs in the modern ocean. A program of coring, dredging, and drilling from the ice cover will be needed to obtain an adequate record of the resulting biota and paleoenvironments of the deep Arctic Basin. In-place samples of the appropriate age are especially needed. Such samples may be within reach of piston-cores on the crests and flanks of the aseismic ridges, but deep coring techniques may be required especially in regions of low relief where older deposits may not be exposed by slumping processes.

Late Tertiary–Early Quaternary Paleoenvironment

Initiation of the present system of bottom-water circulation and renewal in the world ocean, which is driven by cooling at the fringes of the Antarctic ice sheet, probably represents the most drastic paleo-oceanographic revolution of the past 100 m.y. A combination of tectonic and paleoclimatic changes during the Cenozoic produced a perennially ice-covered Arctic sometime between Paleocene or Eocene and Quaternary time. Estimates of the time of onset of the ice cover, however, range from as early as mid-Cenozoic to as late as about 700,000 years ago. Glacial-marine sediments provide evidence for glacial ice-rafting by late Miocene time, but ice-rafting does not necessarily indicate perennial ice cover. Burckle (1981, personal commun.), for example, pointed to a number of late Pliocene–early Pleistocene diatoms that he believed had entered the Atlantic from the North Pacific via the Arctic Ocean, indicating that the Arctic Ocean must have been at least intermittently ice-free during part of the late Cenozoic, or perhaps they were transported around the edges of an ice-covered basin. Uncertainty as to the onset of perennial ice in the Arctic Basin is a problem that is critical to worldwide paleoclimatic and oceanographic models as well as to Arctic marine geology and paleoecology.

Quaternary Sediments and Paleoenvironments of the Arctic Ocean

Thirteen correlatable lithostratigraphic units have been recognized in cores from the central Arctic Basin. The units appear to range in age from late Miocene to Holocene and can be correlated over areas of several hundred thousand square kilometers. The ages have been calculated from paleomagnetic reversal signatures and by the extrapolation of sedimentation rates based on the paleomagnetic data. Unfortunately, the chronostratigraphy of the Arctic Basin sediments is based on only a few radiocarbon and uranium series dates, and at least some of the paleomagnetic data are quite noisy. On the basis of these ages, younger units have apparent sedimentation rates that average 1.14 mm/1,000 yr, and older units have rates that average 0.5 mm/1,000 yr. Pebbles, probably ice-rafted, occur throughout the section. Extrapolation of the sedimentation rate from the Gilbert Polarity Interval 5 reversal indicates that the oldest erratic in undisturbed sediment in the Arctic Basin may have been deposited 5.26 m.y. ago.

Four types of glacial-marine sediment can be characterized by silt-clay histograms in the Arctic Basin. The histograms are similar to those for glacial-marine sediments in non-Arctic marginal seas. Reworking of central Arctic Basin glacial-marine sediment by bottom currents does not appear to have had a major effect on silt-clay distributions. In general, arenaceous lutites and pink-white layers appear to represent times of relatively high sedimentation rates and are interpreted to have been deposited during times of increased glacial ice-rafting. Silty lutites may also be principally ice-rafted sediment, but they are believed to represent times of reduced ice-rafting and lower sedimentation rates.

The geographic distribution of glacial-marine sediment types in the central Arctic Basin suggests that deposition of coarse ice-rafted sediment is dominant in the Alpha–Mendeleev Ridge region. Ice-rafted debris is also deposited on abyssal plains but is there diluted by and interbedded with turbidite deposits. Six intervals of increased glacial ice-rafting during the Pleistocene have been defined in the Amerasia Basin. Correlation with periods of continental glaciation is possible, but correlation between marine and adjacent terrestrial sequences is difficult because Pleistocene climatic fluctuations are better understood in lower latitudes than in the Arctic.

Recent work has raised serious questions about the present dating of some Neogene and Quaternary sediments of the Arctic Basin. The reported sedimentation rates in the Arctic sediment cores are much lower than in other ocean basins even though sediments in the Arctic contain a sizable component of ice-rafted detritus, which should have caused relatively high rates. Sedimentation rates in the Arctic should also have been higher than in other oceans, particularly during the Pleistocene, because the Arctic is a small, almost landlocked ocean with a larger influx of river- and glacier-borne detritus per unit area than other oceans. Preliminary results of amino-acid analyses of material from sediment cores suggest the possible presence of reworked material in the cores. The presence of sizable quantities of sand and pebble-sized erratics in the cores might also have confounded the paleomagnetic measurements. An abundance of *Elphidium* at supposedly early Pleistocene levels in the cores suggests that the cores may represent a much shorter time span than the presently accepted chronostratigraphy would indicate. In view of these uncertainties and because a revision of the chronology would require important changes in our understanding of Arctic geologic and climatic history, the following studies should be given a high priority.

1. A suite of new cores from representative geologic and bathymetric environments should be obtained to provide adequate samples for an intensive program of dating, geochemistry, and physical stratigraphy. Some of the existing cores are largely exhausted, and with the passage of time they are becoming less and less satisfactory for many purposes. Critically needed are cores that are substantially longer than 6 m, the maximum length of existing cores.

2. Additional paleomagnetic studies need to be made and the origin of the noise in the paleomagnetic data determined.

3. Additional geochronological measurements by the amino-acid, improved radiocarbon, uranium series, and perhaps other methods are needed. Their successful application will require newer and larger samples than are presently available.

4. Detailed comparative studies of the biostratigraphy of the Arctic cores and cores from the Norwegian and Greenland Seas, as well as correlative sections in the adjacent continental shelves and onshore, are needed.

5. Provenance studies of the ice-rafted detritus in the Arctic cores will contribute to our understanding of late Cenozoic sedimentation in the Arctic Ocean and of the history of glaciation in the surrounding land masses.

6. Palynological study of the large detrital organic fraction that has been observed in the Arctic cores should extend and refine the dating based on other taxa.

7. Oxygen-isotope studies of Arctic cores and comparison with the large body of such data from other oceans will contribute to the resolution of existing uncertainties in correlation and provide an important body of paleoenvironmental data.

Late Quaternary Paleoenvironment

The modern (interglacial) depositional environment in the Arctic Ocean depends primarily on the structure of the water mass. A low-salinity surface layer is produced by the combined effects of the permanent ice cover, the inflow of low-salinity North Pacific water through the Bering Strait, and the large volume of continental runoff. An inflow of Atlantic water produces a warm intermediate layer, and a cold deep-water mass enters from the Norwegian-Greenland Sea where it is formed. Cores of Quaternary sediment in the Arctic consist of alterations of brownish mud with some foraminiferal tests and gray foraminifer-barren lutite. The brownish mud may represent the Holocene and interglacial parts of the Pleistocene section, and the lutite the late Weichselian and some earlier glacial stages. Important questions include: (1) details of the transition from synglacial to syn-interglacial deposits by oxygen isotope analysis of pelagic and benthic foraminifera, and by palynology; (2) character of sedimentation beneath pack ice in glacial and interglacial periods; and (3) growth, development, and dispersion of biota in the ice-covered Arctic.

Present Sedimentary Environment

The proportion of sediment in the Arctic that is of fluvial and glacial origin is in general higher than in other oceans, and the proportion of biogenic material that is derived from macroorganisms is in general lower. However, we have meager knowledge of sediment accumulation rates and transportation processes in the Arctic and of where erosion, deposition, and bypassing are taking place. Such studies are underway in other oceans—for example, the Benthic Boundary layer studies in the Atlantic and Pacific—but Arctic workers are handicapped by fewer baseline data and inaccessibility. Information about suspended sediment, with emphasis on the location of nepheloid layers that might exist and the measurement of sediment flux, is a particular need. Such studies would also provide a better insight into biologic productivity and the production of biogenic sediments in the Arctic Ocean. A better understanding of the production and diagenesis of biogenic sediments in the Arctic requires measurement of biologic productivity by biological and sediment traps, petrographic and geochemical study of diagenesis in Arctic environments, and determination of the distribution of biogenic materials within the various Arctic lithofacies of the present and former climatic and tectonic regimes.

Physical and biologic postdepositional reworking and retransportation of sediment are especially active on Arctic shelves, and some of these processes, such as deep gouging by sea ice, are unique to polar seas. Research in the Beaufort and Barents Seas, using mainly high-resolution seismic records, side-scan sonograms, and sediment cores, is contributing important insights to sediment dynamics on Arctic shelves, but similarly detailed studies are apparently yet to be done elsewhere. Additional studies of this type and more shallow drilling promise important new understanding of the processes involved.

High-latitude rivers and fjords are potentially important contributors of fluvial and biogenic detritus and nutrients to the Arctic Ocean. The character, dispersal, and biologic effect of these products in the Arctic must be understood before we can formulate quantitative models on Arctic sedimentation. An important study is determination of provenance and relative abundance of man-made and natural airborne detritus in the Arctic Ocean, its ice cover, and the sea floor. A number of specialists can now identify the source areas of many industrial and natural microscopic airborne particles. Freshwater diatoms and opal phytoliths will be especially useful in such studies.

Petroleum exploration and development is now underway on Arctic shelves. It is important, therefore, to determine the physical properties of shelf sediments deposited under Arctic conditions. For example, how do sediments dominated by rock flour react to static and seismic loads, what are the engineering properties of shelf sediments overconsolidated by grounded glaciers, and what is the effect of submarine permafrost and gas hydrate, where present, upon the geotechnical properties of subsea sediments?

Man's Impact on the Arctic

Human activities in the Arctic Ocean, as elsewhere on the globe, can have severe environmental effects. The seasonally expanding and contracting perennial ice cover is largely dependent upon a balance between the influx of lower salinity North Pacific waters into the Arctic Basin through the shallow Bering Strait, the influx of high-salinity Atlantic water through the deep gap between Svalbard and Greenland, and the influx of freshwater from the surrounding continents. Artificial disturbance of this balance, such as massive diversions of Arctic rivers in the Soviet Union for irrigation and other purposes, or blocking the flow of water through Bering Strait, might result in climatic changes that could not be controlled by technical means. The changes might have detrimental consequences for the environment of the Arctic regions, and perhaps for the entire globe.

The effect of burning fossil fuel and the resultant input of large amounts of CO_2 into the atmosphere has been debated extensively. One view is that increased atmospheric CO_2 will lead to a substantial increase in global temperatures, particularly in the polar regions, causing melting of the ice sheets on Greenland and Antarctica and a major rise in sea level. The West Antarctic ice sheet, which is grounded below present sea level, is particularly susceptible to accelerated melting. It is not certain, however, that CO_2 warming will cause a long-term rise in sea level. By some scenarios, a combination of warmer oceans and continuing below-freezing air temperatures above ice sheets could thicken and expand the ice sheets, and eventually lower sea level. Another viewpoint is that the increased input of CO_2 to the atmosphere will be buffered by increased CO_2 intake into the oceans, which would lower their pH and increase the rate at which they dissolve $CaCO_3$. Small increases in dissolved CO_2 in the oceans may therefore drastically raise the carbonate compensation depth (CCD), whose position in the Arctic is yet to be defined. Determining the present position of the CCD and of $CaCO_3$ dissolution patterns in the Arctic Ocean will permit us to estimate the effect upon it of increased levels of CO_2 from the burning of fossil fuels.

By far the most hazardous human activity in the Arctic today is drilling for oil and gas on the polar shelves. In addition to the hazards encountered in offshore exploration and development at lower latitudes, there are severe operational conditions created by the dynamic polar ice pack, Arctic weather, winter darkness, and local subsea permafrost and gas hydrate. As a result, it might not be possible to control blowouts or to clean up oil spills in the Arctic for months, or even years, and spilled oil decomposes slowly in Arctic waters.

EARTH PHYSICS BRANCH CONTRIBUTION NUMBER 982

Selected References

Churkin, Michael, Jr., and Trexler, J. H., 1980, Circum-Arctic plate accretion—Isolating part of a Pacific plate to form the nucleus of the Arctic Basin: Earth and Planetary Science Letters, v. 48, p. 356–362.

Clark, D. L., 1974, Paleontologic response to post-Jurassic crustal plate movements in the Arctic Ocean: Milwaukee Public Museum Special Publication, Biology and Geology, 2, p. 55–76.

Clark, D. L., Whitman, R. R., Morgan, K. A., and Makey, S. D., 1980, Stratigraphy and glacial-marine sediments of the Amerasian Basin, central Arctic Ocean: Geological Society of America Special Paper 181, 57 p.

Coachman, L. K., and Aagaard, K., 1974, Physical oceanography of Arctic and subarctic seas, *in* Y. Herman, ed., Marine geology and oceanography of the Arctic seas: New York, Springer-Verlag, p. 1–71.

Falconer, R.K.H., and Thiede, J., 1980, Geological and geophysical research in eastern Arctic science, *in* P. Gudmansen, ed., Workshop report, Commission for Scientific Research in Greenland: Copenhagen, Denmark, Commission for Scientific Research in Greenland, p. 7–54.

Grantz, A., Eittreim, S., and Dinter, D. A., 1979, Geology and tectonic development of the continental margin north of Alaska, *in* C. E. Keen, ed., Crustal properties across passive margins: Tectonophysics, v. 59, p. 263–291.

Herman, Yvonne, and Hopkins, D. M., 1980, Arctic oceanic climate in late Cenozoic time: Science, v. 209, p. 557–562.

Johnson, G. L., Monahan, D., Gronlie, G., and Sobczak, L., 1979, GEBCO Sheet 5.17 (Arctic): Ottawa, Canada, Canadian Hydrographic Service.

La Brecque, J. L., Kent, D. V., and Cande, S. C., 1977, Revised magnetic polarity time scale for Late Cretaceous and Cenozoic time: Geology, v. 5, p. 330–335.

Nansen, F., 1904, The bathymetrical features of the North Polar Seas, with a discussion of the continental shelves and previous oscillations of the shore-line: The Norwegian North Polar Expedition 1893–1896, Scientific Research, edited by F. Nansen, 4 (13), p. 1–232.

Richardson, N. J., Wimbush, M., and Mayer, L., 1981, Exceptionally strong near-bottom flows on the continental rise of Nova Scotia: Science, v. 213, p. 887–888.

Sweeney, J. F., and Haines, G. V., 1978, Arctic geophysical review—an introduction: Ottawa, Canada, Department of Energy, Mines and Resources, Earth Physics Branch Publications, 45 (4), p. 1–6.

Taylor, P. T., Kovacs, L. C., Vogt, P. R., and Johnson, G. L., 1981, Detailed aeromagnetic investigation of the Arctic Basin, Part II: Journal of Geophysical Research (in press).

Vogt, P. R., Taylor, P. T., Kovacs, L. C., and Johnson, G. L., 1979, Detailed aeromagnetic investigation of the Arctic Basin: Journal of Geophysical Research, v. 84, p. 1071–1089.

The western North Atlantic

Peter R. Vogt
Naval Research Laboratory
Washington, D.C. 20375

Brian E. Tucholke
Woods Hole Oceanographic Institution
Woods Hole, Massachusetts 02543

General Remarks

Goals of Perspectives Chapter

Although available space strongly limits the scope of this essay, we want it to serve variously as a prospectus of subject matter to be covered; an informal "minisynthesis" that geographically and topically delimits synthesis contents; a guide for authors; a definition of *synthesis* problems as well as *scientific* problems; a philosophical perspective; identification of relationships with other volumes; and finally, if we are successful, a stimulus to new research.

Philosophy and Format of D-NAG Volume

It is in a way fortunate that the Geological Society of America was not founded two decades earlier. One can only wonder what kind of syntheses would have resulted for the North America plate (Fig. 1), and particularly for the geology below the Atlantic Ocean, if they had been written during or just *before* the "plate-tectonics revolution." The previous major synthesis of western North Atlantic geology (Emery and Uchupi, 1972) was written a decade ago, shortly *after* the "revolution," and this synthesis remains basically correct, despite a flood of subsequent data. The rate of change in fundamental concepts has slowed recently, and the timing for the D-NAG synthesis project probably is optimal.

Because our western North Atlantic volume will be part of a series of primarily *continental* syntheses, we first note how an ocean basin synthesis differs in the character of data, geology, and approach. The fundamental differences in topography, age, and crustal structure have become common knowledge, so much so that no elaboration is needed here. Few other volumes will address as much territory (more than $5 \times 10^6 km^2$), although the geological time (0 to ?170 m.y.) represents only about 5% of the age of North America's oldest known rocks. We are synthesizing geology essentially unseen by human eyes except for oceanic islands and very limited submersible observations. Thus marine geology has come to depend much more on geophysical techniques and interpretation of a comparatively small collection of samples from dredging, coring, and deep-sea drilling.

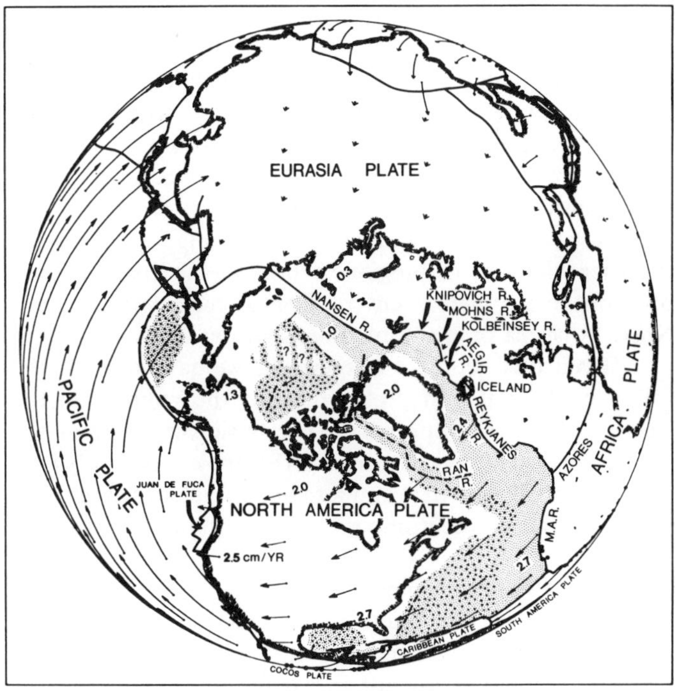

Figure 1. Polar view showing lithospheric plates and absolute motion vectors in a fixed hotspot frame. Oceanic lithosphere included in the North America plate is stippled, with lighter stippling denoting Cenozoic-aged crust. Modified from Minster and Jordan (1980).

Of the many synthesis volumes, the western Atlantic volume perhaps best approaches the uniformitarian ideal. The birth of oceanic crust can be studied along the crest of the Mid-Atlantic Ridge and compared to undisturbed older crust; the ocean crust adjacent to the eastern United States is as old as any known in the ocean basins. Sedimentation processes studied across the present ocean floor are basically the same as observed in older lithofacies even though the Atlantic oceanographic environment has varied greatly in the past. Both sediments and igneous crust are scarcely touched by mid-plate or metamorphic igneous/tectonic processes, and particular lithofacies or average basement characteristics are more homogeneous over areas of 10^5 or 10^6 km^2 than even continental platform sediments, let alone Phanerozoic orogens or the Precambrian shield.

Despite these differences, there also are close links between continental North America and the deep Atlantic. For example: (1) The Mid-Atlantic Ridge has generated a geophysical record of North American plate motion, and it even may help *drive* that motion. (2) Mid-plate seismicity in the western Atlantic and eastern North America may reflect much the same compressional stress field, possibly because of viscous resistance to the westward movement of the plate. (3) A much different paleo–stress field, recorded in Triassic-Jurassic igneous activity and normal faults along the eastern continental margin, prefigured the birth of the modern Atlantic. (4) Subsidence of the Coastal Plain is a syn-Atlantic phenomenon that echoes the subsidence of oceanic crust and results from similar thermo-isostatic processes in the mantle. (5) Possible ophiolites in orogens like the Appalachians represent remnants of Paleozoic oceanic crust and invite comparisons with modern Atlantic crust. (6) The sediments of the deep western Atlantic contain a 170-m.y. record of continental detritus derived under varying climatic regimes and topographic gradients from North America east of the drainage divide (Fig. 2). Assessed together with marine biogenic debris, this sediment probably contains a better environmental history of eastern North America than can be found ashore.

Given this geologic framework, we have devised the following format and philosophy for the volume on the western North Atlantic basin. The basic simplicity of the crustal section allows us to divide the book subequally into an igneous-tectonic part and a sediment-paleoceanography part. The igneous-tectonic part will introduce the volume because these processes have influenced sediment distribution and paleoenvironment, rather than vice versa. About 15% of the text is devoted to an introduction and to synthesis perspectives, as well as to several important but miscellaneous topics that do not fit easily elsewhere. Three of these topics (History of Geological Exploration, Bathymetry, and the Geomagnetic Time Scale) follow our Introduction. Three others (Resources, Engineering Properties and Hazards, and Law of the Sea) precede the Conclusion.

The philosophy adopted here is that our D-NAG volume should be useful to both continental and marine geologists and should summarize both "known" and "problematic" aspects of the western Atlantic geologic framework. The "known" geology will emphasize the major advance of the past 20 years and how these advances have modified earlier thinking. Our treatment of "unknown" geology can be divided into (1) clear definition of major problems that cannot at this time be resolved, but for which research strategies can be conceived, and (2) the resolution, via the D-NAG synthesis vehicle, of some currently perceived problems. Although we expect most of the chapters to advance the state of knowledge, the synthesis volume will only achieve its potential if it is "greater than the sum of its parts." This depends on our insight to recognize important problems and then coordinate the efforts of individual authors to resolve them.

The geographic limits of our synthesis volume will depend to some degree on the topic. For example, in the north and west the continental slopes of eastern North America, Greenland, and Iceland provide natural limits for discussions of downslope sedimentation; on the other hand, plate tectonics and crustal studies suggest western limits at the continent-ocean crustal transition but demand consideration of Cenozoic ocean basins north of Iceland and in the Arctic (Fig. 1). The southern limit for most chapters extends from Cuba east to the Antilles island arc, and eastward at latitude 10 °N to the Mid-Atlantic Ridge (MAR) crest. This latitudinal boundary is somewhat arbitrary because, although it is known that North and South America are not part of the same plate, the relative motion between the two plates is slow and poorly constrained (Minster and Jordan, 1980). The boundary may extend along the Barracuda Ridge and intersect the MAR crest near 13.7 °N. The eastern limit is provided by the active accreting plate boundary that extends north through the Azores triple junction to Iceland and farther along the Nansen Ridge (Fig. 1). There will be inevitable overlaps with D-NAG volumes dealing with the U.S. Atlantic margin, Greenland, and the Arctic, and with the volume on the continental margin of Eastern Canada. However, we think such overlaps strengthen the D-NAG project as a whole. Coordination among volume leaders should take advantage of differing perspectives (and authors) in the areas of overlap.

Introductory Topics

Bathymetry

Field geology, including marine geology, properly begins with a topographic map, and a chapter on bathymetry therefore leads in our volume. If anything, bathymetric

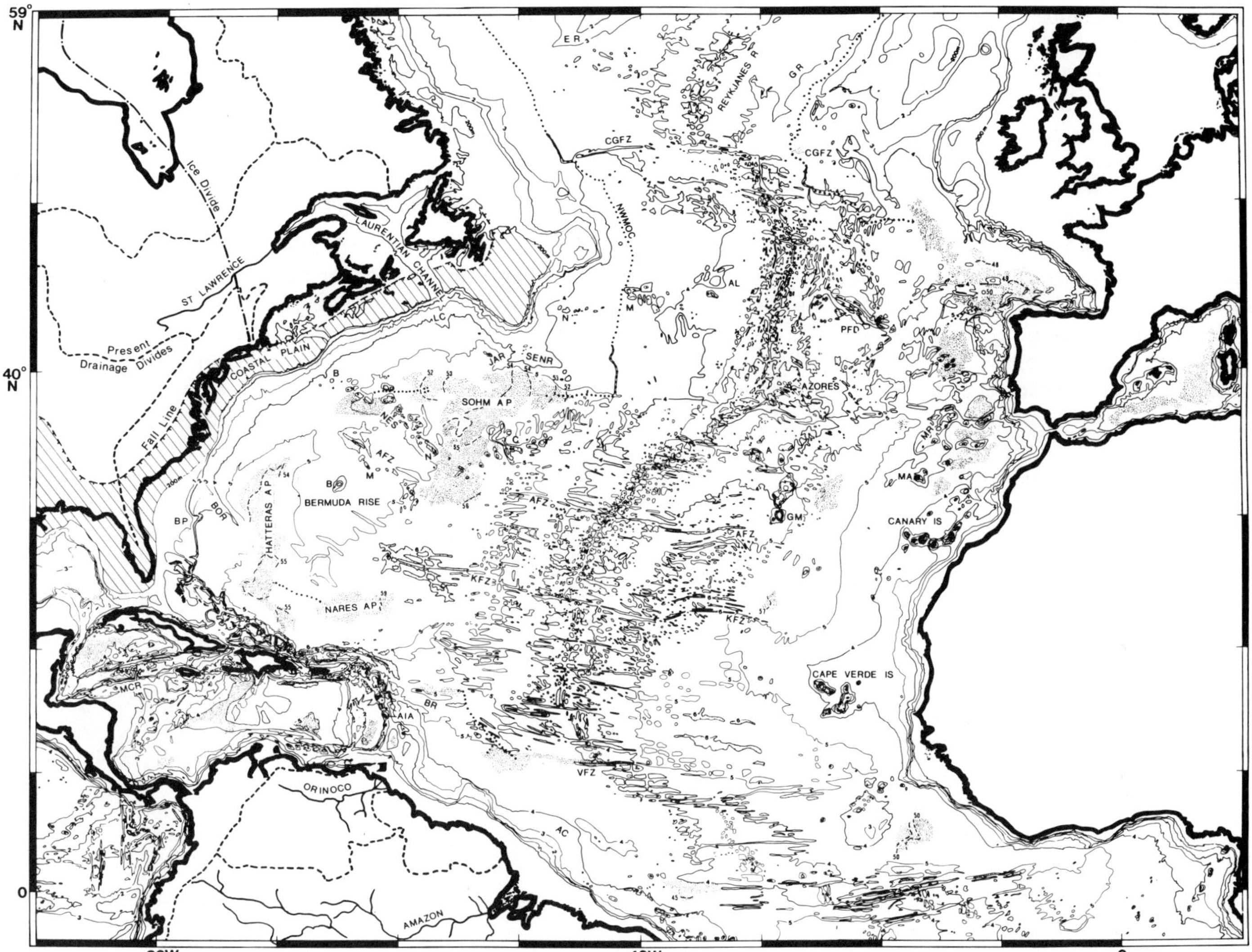

Figure 2. Bathymetry of the North Atlantic (Perry and others, 1981; Vogt and Perry, 1982) at 1-km-contour interval. Stippled areas are abyssal plains; dashed lines show plain contours in hundreds of meters. Mid-ocean channels are dotted. Atlantic Coastal Plain, major rivers and present drainage divides, and probable ice drainage at 20,000 yr B.P. added to indicate source areas for western Atlantic late Neogene terrigenous sediment. Place names clockwise from upper right: GR, Gardar Ridge; CGFZ, Charlie-Gibbs Fracture Zone; PFD, Peake-Freen Deeps; MR, Madeira Rise; MA, Madeira Island; A, Atlantis Seamount; GM, Great Meteor Smt.; AFZ, Atlantis F.Z.; KFZ, Kane F. Z.; VFZ, Vema F. Z.; BR, Barracuda Ridge; AC, Amazon Cone; AIA, Antilles Island Arc; MCR, Mid-Cayman Rise; BOR, Blake Outer Ridge; B, Bermuda; M, Muir Smt.; C, Corner Smts.; N, Nashville Smt.; NES, New England Smts.; B, Bear Smt.; LC, Laurentian Cone; JAR, J-Anomaly Ridge; SENR, South-East Newfoundland Ridge; AL, Altair Smt.; MI, Milne Smts.; N, Newfoundland Smts.; NWMOC, Northwest Mid-Ocean Canyon; ER, Eirik Ridge.

charts play an even more crucial role in marine geology than topographic charts do on land, because soundings are often the *only* data at hand. Submarine erosion of hard rocks is virtually nonexistent; therefore volcanic and tectonic topography (e.g., seamounts, fracture valleys, and ridges) may persist almost unchanged for 10^7 or 10^8 years unless buried by sediment. The characteristic sea-floor shapes of sedimentary features such as current-deposited "outer ridges," deep-sea channels, abyssal fans, or abyssal plains often suggest the associated depositional processes. Thus morphology alone may define a substantial part of the basin's geologic origin and evolution.

Synthesizing the current state of bathymetric knowledge immediately highlights the fact that the data base is highly inhomogeneous; detailed surveys of 10^4 to 10^5 km^2 are scattered irregularly in an unsystematic "matrix" of tracks, and uncharted areas of 10^3 to 10^4 km^2 remain between the sounding lines. Unfortunately the topographically roughest areas (i.e., crest and flanks of the mid-ocean ridge) are also the more poorly surveyed. The unevenness of data density requires that "geological experience" be used to contour soundings, but such wisdom also introduces bias. Large-scale trends (> 100 km) now are generally recognized (e.g., "transform faults" and "ridge-crest" trends). However, detailed surveys show that deviations such as oblique trends may be extremely important.

A variety of high-resolution techniques (e.g., multibeam echosounding, deep-towed acoustics systems, submersible observations, and large-area bottom photography) have surveyed even smaller areas within financial and time constraints. However, these bits of information are very important to our understanding and "prediction" of unsurveyed terrain. In our synthesis we will complement a large-scale, generalized bathymetric map of the basin with charts of selected small areas that are well controlled by higher resolution data. At all scales it will be important to distinguish bathymetric measurements from their interpretation, particularly in contour charts.

The Geomagnetic Reversal Sequence

A separate chapter will be devoted to the Earth's magnetic reversal history for the time since the initial opening of the North Atlantic. Because reversal time scales impact on diverse fields of research, this chapter will be an important foundation for other chapters in our volume and other D-NAG volumes. Reversals of the geomagnetic field from Middle Jurassic time to the present are recorded as magnetic anomalies in the western North Atlantic (Figs. 3, 4) and often as polarity changes in weakly magnetized pelagic

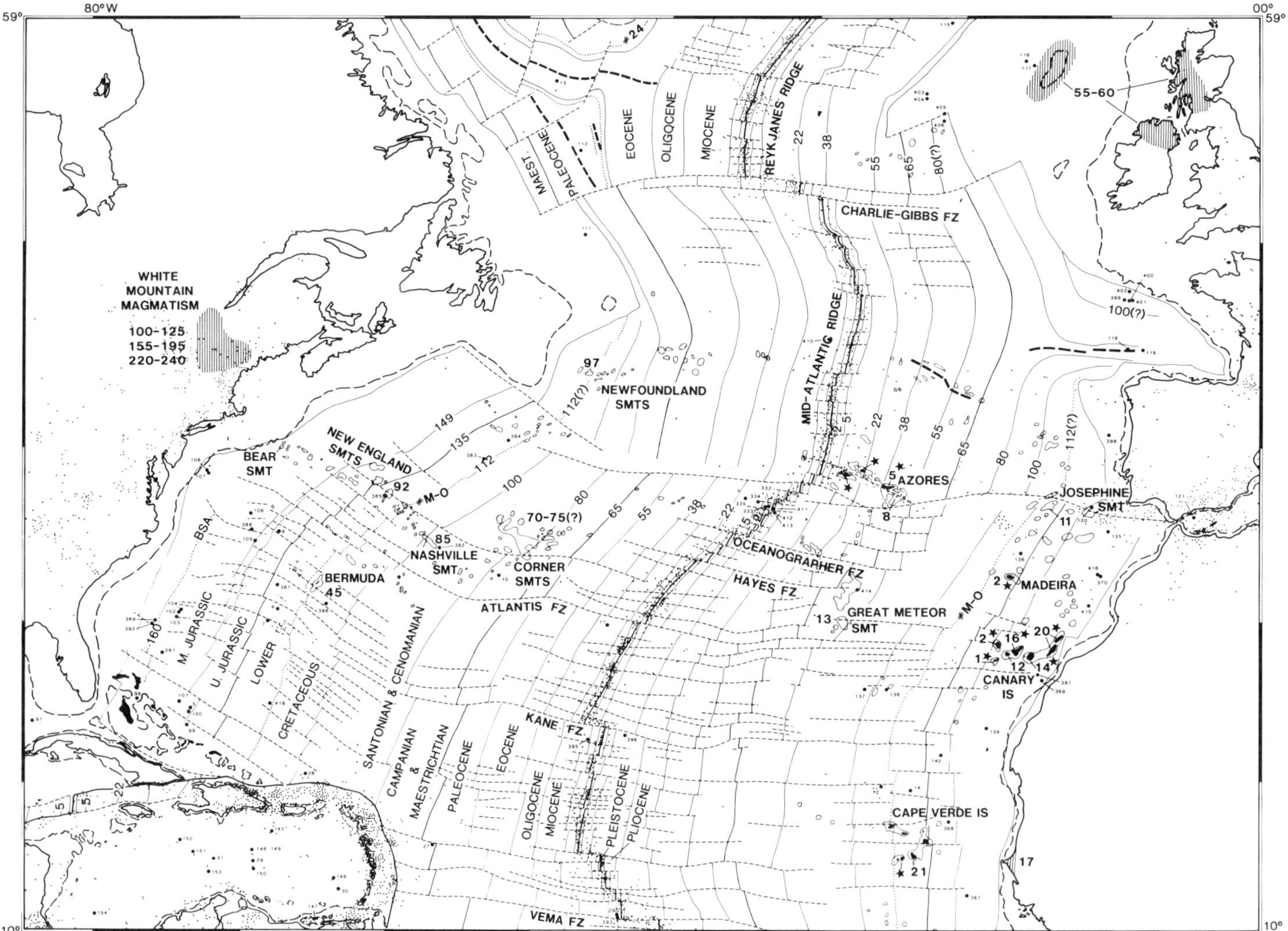

Figure 3. Crustal ages (m.y. contours) in North Atlantic (Vogt and Perry, 1982). Dots show earthquake epicenters (1961–1979); stippling indicates seamounts; vertical ruling denotes continental magmatism possibly related to volcanism on adjacent oceanic crust. Maximum ages given, where known, for volcanic rocks on seamounts and islands; stars denote historical volcanism.

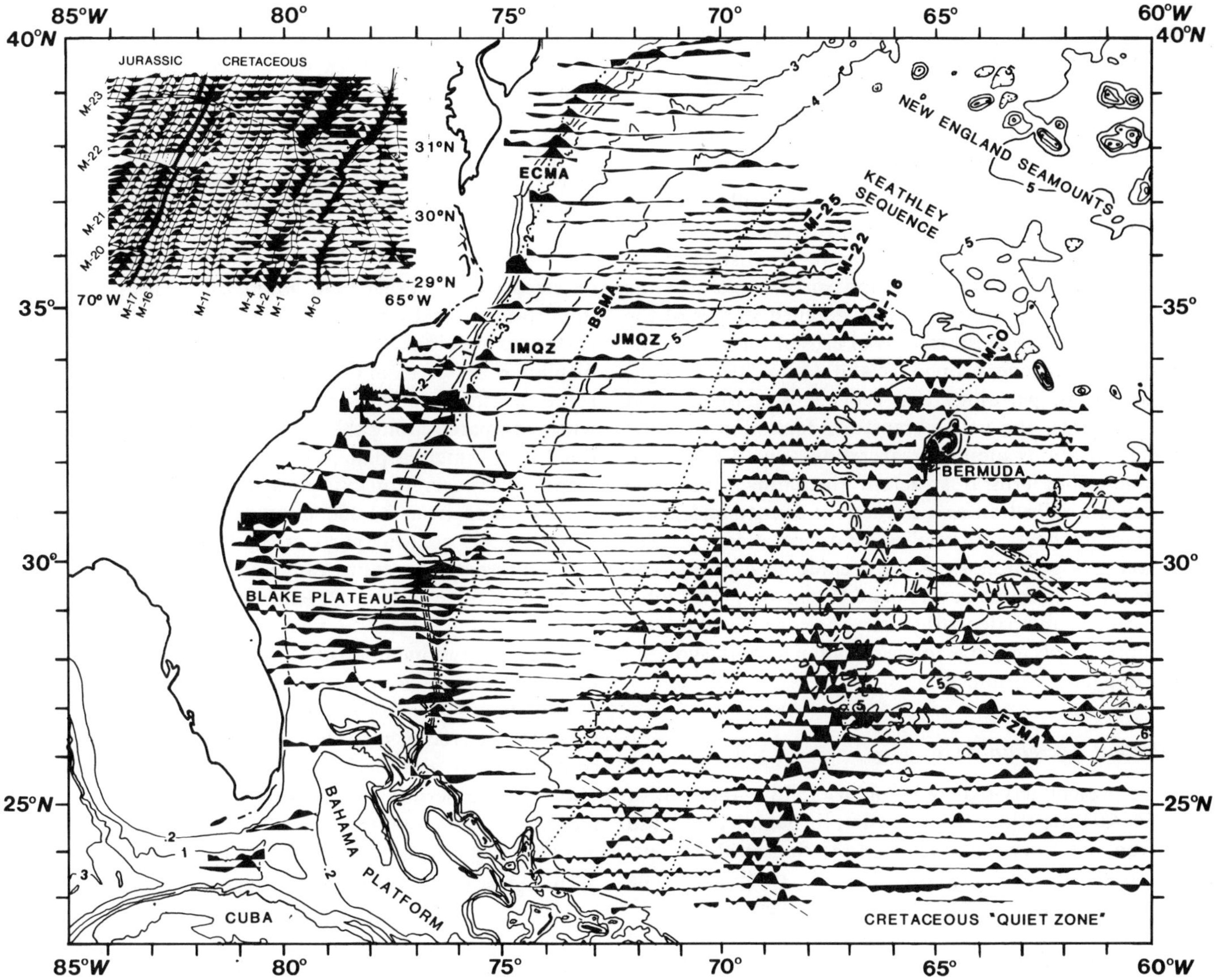

Figure 4. Residual magnetic anomaly profiles collected by USNS Keathley in western North Atlantic, showing major magnetic/tectonic crustal provinces. ECMA, East Coast Magnetic Anomaly; IMQZ, Inner Magnetic Quiet Zone; BSMA, Blake Spur Magnetic Anomaly; JMQZ, Jurassic Magnetic Quiet Zone (includes IMQZ). FZMA, Fracture Zone Magnetic Anomalies (dashed). Inset in upper left: Magnetic anomaly details in box-outlined area southwest of Bermuda. Modified from Vogt and Einwich (1979).

sediments. The sedimentary magnetic record is more fragmentary than the anomaly sequence but can have higher resolution. For example, several short reversed intervals were detected in Albian sediments cored at DSDP Site 386 near Bermuda, but no corresponding marine magnetic anomalies have so far been identified in any ocean.

The magnetic reversal sequence is central to the exposition of plate tectonics in the North Atlantic. Calculations of spreading rates and times of plate acceleration depend critically on the correctness of the reversal time scale. Because higher Pacific spreading rates provide a higher-resolution, lower-noise record of reversals, Atlantic magnetic data have not been used to erect the reversal sequence. However, the North Atlantic has contributed many of the biostratigraphic basement ages used in *calibrating* the reversal sequence, and it has provided sedimentary sections where sediment magnetization has been used for precise age dating, interhole correlation, and information on short-period field behavior ($\sim 10^4$ yr) not resolved by magnetic anomaly data. New reversal time scales are published every year or two as new data are acquired. Our synthesis will be an opportunity to update and combine such time scales, but no significant improvements can be expected without a large input of new data. The many technological and geological advances of the last decade justify a field (drilling) program designed specifically to develop a new, high-precision reversal time scale.

Plate Boundary Processes, Regional Crust/Mantle Properties, and Plate Kinematics

The igneous-tectonic part of the volume will be divided into three groups of chapters in progression from the present crustal accretion axis, to the regional state of the oceanic crust and lithosphere, and on to plate-tectonic evolution. The first group includes detailed studies of the Mid-Atlantic Ridge crest, models of crustal formation, the nature of crustal generation in Iceland, seismicity and fault-plate solutions, petrology and geochemistry (major and minor element), the active transform-fault domain, and regional variations along the accretion axis. The second group describes crustal structure (including sediments) from seismic reflection and refraction; gravity and geoid anomalies; magnetic anomalies; and heat flow, present state, and aging processes of the oceanic lithosphere. The third group of chapters covers plate kinematics, "absolute" plate motions, and paleobathymetry/paleogeography. This last chapter forms a bridge to the second part of the volume on ocean paleoenvironment.

The Present Accretion Axis

The present active plate boundary follows the crest of the Mid-Atlantic Ridge, its extension through Iceland and the Greenland-Norwegian Sea, and the Nansen Ridge cutting through the Eurasia Basin of the Arctic (Fig. 1). The total length of this accreting plate boundary between 13 °N, 44.5 °W and 68.5 °N, 133 °E, is about 14,100 km, of which 2,800 km (20%) consists of transform faults. The accretion presently occurs about two poles of rotation, one near 61.9 °N, 143.1 °E for relative motion between the North America (NOAM) and Eurasia (EURA) plates north of the Azores triple junction, and one near 80.4 °N, 56.4 °E for NOAM-Africa (AFRC) relative motion. Except for minor growth at the Mid-Cayman Rise, accretion to the North America plate occurs along the MAR and its Arctic extension. NOAM has been growing at about 0.1 km^2/yr on the average for the past few million years. Loss of NOAM plate area by subduction under the Caribbean arc is an order of magnitude less than growth.

Considering the entire accreting boundary as a unit is desirable for regional synthesis purposes. For example, to test the dependence of accretion processes on spreading rate one should examine a large range of rates. Older crust in the western Atlantic that formed at rates of about 1 to 4 cm/yr can then be examined in light of these rate-dependent processes. The NOAM-EURA boundary covers the entire "lower" end of globally observed present spreading half-rates, from 1.20 cm/yr near the Azores to less than 0.5 cm/yr along the southeastern Nansen Ridge. Half-rates increase southward along the NOAM-AFRC boundary and reach 1.6 cm/yr at 10 °N.

Treating the accreting boundary as a unit also is desirable from a geodynamic viewpoint. Iceland and the Azores, as prominent regional topographic highs on the MAR, are logical *geographic* limits for dividing the plate boundary, but these features and their associated geophysical and geochemical anomalies should for synthesis purposes be considered as *centers* rather than *boundaries*. This is especially important if Iceland and the Azores are "hotspots"—surface expressions of ascending mantle convection centers or plumes. Aside from changes in spreading rate and plate rotation latitude, other parameters also exhibit large ranges when compared along the entire eastern accreting edge of the North America plate: (1) Altitude of spreading axis: Depth of the accretion axis (away from fracture zone influences) varies from +1,300 m in southeastern Iceland to below −5,300 m along parts of the Nansen Ridge. This in fact represents the entire range found along the present mid-ocean ridge system. (2) Ridge crest morphology: Most slow-spreading ridges (<2.5 cm/yr half-rate) exhibit rough topography and median rift valleys, but these features are underdeveloped near the Azores (37.5°–39.5 °N) and within about 900 km of Iceland. (3) Plate-boundary geometry: Most accreting plate boundaries consist of spreading centers (axes) and nearly orthogonal transform faults. However, numerous oblique spreading centers have been identified along the MAR both south of Iceland (23°–27 °N, 47°–51 °N, 60°–64 °N) and to the north (71°–64 °N). Although some of these oblique axes make average angles as small as 15° to 30° with the direction of relative plate motion, the smaller-scale morphology of oblique rifts may be dominated by *en echelon* fissure eruptions less oblique to plate motion. (4) Plate boundary sedimentation: During the initial stages of spreading between two continents, a newly developed accretion axis generally acquires and retains a substantial elevation above the adjacent basins. This isolates even a deep rift valley along the accretion axis from all but a light dusting of pelagic sediments. However, there are exceptions, such as the large terrigenous input from the Amazon Cone into the Vema Fracture Valley at 11 °N and the burial of parts of the Knipovich and Nansen ridges by terrigenous debris in the Arctic. A study of the accretion process operating in an area of heavy terrigenous sedimentation has important ramifications for understanding the early evolution of rifted ocean basins because the first-formed oceanic crust, now commonly buried 5 km or more below the continental slope/rise, probably was exposed to a similar environment.

The preceding regional view of the accretion axis will be contained within only one chapter of our volume, but its perspective of establishing relationships among a variety of parameters at the larger scale will apply to associated chapters on the accretion axis. These chapters will deal mostly with the Mid-Atlantic Ridge and its Arctic extension. We will offer in-depth discussions of observations and models for crustal formation on Iceland and more typical MAR segments studied in detail by shipboard and submersible surveys (e.g., the FAMOUS-AMAR areas at 36°–37 °N). The regional seismicity and fault-plane solutions of the MAR will be examined and related to these models, and a separate chapter will treat the structure and seismicity of the active transform-fault domain. Two other chapters will discuss the petrology and geochemistry (major, minor, and trace element) of igneous, metamorphic, and hydrothermal samples from the immediate spreading axis where unaltered outcrops are common and sediment thickness is insignificant. Results of all these studies will be generalized and extrapolated to other, less well-known parts of the plate boundary and to older crust. In both instances care must be taken to determine what is normal (or typical) and what is anomalous. When the entire Mid-Oceanic Ridge is considered—particularly the slowly spreading parts—the MAR from 15° to 30 °N is in many respects the most typical and should therefore be used as a point of departure in regional comparisons.

Regional Geology and Geophysics

1. Gravity and Altimetry. The D-NAG synthesis of gravity and altimetry must draw on the results from other oceans and address fundamental topics ranging through basement topography, plate response to loading, aging of the lithosphere, and sublithospheric processes such as mantle convection.

The acceleration of gravity has been routinely measured on ships for years; gravity now compares with seismic reflection in terms of areal coverage. The ideal ocean surface (without currents or meteorological effects) is a potential surface whose undulations can be derived from the gravity field by the Stokes Integral. With the advent of geodesic satellites (GEOS; SEASAT) it became possible to map the ocean surface—and after suitable corrections, the potential surface—by measuring the reflection time of radar pulses bounced off the sea surface. Thus there are now two independent approaches to measuring the Earth's gravity field at sea.

The spectrum of gravity (and geoid) anomalies observed in oceanic areas, including the Atlantic, divides into three domains: short (30 to 300 km), intermediate (300 to 3,000 km), and long (>3,000 km), with overlaps at the boundaries.

Short-wavelength anomalies reflect density contrasts associated with sea-floor topography and to a lesser extent buried basement topography and crustal/upper mantle structures. In the past, gravity measurements were often used to help constrain seismic models of crustal structure. The data have also been used to study the flexural response of the lithosphere to a load subsequently imposed; gravity anomalies over a seamount formed on young crust (e.g., the Azores) differ from those over a seamount like Bermuda that was formed on crust already many tens of millions of years old. A third application lies in the near future when measurement of gravity anomalies by aircraft will be accurate enough to delineate major topographic (or basement) features in acoustically uncharted areas. Geoid (ocean-surface) anomalies measured by the SEASAT radar altimeter can be applied in the same three ways as gravity data. The SEASAT data set, however, is not dense enough to permit reliable contouring of the short-wavelength class of anomalies. The short-wavelength resolution of shipborne and future airborne gravity data is somewhat better than SEASAT altimetry, although major Atlantic fracture zones are readily resolved in the altimetric data.

Intermediate-wavelength gravity and geoid highs are associated with features like the Mid-Atlantic Ridge, Azores Plateau, and Bermuda Rise. Such anomalies reflect lithosphere, asthenosphere, and perhaps deeper density contrasts of thermal or compositional origin. A regular decrease of free-air gravity with crustal age has been observed for the first 40 m.y., and a corresponding effect is described for the geoid. Combined with basement depth, these data help constrain such parameters as upper mantle temperature, thermal diffusivity, and depth of compensation.

Long-wavelength anomalies, such as the gravity/geoid low that dominates the western Atlantic, are more mysterious because surface features or other geophysical anomalies are not obviously correlated with most of them. Such anomalies may relate to asthenosphere or deeper mantle processes (such as convection), which in turn may bear on the driving force of plate tectonics. The striking difference in regional gravity anomalies between the eastern and western North Atlantic may relate to two other observations: (1) the eastern Atlantic has been more influenced by mid-plate ("hotspot") volcanism, and (2) the western Atlantic lithosphere is moving relatively much faster in an "absolute" sense (Fig. 1).

2. Magnetic Anomalies. Unlike other marine geophysical data, the magnetic field can be readily measured by aircraft as well as by ships. When in the late 1960s linear anomalies of the total field strength came to be recognized as formed by sea-floor spreading and geomagnetic reversals, magnetics materialized as a kind of geophysical Cinderella. In our D-NAG volume, the chapter on geomagnetic reversals and two chapters on plate kinematics will discuss these now-classic applications, but an additional chapter, synthesizing the entire western Atlantic magnetic anomaly data base, will emphasize the following other kinds of information that can be extracted from such data:

Even in the plate kinematic area, both magnetic lineations and seamount-type anomalies constrain the location of the virtual geomagnetic pole (VGP) "seen" by the cooling igneous rock bodies. Magnetic anomalies therefore help constrain the absolute motion of the North America plate from Middle Jurassic time to the present. If the apparent polar wander path is assumed, VGP information from an undated feature such as a seamount helps fix its age. Analyses of seamount magnetization can also exploit Poisson's Theorem, which relates the magnetic and gravity anomalies of homogeneous bodies. Magnetic anomaly amplitudes must also reflect the paleomagnetic dipole strength, but such analysis has not been attempted.

Magnetic anomalies constrain structural and petrologic-mineralogic models of the oceanic crust. The synthesis will attempt to relate these geologic effects to variables such as spreading rate, crustal age, fracture spacing, proximity to hotspots, and crustal age. The local variability ("garbling") of the lineation pattern also provides statistical information on the texture of the source layer, that is, the temporal and spatial pattern of igneous emplacement and deformation in the accretion zone. Even when reduced to the way they would look at the North Magnetic Pole, linear anomalies may retain some skewness related to magnetization in the gabbroic layer, which may contribute one-third of the observed anomaly amplitude. Between the gabbro and the ~1-km-thick highly magnetized pillow basalt, about 1 km of essentially nonmagnetic greenschist facies is presumed to exist. Geographic variations in the thickness or magnetization of these units might be "seen" by magnetic anomaly data.

Over certain parts of the oceanic crust—notably near some Pacific hotspots and along fracture zones—anomaly amplitudes are enhanced to several times typical values. Such enhancements, not yet identified in the western North Atlantic except possibly in the J-anomaly Ridge area (Fig. 2), apparently reflect highly fractionated FeTi-type basalts of high remanent magnetization. (However, in the FAMOUS rift valley south of the Azores, fine-grained olivine basalts are the most magnetized.) Anomalously low amplitudes occur along some continental margins and in parts of the Greenland-Norwegian Sea; the association of many such "magnetic smooth zones" with thick sediment overburden suggests erasure of crustal magnetization by thermal effects, perhaps through accelerated maghematization. The latter process is often invoked to explain the rapid magnetization decline in the first few hundred thousand years after cooling. Still other factors may influence magnetization when extrusion occurs in shallow water (e.g., explosive brecciation and loss of volatiles).

3. Heat Flow, Age-Depth Anomalies, and Aging of the Lithosphere. The oceanic lithosphere being accreted to the trailing edge of the North America plate cools, subsides, and thickens as it ages. Although the average depth of newly formed oceanic crust may vary from –4 km (Nansen Ridge) to more than +1 km (Iceland), it is mostly in the range of –2.5 to –3 km. The subsidence thereafter, essentially independent of spreading rate or initial depth, proceeds as the square root of crustal age to about 80 m.y. Still debated is the behavior after 80 m.y.: Does the subsidence exponentially approach a constant depth, or does the observed flattening simply reflect the progressively greater chance of subsequent exposure of the plate to thermal anomalies? The central North Atlantic, underlain by crust from 0 to more than 170 m.y. in age, is an ideal laboratory for studying how slowly accreted (~1 to 2 cm/yr) oceanic lithosphere changes as it ages. This problem is interdisciplinary in that any aging model for the oceanic lithosphere must be consistent with thermal models of the crust and upper mantle, heat-flow measurements and attendant problems of hydrothermal circulation, experimental rock mechanics, petrologic models, gravity and bathymetric surveys, and magnetotelluric data, as well as with patterns of seismicity and focal mechanisms.

A fundamental problem for all such studies is to distinguish true aging effects from (a) variations in time and space of the *initial* plate properties and (b) later volcanic or sedimentary loading, and igneous, tectonic, or thermal "overprinting" of the original crust. Since only this subsequent history is unique to the North America plate (or to any particular oceanic area), a provincial discussion of the aging process is not sensible. Instead, data from all oceanic parts of the North America plate should be combined with data from all oceans in a search for dependence on age and other parameters. More specifically, the eastern and western North Atlantic should be compared since one is essentially a mirror image of the other. Any differences in crust or mantle properties might then relate to their different subsequent histories, including differences in absolute motion, which in turn may relate to asthenosphere properties, convection, and intraplate stress.

Another problem arises simply in the use of the "lithosphere" concept. Do we distinguish the bottom boundary of this aging "plate" by thermal, mechanical, or other parameters? Different approaches may produce different, although not necessarily incompatible, estimates of plate thickness and properties. Furthermore, it seems clear that the oceanic lithosphere is merely the upper boundary layer in a mantle convection scheme whose configuration at depth is at present still conjectural. The asthenosphere below the North American lithosphere, an unknown thickness of still deeper mantle, and anomalies such as possible plumes or blobs rising from the deeper mantle all must be considered as related to the aging or evolution of this plate.

4. Structure, Seismic Velocity Structure, and Thickness of Ocean Crust and Sediments. The only crustal aging effect thus far resolved in seismic refraction measurements from the central North Atlantic is a gradual increase in compressional (P) wave speed in the uppermost kilometer of igneous crust; after about 40 m.y. a constant value is reached, with much small-scale scatter. With the exception of anomalously thick crust like that beneath the aseismic Iceland-Faeroe Ridge, crustal thickness remains constant at about 4.8 km, with considerable local scatter, out to about 150 m.y. where thickness appears to increase by about 2 km. (Houtz, 1980). Given the fact that much of the scatter may be due to topographic effects that cannot be removed from present refraction/sonobuoy data, we probably can only approximate the true velocity/thickness distribution of oceanic crust in our synthesis. On the other hand, the velocity structure in sediments can be more accurately determined because of the lesser topographic complexity, closer proximity to measuring devices, and larger data set (including multichannel seismic reflection). This velocity information, when applied to sediment "thickness" (in reflection time) picked along seismic reflection profiles available for the western North Atlantic, allows a reasonably accurate portrayal of sediment distribution patterns and their relation to plate evolution, basement topography, sediment sources, bottom currents, and other factors (Fig. 5). The same data set also provides a remarkably complete picture of depth and surface structure of the igneous crust to about 160 m.y. age. Multichannel methods now permit continuous reflection profiling of the M-discontinuity, at least where basement topography is not too rough. In our

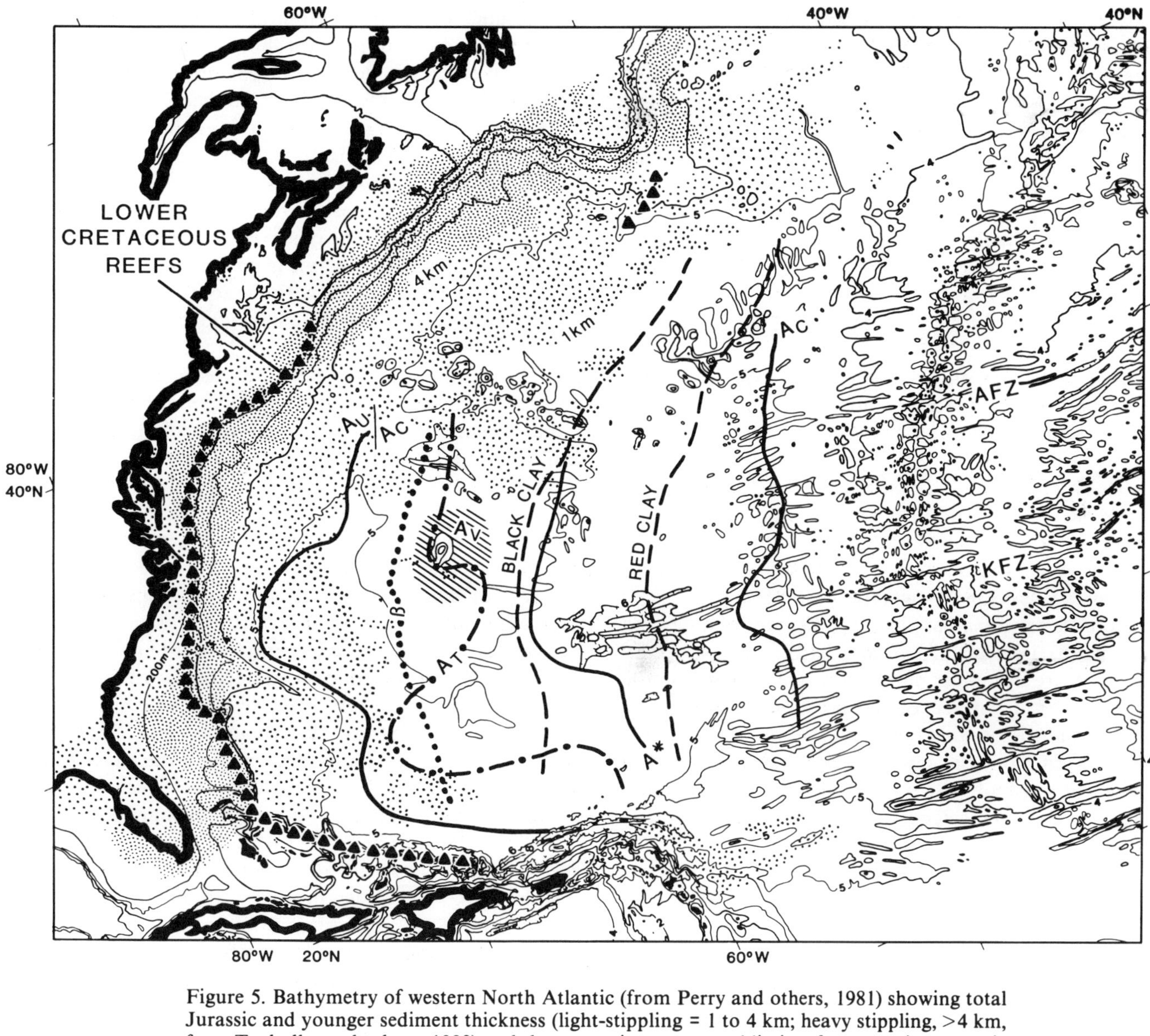

Figure 5. Bathymetry of western North Atlantic (from Perry and others, 1981) showing total Jurassic and younger sediment thickness (light-stippling = 1 to 4 km; heavy stippling, >4 km, from Tucholke and others, 1982) and the approximate seaward limits of some major acoustic reflecting horizons (solid lines) and lithofacies (dashed) from Tucholke and Mountain (1979).

synthesis we will integrate this basement structural information with regional data from other geologic and geophysical disciplines to form a clearer picture of the present crustal configuration and its plate kinematic evolution. This synthesis, in turn, forms a sound basis for later examination of sedimentation patterns and associated ocean paleoenvironment.

*5. **Crustal Homogeneity.*** As just discussed, seismic reflection and refraction data have long provided *regional* averages of crustal structure; more recently such data have revealed apparent aging trends in the crust. (As with other parameters, care must be taken to distinguish the effects of aging on a given parcel of crust from the variations, in time and space, of properties acquired *initially* at the spreading axis.) Inasmuch as striking *local* (10- to 100-km scale) variations in basement topography are well known, it is important to investigate possible variations in crustal structure at this scale. Among the local *apparent* variations in velocity structure and crustal thickness, some variations may be real, for example, the anomalously shallow mantle beneath several fracture zone traces in the North Atlantic (Detrick and Purdy, 1980). To the extent that crust is pervaded by fracture zones, spaced as little as 30 km apart, the crust and upper mantle are likely to be inhomogeneous at this scale. At other scales (e.g., between fracture zones or from one crustal "ribbon" to another) the lateral homogeneity (Purdy, 1981) and velocity anisotropy (e.g., Stephen, 1981) are beginning to be studied. A search for *mantle* velocity anisotropy in the Atlantic is another important goal, particularly since this may relate to flow orientation of olivine crystals and thus evidence of past flow under the accretion axis.

*6. **Petrology and Geochemistry of Pre-1-M.Y. Crust.*** Extrusive rocks of basaltic (tholeiitic) composition dominate North Atlantic basement lithology. Such rocks probably persist to depths of 1 to 2 km below the basement because the deepest DSDP penetration (583 m at Site 332) showed only a slight increase in the percentage of dikes and sills at depth. In Iceland's deepest drill hole, subaerial basalt lavas are common more than 4,000 m below the original uneroded surface of the crust. Composition of deeper parts of the oceanic crust and the top of the mantle can be inferred from the seismic velocity structure, from analogies with ophiolite sequences on the continent, and from plutonic rocks such as serpentinized peridotites (altered mantle?) and gabbros (lower crust?) dredged in the major Atlantic fracture zones or recovered at anomalously shallow (50 m) sub-basement depth at some DSDP sites (e.g., 334 and 395). The "ophiolite connection" ties the North Atlantic volume to others in the D-NAG series, particularly the Appalachian volume, since obducted remnants of an earlier Atlantic Ocean (the "Iapetus Ocean") apparently survive there.

A separate chapter will be devoted to the composition of "older" (>1 m.y.) basement samples, most of which are of DSDP-IPOD origin. Two important sets of problems should be considered: (1) What are the nature and extent of chemical alteration (principally as a result of low-temperature reactions with sea water), and how does this alteration relate to physical properties (e.g., sound velocity) of the upper crust and to the composition and long-term chemical budget of the ocean? (2) What are the chemical and petrologic characteristics of the unaltered rock, and how do they compare with the composition of rocks recently erupted at the MAR axis?

The first problem is not specifically Atlantic; data from that ocean thus could be combined with others from the world ocean. Factors such as variable degrees of crustal "sealing" by sediment cover and the possibly more "weathering"-prone, topographically rough Atlantic crust formed by slow spreading rates should be considered.

The second problem becomes more interesting for purposes of basin-wide synthesis if there are systematic variations in average composition over millions of years and/or hundreds of kilometers. One hypothesis that can be tested is whether composition is a function of spreading rate, for example increased fractional crystallization at higher spreading rates. The observed southward increase of Ti and Na along the MAR axis from the Azores area to the equator has been correlated with the southward increase in spreading rate. Spreading-rate dependence, if any, should be sought globally, but can be tested in the western North Atlantic by comparing the "faster" Jurassic, Middle to Late Cretaceous, and early Cenozoic basalts with those retrieved in the "slower" Early Cretaceous and middle Cenozoic crust. A second testable hypothesis is that the Large Ion Lithophile (LIL)–enriched centers on the MAR at 35°, 39°, 45°, and 63°N represent stationary mantle plumes. If such plumes maintain their position at the plate boundary and do not change in composition, a transect along the trajectory of "absolute" (plate-mantle) motion should yield basalts of similar LIL enrichment. (It would be misleading to conduct such transects along trajectories of *relative* plate motion.) A similar test may be applied to possible mantle-plume manifestations such as the New England Seamounts.

*7. **Mid-Plate Stress, Deformation, and Seismicity.*** Most earthquake seismicity occurs along the edges of the North America plate (Figs. 1, 3) and in the western cordilleras, but seismic events are by no means uncommon in many parts of the plate interior. The nature of the stresses responsible for intraplate seismicity is poorly understood. Many possible sources have been proposed, for example, deglaciation, sediment loading, differential cooling or underplating of the lithosphere, drag at the base of the plate, and compression transmitted from "ridge-push" forces acting along the Mid-Atlantic Ridge.

Mid-plate seismicity appears much less common in older oceanic lithosphere than in the adjacent continental portion of the plate (Fig. 3). However, the 1978 occurrence of a magnitude 6 event on the Bermuda Rise, plus other data, suggests that seismicity in the western North Atlantic is comparable to the activity in eastern North America. Western North Atlantic seismicity possibly is related to continental seismicity in several ways: (1) Differential cooling of oceanic and continental lithosphere, loading by continent-derived sediments in oceanic depocenters like the Laurentian Fan, and loading/unloading by continental ice sheets that extended to the shelf edges may all be important sources of stress. (2) Oceanic transform or other fracture zones may have continental extensions that structurally segment the margins and act as zones of weakness. (3) Oceanic intraplate igneous belts (e.g., the New England Seamount chain) also may lie on the seaward extensions of seismically active continental fracture zones. (The alternative view, that they are hotspot traces, will be discussed later.) (4) Although there is evidence for some regional variation in the stress field (e.g., between the mid-continent and eastern parts of the U.S.; Zoback and Zoback, 1981), much of the North America plate east of the Rocky Mountains appears to be under large horizontal compressive stress oriented roughly eastward, that is, anti-parallel to the present direction of absolute plate motion. This would be consistent either with "ridge-push" forces developed by the Mid-Atlantic Ridge or with asthenospheric drag experienced by the underside of the North America plate. In either case a plate-wide treatment of seismicity and stress is indicated. Placing this treatment in the North Atlantic synthesis presumes that mid-plate stress is closely related to absolute plate motion and/or to "ridge-push" forces.

Plate Kinematics

According to the plate-tectonics paradigm, the North America plate moves with respect to its neighbors and with respect to "fixed" coordinate systems such as the Earth's rotation axis, the mean geomagnetic dipole, or hotspots that are presumed to originate in the less mobile mantle below the asthenosphere. An in-depth discussion of these "relative" and "absolute" motions is a key ingredient of the D-NAG synthesis. Although the threads of plate tectonics will be interwoven throughout the D-NAG volumes, it seems most appropriate to discuss the post-Triassic motions of the North America plate in the western North Atlantic volume. There are two good reasons for this: (1) The Mid-Atlantic Ridge best constrains the present motions of the plate with respect to the Eurasia and Africa plates and, by vector additions, with other plates (Minster and Jordan, 1980). (2) A continuous history of relative motion between NOAM and AFRC since their separation in Early Jurassic times is recorded in the central Atlantic crust south of the Grand Banks–Gibraltar line; to the north the oceanic crust carries a record of mid-Cretaceous and later plate motions involving NOAM, EURA, and additional plates that had a temporary existence. The principal kinematic data (e.g., Fig. 3) are transform fault traces and magnetic lineations (crustal isochrons).

1. Relative Plate Motion. The D-NAG treatment of relative plate kinematics in the western North Atlantic will consist of two chapters: (1) a kinematic model that attempts to recreate the history of motion between AFRC and NOAM to form the central North Atlantic, and (2) a kinematic model that treats the more complex plate motions north of the Azores, involving the NOAM, EURA, and various lesser plates, such as Greenland, which once existed along the NOAM-EURA boundary region.

The treatment of plate motions north of the Azores must consider the opening of the Labrador Sea and Eurasia basins, but detailed discussions of these northern arms of the North Atlantic rift will be found in the D-NAG volumes on the Arctic Ocean and the eastern margin of Canada.

A plate kinematic history of the North America plate actually is meaningful only for Cenozoic and perhaps Late Cretaceous time. During Jurassic and Early Cretaceous times the proto-MAR formed the break between the two supercontinents, Gondwanaland to the south and Laurasia to the north. Only about 110 to 120 m.y. B.P. did the MAR accreting plate boundary begin its episodic drive north of the Grand Banks into the Laurasian heartland following a chain of epicontinental seas that had developed along the Caledonian orogenic structure. The subsequent development of this trailing plate margin will be a central theme in our volume.

Kinematic models north and south of the Azores are not independent; some overlap is inevitable but not undesirable. Although plate motions can be summarized in a table of rotation angles and poles, the D-NAG synthesis will demand geographic reconstructions for various important and well-constrained times. "Well-constrained" refers to widely mapped and easily identifiable magnetic lineations (e.g., anomalies 5, 6, 13, 21, 22, 24, and 34) and well-mapped continuous transform fault traces with large offsets. "Important" refers to tectonic events or turning points. The opening of new oceanic rifts and major changes in plate rotation pole and rotation rate are important tectonic events. One problem is to establish the time and magnitude of such events even when they fall somewhere between key magnetic lineations. In the western North Atlantic this is especially difficult because 45% of the ocean crust is represented by the Jurassic Magnetic Quiet Zone (JMQZ) and the Cretaceous normal (Mercanton) interval (Figs. 3, 4).

A problem with many published plate kinematic models is the lack of adequate error limits. As a consequence, it is often impossible to tell whether two reconstructions (or one reconstruction and some independent data) are in significant conflict, given random errors in anomaly identifications, and so forth. The analysis of present plate motions by Minster and Jordan (1980) should set a standard for the kind of "statistical objectivity" desired in future plate reconstructions.

2. Absolute Plate Motion and Hotspot Traces. The absolute plate motion of NOAM, presently 0.25° m.y. about a pole at 58.3 °S, 40.7 °W, may have been recorded by hotspot traces, that is, belts of mid-plate volcanism such as the Hawaii-Emperor Chain in the Pacific. For example, Morgan (1982) considers the igneous belt stretching from the White Mountains along the New England Seamounts (Fig. 2) as the trace of Cretaceous absolute motion of the North America plate. Other possible hotspot traces at least partly on the oceanic parts of this plate are represented by the Greenland-Scotland Ridge, the Newfoundland Seamounts, Muir and Caryn Seamounts, Bermuda, and perhaps the Bahama Banks (Figs. 2, 3). According to the hotspot concept, such traces may cross from continent to ocean basin (e.g., from the White Mountains magma centers to the New England Seamounts; Fig. 3), or they may be entirely oceanic (e.g., the Canary Islands on the Africa plate) or continental (e.g., the Yellowstone-Snake River trace). Hotspot traces may also cross plate boundaries, but if a hotspot remains near a plate boundary it will simultaneously produce a trace on each plate. The Azores hotspot, for example, lies close to the NOAM-AFRC-EURA triple junction and has caused volcanoes to erupt on all three plates (Fig. 3).

The above kinematic complexities demand a unified, plate-wide treatment of hotspot traces in relation to absolute motions of the North America plate. Moreover, as with relative plate motions, it is also necessary to consider hotspot traces developed on neighboring plates. (However, detailed discussions of North American mid-plate magmatism properly belong in other D-NAG volumes.) A significant difficulty in our synthesis is the lack of reliable and unambiguous age dates on magmatic features. For example, there are only a handful of age determinations on the New England Seamounts, not enough to test the sort of age gradient (a few centimeters per year younging to the southeast) that would be compatible with the fixed-hotspot hypothesis. Moreover, in the absence of deep drill holes fully penetrating volcanic edifices, the age of onset of volcanism is not definitely known for any island or seamount in the North Atlantic. If most of an edifice was constructed in a brief interval of time—1 to 2 m.y. or less—it would not be so critical to sample the first eruptions (Vogt and Tucholke, 1979). However, the evidence for an initial, rapid, edifice-building stage still is very limited for Atlantic volcanoes. Simultaneous or episodic volcanism along a fracture or progressive volcanism along a propagating fracture also remain viable alternatives to the hotspot hypothesis. To the extent that fracturing plays a role in intra-plate volcanism, the geographic pattern and timing of this volcanic activity will reflect both the preexistence of fractures and the state of stress of the lithosphere at the time of magmatism. Although seamount chains erupted according to "fracture" mechanisms would not record the direction of absolute plate motion, they would still retain an important plate-tectonics relevance.

3. Paleobathymetry and Paleogeography. The past distribution and elevation of crustal topographic elements are a direct consequence of the plate kinematic evolution of the North Atlantic. By exerting a profound influence on ocean circulation, sedimentation, and climate, "paleobathymetry" deserves a thorough treatment in our volume. Although the first-order topography of the sea floor can be derived from generalized empirical treatments, there are many departures from this norm. Prominent examples include the effects of sediment accumulation and erosion, possible variations in subsidence rate due to varying crustal thickness or proximity to hotspots, and subsequent depth changes such as formation of the Bermuda Rise. A comprehensive synthesis will have to investigate the various phenomena.

In addition, the distribution of subaerial topography, the subsidence history of the continental margin, and the interplay of margin flooding and sea-level fluctuations must be documented. Their relations to plate kinematic evolution can then be tested, thus allowing significant improvement in our application of paleobathymetry and paleogeography to studying the ocean paleoenvironment.

Sedimentary Record

The chapter on paleobathymetry and paleogeography provides the link between the igneous-tectonic part of our D-NAG volume and the sedimentary part. The sedimentary part is in turn divided into discussions of surficial sedimentation, biofacies, and ocean paleoenvironment. The latter topic includes seismic stratigraphy, lithostratigraphy, and facies analysis.

About five years will have elapsed between the publication of Volume 3 of the Maurice Ewing Series, *Deep Drilling Results in the Atlantic Ocean* (Talwani and others, 1979), and the forthcoming D-NAG volume. The newer synthesis, although probably not dramatically different except for its geographic emphasis, will be able to incorporate new multichannel seismic data, especially over Jurassic ocean crust; the results of recent IPOD drilling; and the added wisdom and insight developed over this period. Improved correlation between deep-sea and shallow-water facies also may be possible, based on recent exploratory drilling and seismic work along the U.S., Canadian, and Greenland continental margins.

Surficial Sedimentation

A group of chapters will synthesize late Pleistocene–Holocene sediment characteristics, including echo-character, in relation to depositional *processes,* in order to form a foundation for subsequent discussions of ocean paleoenvironment. The chapters will discuss four principal processes (or regimes) of deep-sea sedimentation: mass movements, suspension (turbidity) flows, bottom-current effects, and pelagic/aeolian sedimentation. The first two processes are related, since slumps and slides may generate turbidity flows. The first three involve both erosion and deposition.

All syntheses should relate observations of surficial sediments to in situ measurements of present-day processes (e.g., current velocity, pelagic fertility, atmospheric dust, etc.) and to models of such processes. Two major problems occur in relating measurable processes to the observed sedimentary conditions: (1) the Quaternary and late Pliocene climatic fluctuations and (2) the episodic nature of some processes (e.g., a turbidity current has never been directly observed). The effects of climatic fluctuations on the biostratigraphy and composition of calcareous pelagic/hemipelagic sediments have been particularly striking at intermediate latitudes such as the western North Atlantic. Indeed, North Atlantic sediment cores above the calcite compensation depth provide one of the best records of Pleistocene glacial-interglacial climatic cycles. These are now generally believed to be forced by small insolation changes of astronomical origin, greatly amplified by oceanographic and meteorological feedback effects. The D-NAG *Quaternary: North America* volume will deal with some of these topics, but we still are confronted with the problem of separating Holocene (and previous interglacial) effects from those inherited from full-glacial or transitional regimes. For example, are such features as the Northwest Atlantic Mid-Ocean Channel (Fig. 2) or giant ripples and furrows on the Blake-Bahama Outer Ridge related to Holocene conditions, or are they relict features formed during the Wisconsin glacial stage? Episodic events, some of which depend on factors such as slope stability, currently are even less tractable problems.

Biofacies

Pelagic and occasionally terrestrial biogenic debris composes a significant fraction of the marine sedimentary record in the western North Atlantic. These biofacies are important for reconstructing paleoceanographic conditions in surface and deep waters, for biostratigraphic age control, and often for study of postdepositional diagenesis. The most common biogenic components are planktonic and benthonic foraminifera, calcareous nannofossils, and radiolarians. Less common or stratigraphically more restricted are diatoms, palynomorphs, icthyoliths (fish teeth), silicoflagellates, crinoids, ammonites, ostracods, rudists, and coralline algae.

Our knowledge of these fossils in the pre-Pliocene sedimentary record is derived almost exclusively from Deep Sea Drilling Project boreholes, which unfortunately are few (43), far between, and poorly distributed for attaining a regional perspective. However, there is no question that these data have been of inestimable value in the study of Atlantic paleoceanography. The D-NAG synthesis is an appropriate vehicle for summary of the diverse biofacies in the western Atlantic. Obviously it will not be possible to give a complete review of Mesozoic and Cenozoic marine biostratigraphy, which is properly a global problem, within the limited confines of one section of the D-NAG volume. However, description of the distribution of Atlantic biofacies in space and time and their impact on development of biostratigraphic usage in the global oceans is a tractable and worthwhile goal. Three chapters in our volume will describe Neogene, Paleogene, and Mesozoic biofacies, and a summary chapter will deal with the paleobiogeography of the North Atlantic.

Ocean Paleoenvironment

We do not mean to downplay the scientific challenge presented by postdepositional processes (e.g., formation of the mid-Eocene cherts) or the physical, chemical, and engineering properties of deep-sea sediments; both topics will be addressed in the synthesis volume. Yet, it is the oceanic environment that prevailed in the Atlantic *at the time* sediments were (or were not) deposited that must take center stage in exposition of the sediment history. We call this topic "ocean paleoenvironment" rather than "paleoceanography" to emphasize that it is not only ancient ocean currents, temperature, water chemistry, fertility, and biogeography that capture the workings of the system but also the shape and character of the sea floor below; winds, insolation, clouds, and rainfall from above; and runoff and detritus from adjacent continents and perhaps interior sources like island volcanoes and reefs. Drawing as it does from disciplines as diverse as plate kinematics, biostratigraphy, isotope geochemistry, and facies analysis, the synthesis of oceanic paleoenvironment (Neogene, Paleogene, Cretaceous, and Jurassic chapters) is placed near the end of our volume. These chapters present not only the greatest scientific challenges but also the greatest problems in synthesis.

A significant part (~150 m.y.) of the ?170-m.y. stratigraphic record of the western North Atlantic has been sampled, however sparsely, by deep-sea drilling. Studies of the lithofacies, biofacies, and geochemical characteristics of these sediments have resolved the general depositional environments since Late Jurassic time (Table 1) and have attempted to explain their remarkable diversity. Further-

TABLE 1. PRINCIPAL LITHOFACIES AND PALEOENVIRONMENTS IN THE WESTERN NORTH ATLANTIC

Unit Name	Age	Principal Facies	Depositional Environment	Principal Correlative Reflectors
Blake Ridge Formation	Holocene-Middle Eocene	Brown and green-gray hemipelagic silty clay, turbidites	Bottom-current/turbitidy-current control on deposition of largely terrigenous debris at and below CCD	Reflector X near middle Miocene
Great Abaco Member	Miocene	Intraclastic chalk	Mass-flow deposit in Blake-Bahama Basin, from adjacent carbonate banks	Reflector M at top
-	Holocene-Middle Eocene	Calcareous ooze on MAR flank-laterally correlative with Blake Ridge Formation	Pelagic, above CCD	-
-	Oligocene-Early Eocene	Volcaniclastic turbidites surrounding Bermuda	Turbidity-current flow from Bermuda pedestal	Horizon A^T
Bermuda Rise Formation	Early-Middle Eocene	Siliceous mudstone, chert	Hemipelagic with high biosiliceous productivity	Horizon A^C at top
Plantagenet Formation	Paleocene-Turonian	Multicolored pelagic shales	Pelagic, well below CCD	-
Crescent Peaks Member	Early Maestrichtian	Calcareous ooze/chalk	Pelagic, above CCD	Horizon A*
Hatteras Formation	Cenomanian-Barremian	Black and green-gray carbonaceous shales	Hemipelagic deposits well below CCD, under intermittently anoxic conditions	-
Blake-Bahama Formation	Barremian-Early Tithonian	Gray-white cherty limestone, chalk, marl	Pelagic, above CCD. Some turbidity current influence	Horizon β at top
Cat Gap Formation	Early Tithonian-Oxfordian	Reddish and green-gray argillaceous limestone/chalk	Pelagic, near CCD. Some turbidity current influence	Horizon J_1 or C at top
	Oxfordian-Callovian and older?	Dark shale radiolarian shale, pelletal limestone	Pelagic, near and below CCD. Intermittently anoxic. Some turbidity current influence	Horizon D?
	Pre-Callovian	Not yet drilled. Probably includes terrigeneous, pelagic, and evaporite facies	Variable	Horizons J_2, J_3

Note: A major unconformity (= Horizon A^u), eroded along the deep western margin by Oligocene bottom currents, truncates Eocene to Barremian facies.

Table based on: Jansa and others (1979), Tucholke (1979), Tucholke and Mountain (1979), Bryan and others (1980), Klitgord and Grow (1980) and Scientific Party (1981).

more, studies of seismic stratigraphy and its correlation to lithofacies at the boreholes have greatly amplified our knowledge of the extent of lithofacies in the basin (Fig. 5) and our ability to interpret the depositional environments.

However, these and future studies are faced with two basic problems. First is the fact that the Atlantic is and always has been part of the world ocean and cannot even be approximated as a separate system. Perhaps the best example of this fact is the flows of deep and bottom water masses that have profoundly affected Cenozoic sedimentation patterns, especially along the western boundary of the basin. The deepest water mass is Antarctic Bottom Water which is derived largely from the Weddell Sea and flows through the western South Atlantic into the North American Basin. Above it, the North Atlantic Deep Water originates largely in the Norwegian-Greenland Sea, and then follows a tortuous, contour-following path southward along the western basin margin. Obviously tectonic and climatic changes in the Arctic/Antarctic that otherwise had little or no direct impact on the North Atlantic could have had (and did have) a dramatic effect on its abyssal circulation/sedimentation patterns.

This leads us to a second basic problem in reconstructing ocean paleoenvironment, namely, what are the exact nature and hierarchy of cause and effect relations that control paleoenvironment? For a problem like the Eocene "blizzard" of biogenic silica in the North Atlantic, two of several possible causal chains look like this: (1) tectonic breaching of Arctic landmasses → influx of Arctic bottom water to the North Atlantic → enhanced upwelling of nutrient-rich water → increased biosiliceous productivity (Berggren and Hollister, 1974), and (2) increased plate-tectonic activity → increased subaerial and submarine volcanism → increased silica and nutrients → enhanced productivity and preservation (e.g., Mattson and Pessagno, 1971; Hermann, 1972). At the head of these and many other cause-effect chains is plate tectonics, which in turn is driven by or related to processes in the sublithospheric mantle.

A separate set of cause-effect chains is set up by astronomical (e.g., Milankovitch) driving forces and by catastrophes of extraterrestrial origin (e.g., asteroid impact as the cause of the end-Cretaceous extinctions). Understanding the constructive/destructive interference of all these event chains is a challenge that will extend through and well beyond our synthesis efforts.

Man and the Sea

Human pressure toward potential use of geology beneath the ocean, including the western North Atlantic, perhaps has never been so great as now. Offshore exploration for fossil fuels is both intensifying and extending toward deeper water. The search for economically recoverable deposits of manganese and phosphorite nodules is expanding. Recent discoveries of massive sulfide deposits at the crest of the Pacific Rise has stimulated speculation on the possibility of similar deposits along other, more slowly accreting plate boundaries such as at the Mid-Atlantic Ridge, or over oceanic crust that was formed at high spreading rates and is now locked in the interior of the North America plate. Knowledge of the stability and engineering properties of marine sediments accrues added importance as platforms and instruments are anchored on the seabed. The sea floor is a source of nutrients and home to bottom-dwelling organisms that directly or indirectly relate to man's finned-fish and shellfish harvests. And philosophical, political, and economic factors all are being scrutinized as "Law-of-the-Sea" negotiations attempt to define man's future access to the oceans and sea floor.

Addressing these topics are three chapters near the close of our volume: "Resources," "Engineering Properties and Hazards," and "Law of the Sea and Man's Use of Ocean Geology." Geologic knowledge of the western North Atlantic ocean basin has progressed from virtually nonexistent to sophisticated (albeit incomplete) in GSA's 100 years. It is perhaps fitting that our volume, the first volume projected for completion in the D-NAG series, will have the opportunity to assess how successful we have been in application of this knowledge to the benefit of mankind.

Acknowledgments

The perspectives developed in this essay have been fostered over the years by funding from the Oceanography Section of the National Science Foundation and from the Office of Naval Research. Their continuing support of basic science is largely responsible for the progress in oceanography that will be synthesized during the Decade of North American Geology, and we gratefully acknowledge that debt. We also thank Irene Jewett, Carol Rockelli, and Tracie Williamson for assistance, and J. Ewing and E. Uchupi for reviewing the manuscript.

Woods Hole Oceanographic Institution Contribution No. 5066.

References Cited

Berggren, W. A., and Hollister, D. C., 1974, Paleogeography, paleobiogeography and the history of ocean circulation in the Atlantic Ocean, *in* Hay, W. W. (ed.), Studies in paleo-oceanography: SEPM Special Publication 20, p. 126–186.

Detrick, R. S., Jr., and Purdy, G. M., 1980, The crustal structure of the Kane fracture zone from seismic refraction studies: Journal of Geophysical Research, v. 85, p. 3759–3777.

Emery, K. O., and Uchupi, E., 1972, Western North Atlantic Ocean: Topography, rocks, structure, water, life, and sediments: American Association of Petroleum Geologists, Memoir 17, 532 p.

Hermann, Y., 1972, Origin of deep-sea cherts in the North Atlantic: Nature, v. 238, p. 392–393.

Houtz, R. E., 1980, Crustal structure of the North Atlantic on the basis of large-airgun-sonobouy data: Geological Society of America Bulletin, v. 91, p. 406–413.

Mattson, P. H., and Pessagno, E. A., Jr., 1971, Caribbean Eocene volcanism and the extent of Horizon A: Science, v. 174, p. 138–139.

Minster, J. B., and Jordan, T. H., 1980, Present-day plate motions: A Summary, *in* Soma mechanism and earthquake prediction: Editions du Centre National de la Recherche Scientifique Paris, p. 190–124.

Morgan, W. J., 1982, Hotspot tracks and the opening of the Atlantic and Indian oceans, *in* Emiliani, C., ed., The Sea: Wiley and Sons (in press).

Perry, R. H., Fleming, H. S., Vogt, P. R., Cherkis, N. Z., Feden, R. H., Thiede, J., Strand, J. E., and Collette, B. J., 1981, North Atlantic Ocean: Bathymetry and plate tectonics evolution: Geological Society of America, Map and Chart Series MC-35.

Purdy, G. M., 1981, The seismic structure of 140 m.y. old crust in the western central Atlantic Ocean [abs.] EOS, v. 62, p. 953.

Stephen, R. A., 1981, Seismic anisotrophy observed in upper oceanic crust: Geophysical Research Letters, v. 8, p. 865–868.

Talwani, M., Hay, W., and Ryan, W.B.F. (eds.), 1979, Deep drilling results in the Atlantic Ocean: Continental margins and paleoenvironment: American Geophysical Union, 437 p.

Tucholke, B. E., 1979, Relationships between acoustic stratigraphy and lithostratigraphy in the western North Atlantic basin, *in* Tucholke, B. E., and Vogt, P. R. (eds), Initial reports of the Deep Sea Drilling Project: U.S. Government Printing Office, v. 43, p. 827–846.

Tucholke, B. E., and Mountain, G. S., 1979, Seismic stratigraphy, lithostratigraphy and paleosedimentation patterns in the North American Basin, *in* Talwani, M., Hay, W., and Ryan, W.B.F. (eds), Deep drilling results in the Atlantic Ocean: Continental Margins and Paleoenvironment: American Geophysical Union, M. Ewing Series, v. 3, p. 58–86.

Tucholke, B. E., Houtz, R. E., and Ludwig, W. J., 1982, Map of sediment thickness and depth to basement in the western North Atlantic, American Association of Petroleum Geologists Bulletin (in press).

Vogt, P. R., and Einwich, A. M., 1979, Magnetic anomalies and sea-floor spreading in the western North Atlantic, and a revised calibration of the Keathley (M) geomagnetic reversal chronology, *in* Tucholke, B., and Vogt, P. R., et al., Initial reports of the Deep Sea Drilling Project: U.S. Government Printing Office, v. 43, p. 857–876.

Vogt, P. R., and Tucholke, B. E., 1979, The New England Seamounts: Testing origins, *in* Tucholke, B. E., Vogt, P. R., et al., Initial reports of the Deep Sea Drilling Project: U.S. Government Printing Office, v. 43, p. 847–856.

Vogt, P. R., and Perry, R. K., 1982, North Atlantic Ocean: Bathymetry and plate tectonics evolution, text to accompany Map and Chart Series MC-35: Geological Society of America (in press).

Zoback, M. L., and Zoback, M., 1980, State of stress in the conterminous United States: Journal of Geophysical Research, v. 85, p. 6113–6156.

The Continental Margin: Eastern Canada

Richard T. Haworth
Atlantic Geoscience Centre
Geological Survey of Canada
Bedford Institute of Oceanography
P.O. Box 1006
Dartmouth, Nova Scotia B2Y 4A2

Introduction

Over 80% of the continental margin of eastern Canada is underlain by a seaward-thickening wedge of Tertiary and Mesozoic sedimentary rocks, a most attractive exploration target for oil and gas. The first offshore wells were drilled on the Grand Banks in 1966, just before publication of the 1970 edition of the *Geology and Economic Minerals of Canada* in which a mere 2,000 words attested to the fact that "the Mesozoic record of southeastern Canada [was] scant." By 1981, a total of 158 wells had been drilled, significant shows of gas and/or oil had been recorded in 22 of the wells, and production appeared to be imminent in the Sable Island and northern Grand Banks areas. With the contemplation of production facilities comes a requirement for extensive information regarding the nature of the seabed and its stability, environmental hazards such as icebergs and their seabed scours, and the susceptibility of the coastal environment to the pollution created by marine oil spills. Thus, economic and environmental pressures have been largely responsible for the rapid accumulation of knowledge about the continental margin of eastern Canada.

Each section of this report is a precis of each chapter of the synthesis volume, *The Continental Margin: Eastern Canada.* Data and interpretations will be displayed in the text on standard 1:10 million scale, page-size maps (Figs. 1, 2, 3) as background for discussion of the entire margin. A comprehensive bibliography will also be included. Contributions to the volume will be organized so as to make the clearest possible distinction between fact and interpretation, and in the latter case, present the different sides of controversial questions.

Physiographic Provinces

The continental shelf adjacent to Nova Scotia and Newfoundland is generally covered by less than 200 m of water. The shelf edge is well marked and cut by numerous canyons leading down the continental slope, but the base of the continental rise is not so clearly defined. The Laurentian Fan extends to water depths of over 5,000 m at the mouth of the glacially deepened Laurentian Channel. The Grand Banks of Newfoundland have less relief than the Scotian Shelf and Georges Bank, except close to shore where the mouths of fjords are overdeepened. The Newfoundland Basin, southeast of the Grand Banks, is bounded to the north by Flemish Cap and to the south by the southeast Newfoundland Ridge. The Newfoundland Seamounts follow an east-west line roughly dividing that basin. Flemish Cap is separated from the Grand Banks by a narrow, deep channel. Orphan Knoll, an isolated bathymetric high although smaller and deeper than Flemish Cap, lies at the intersection of the Charlie Fracture Zone and the Canadian margin. The boundary between the Grand Banks and the deep-water area southwest of Orphan Knoll is marked by slopes that are much less steep than the continental edges elsewhere off Atlantic Canada. The Northeast Newfoundland Shelf has highly irregular relief, and the shelf break lies at the abnormal depth of about 300 m. The Labrador Shelf is characterized by a series of banks separated from each other and the mainland by deep channels. Evidence for iceberg scouring is seen throughout the Atlantic margin. Hudson Strait separates the Labrador Shelf from the deeper water continental shelf southeast of Baffin Island which trends northward into the Davis Strait sill separating Baffin Bay from the Labrador Sea. Physiographic distinctions between shelf and slope become less well pronounced in this area. The Labrador Sea becomes progressively shallower throughout the northern third of its length with a much sharper transition from narrow shelf to deep ocean floor on the Greenland side than on the Labrador margin. Baffin Bay has maximum water depths of 2 km, the shelf on the Greenland margin in general being substantially wider and deeper than on the Canadian margin. Lancaster Sound is a broad channel that leads westward into the Arctic Islands from the northern end of Baffin Bay. Nares Strait, between the Arctic Islands and Greenland, is the other major physiographic feature at the northern limit of the area to be discussed.

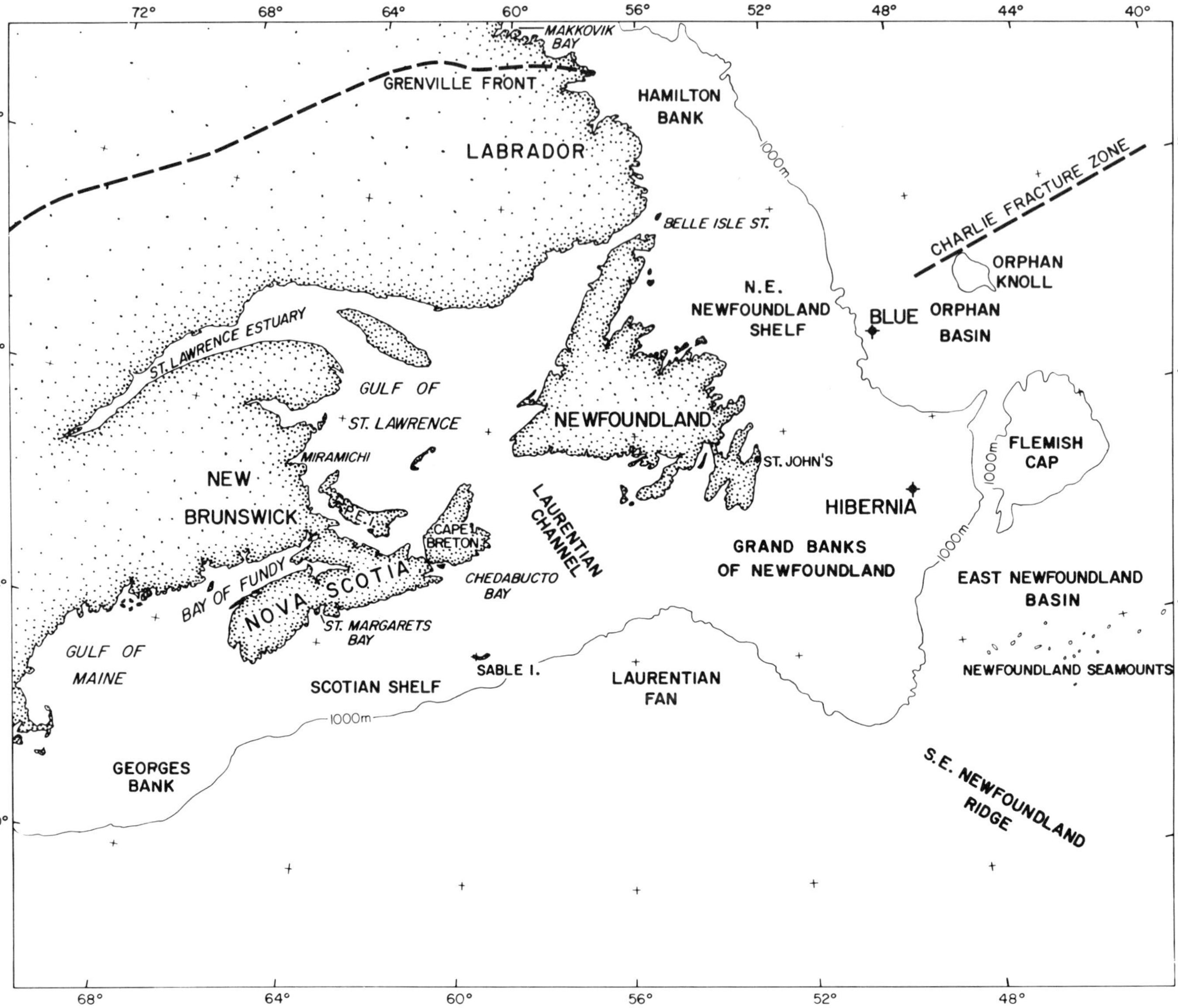

Figure 1. The continental margin of eastern Canada: The Atlantic margin.

Each of these physiographic features has a tectonic significance that will be dealt with in subsequent sections.

History of Exploration

The earliest recorded exploration of the eastern Canadian margin took place in the late fifteenth and early sixteenth centuries, and farther north during the late sixteenth and early seventeenth centuries.

However, detailed exploration only came with the requirement to lay submarine telegraph cables across the Atlantic in the nineteenth and twentieth centuries. Such dramatic events as the cutting of these cables by turbidity currents in the Laurentian Fan area, as part of the 1929 Grand Banks earthquake, and World War II provided an even greater impetus to submarine investigations of the margin. Lamont-Doherty Geological Observatory, in particular, provided information about the gross crustal structure of the Canadian margin, which was supplemented during the 1960s by the work of Canadian scientists. At the same time, industry began geophysical exploration of the offshore sedimentary basins for structures that might be conducive to hydrocarbon entrapment. Apart from minor oil and gas shows, the initial drilling results from the Scotian Shelf and Grand Banks were discouraging, and gas finds in the hostile environment of the Labrador Shelf were discounted; therefore, exploration activity waned during the mid 1970s. The discovery in 1979 of the oil potential of the Hibernia structure revitalized exploration activity and necessitated increased investigation of the processes that

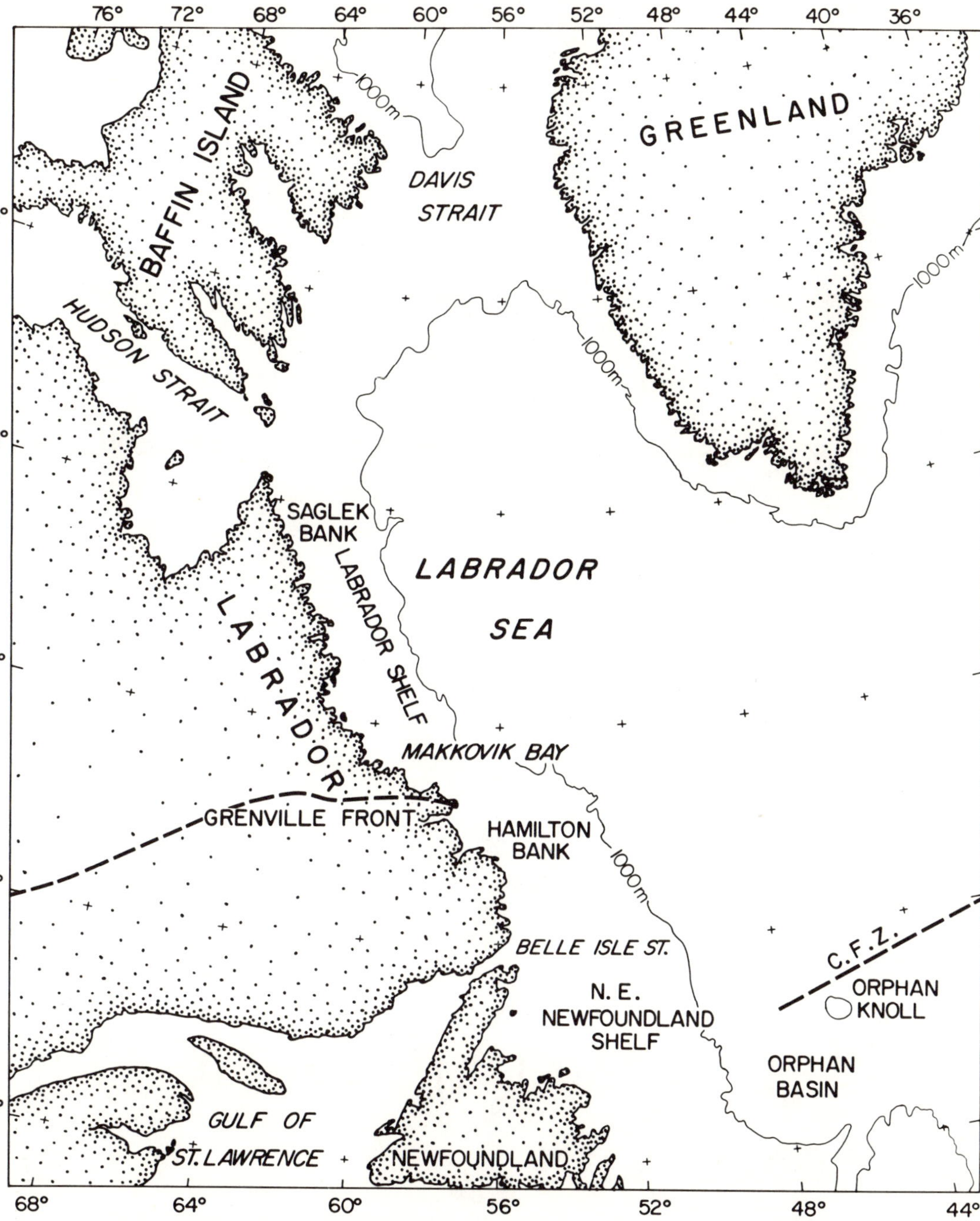

Figure 2. The continental margin of eastern Canada: Labrador Sea.

shape and modify the offshore areas. Most of the vast increase in our knowledge of the continental margin of eastern Canada can be directly attributed to this increased commercial and government interest.

Tectonic Framework

The continental margins of eastern Canada are passive margins whose history is directly related to the opening of the North Atlantic Ocean, Labrador Sea, and Baffin Bay. The evolution of these margins is generally considered to consist of an early rifting stage in which the continental lithosphere was heated and thinned, followed by a spreading stage during which continents moved apart as sea floor was generated between them. Two hundred million years ago, North America, Greenland, Europe, and Africa were consolidated in one huge continent—Pangea. Red-bed deposition in grabens dominated Permian-Triassic time. As rifting progressed, the grabens were flooded and evaporites were deposited in a restricted zone running from the Bay of

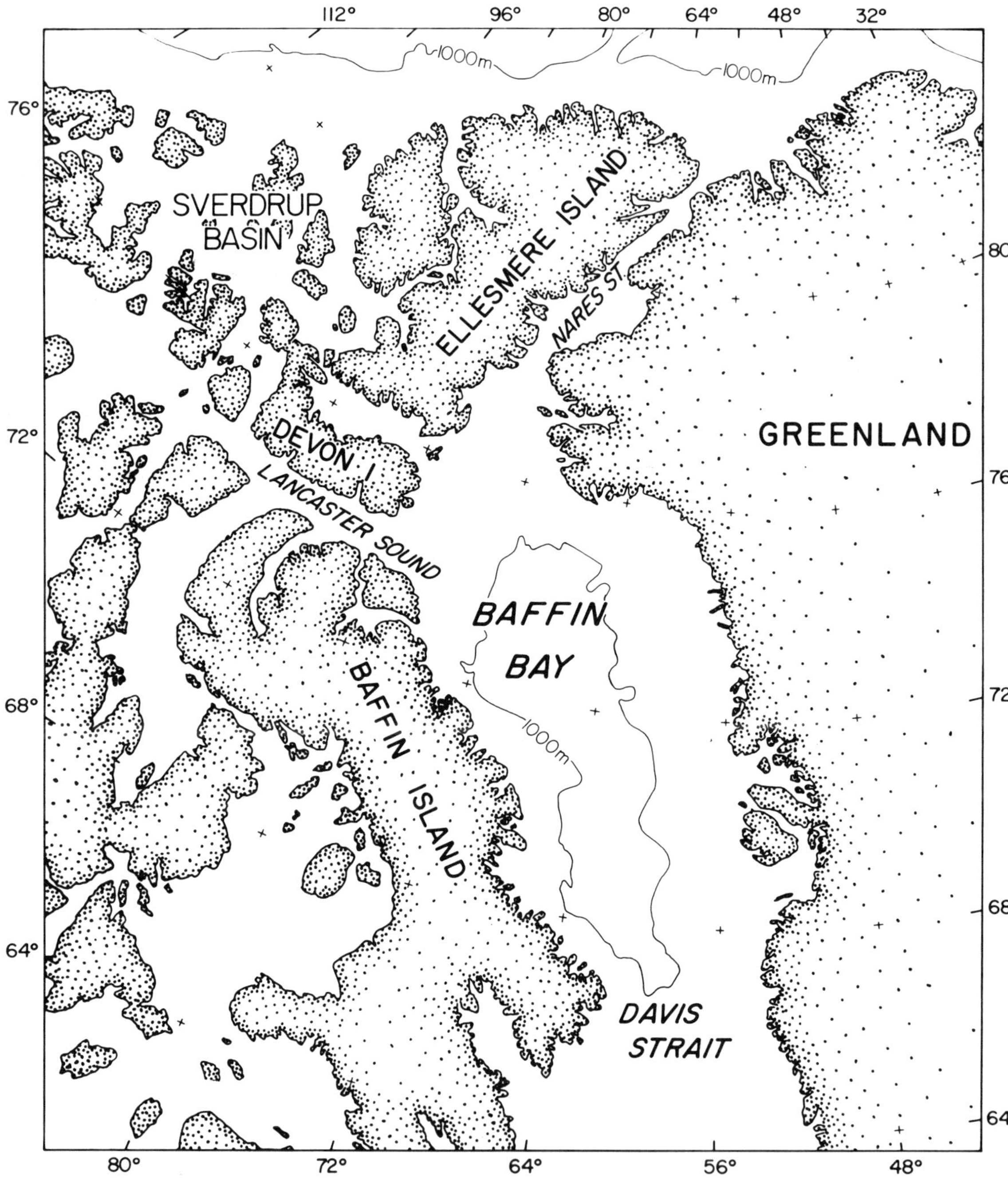

Figure 3. The continental margin of eastern Canada: Baffin Bay and the eastern Arctic.

Fundy across the Scotian Shelf. However, the actual separation began between Early and Middle Jurassic time at a different locality and produced a rifted margin at the southeastern edge of the Scotian Shelf and a transform margin along the southwestern edge of the Grand Banks.

In the Labrador Sea, rifting began in Late Jurassic or Early Cretaceous time, with complete separation in the Late Cretaceous and sea-floor spreading between Greenland and North America continuing until the early Oligocene. Some authors, however, maintain that vertical rather than lateral crustal displacements are necessary to circumvent problems with palinspastic reconstructions of Labrador Sea and Baffin Bay. These problems are most severe for sea-floor spreading hypotheses if less than 25 km of strike-slip (transform) motion occurred along Nares Strait during the formation of Baffin Bay. However, neither crustal extension, mantle upwelling, nor magmatic intrusion as alternatives to sea-floor spreading can account for both the thin crust in central Baffin Bay and the lack of motion along Nares Strait without motion between the Arctic Islands.

Quantitative models for the evolution of passive margins are quite successful in explaining their gross subsidence and sedimentary history, but locally the evolution is not so simple. For example, the crust in Davis Strait is atypical of either oceanic or continental areas. The area has abnormal elevation, and Paleocene basalts apparently underlie much of the strait. It is almost impossible to quantify the tectonic evolution of this area, despite its intimate relation to the relative evolution of Labrador Sea and Baffin Bay. Similar complications occur in the vicinity of the Grand Banks, which were the northern limit of the Jurassic opening between Africa and North America and the western margin for subsequent opening between Iberia and North America. South of the landfall of the Charlie Fracture Zone, Orphan Knoll and Flemish Cap are continental fragments separated from the shelf by deep-water basins underlain by atypically thin continental crust. The Avalon Basin, within which the Hibernia structure was found, plunges northward into that complicated area. Similar structures on the east Atlantic margin (with which the Grand Banks were once contiguous) are also the focus of intense investigation.

Geological and Geophysical Data

Data describing the Atlantic margin of Canada have primarily come from geophysical surveys by industry, university, and government laboratories and geological analysis of samples obtained from commercial exploratory wells offshore. Micropaleontological and lithological analyses are routinely carried out on well cuttings and used in the postulation of paleoenvironments. Physical properties such as conductivity, radioactive heat content, and velocity are selectively measured to provide control in geophysical models. Published models are primarily based on a regional (~80 km) grid of industrial multichannel seismic reflection sections on the continental shelves south of Davis Strait, and less systematic single-channel profiles of the margin and adjacent oceanic crust. Seismic refraction profiles have also been obtained at selected locations. Single-channel and high-resolution seismic reflection data and samples have also been collected for the purpose of mapping the distribution of bedrock and surficial geological units on the continental shelf with a detail not possible from the nonconfidential multichannel data. Gravity, magnetic, and bathymetric data have been collected along lines spaced 9 to 36 km apart for the entire margin south of Baffin Bay. Inshore sampling and aerial photography have provided at least reconnaissance coverage and classification of coastal environments throughout eastern Canada.

Each of these parameters or derivatives will be displayed at scales of 1:2 million, 1:5 million, or 1:10 million as appropriate.

Precambrian and Paleozoic Evolution

Basement rocks underlying the Mesozoic-Cenozoic sedimentary section of the present continental margin are up to 3.7 Ga old. The Baffin Island coast and northernmost Labrador are part of the Churchill province dominated by structures developed during the Hudsonian orogeny (1,700 Ma). Central Labrador is part of the younger Nain province (1,400 Ma). The boundary between these two metamorphic provinces is oblique to the margin, with a trend similar to that of major structures in Davis Strait, and of geophysical anomalies of the northern Labrador Shelf. The boundary between the Nain province and the Grenville province of southern Labrador (affected by the 1,000 Ma Grenville orogeny) is almost orthogonal to the present coast line. It can be traced offshore to a major offset of the continental margin adjacent to a change in the magnetic lineation pattern of the Labrador Sea. Accordingly, the boundaries between the Precambrian structural provinces seem to have had a significant effect on the structure of the present margin and the development of adjacent oceanic structure.

The boundary between the Grenville province and the Appalachian province follows the St. Lawrence Estuary, curves through the centre of the Gulf of St. Lawrence, and emerges onto the continental margin through the Strait of Belle Isle. The area south of this line was affected by the Taconic (450 Ma), Acadian (350 Ma), and Maritime equivalents of the Alleghenian (315 Ma) orogenies. The Appalachian orogen comprises several zones, each of which has been interpreted to represent an element of an ocean that developed during late Precambrian time and was destroyed by closure during early Paleozoic time. Although much of the area was subsequently covered by Mesozoic and Cenozoic sedimentary rocks deposited as a result of the opening of the present seaways, the preexisting structure has had a significant effect on the post-Paleozoic evolution. For example, the general northeasterly structural grain of the Precambrian "Avalon" zone in western Newfoundland is mimicked by that of the Mesozoic section on the Grand Banks, but all such trends are truncated obliquely by the present continental margin.

Devonian intrusions, related to the Acadian Orogeny, pervade the terrestrial part of the Atlantic margin and can be traced offshore. Upper Devonian to Permian sedimentary rocks, deposited in a zone from the Bay of Fundy to Northeast Newfoundland, contain coal and minor oil deposits in New Brunswick and coal onshore and offshore in Nova Scotia. Lower Paleozoic strata are also found along the Labrador Shelf and the Baffin Island Shelf. The relationship of these occurrences to each other, to those on land, and to those in marine areas farther south is not presently known.

Early History of the North Atlantic

Information about the initial phases of opening of the North Atlantic is available from three independent sources: stratigraphy, crustal structure, and "oceanic" magnetic lineations. These data indicate a sequence of events defining the evolutionary process.

Stratigraphic similarities have been established between early Mesozoic sedimentary basins on the North American margin (the Georges Bank, Scotian, Flemish, and East Newfoundland basins) and those on the European and African margins (the Tarfaya, Essaouiran, Lusitanian, Aquitaine, and Celtic Sea basins). In addition to the lithologic, petrological, and paleontological comparisons between cuttings from commercial exploratory wells in both of these areas, the cores obtained from the International Program of Ocean Drilling (IPOD) holes adjacent to the continental margin have provided critical comparisons between the oldest sediments deposited on the narrow zone of oceanic crust between the newly separated continents. Such stratigraphic information, primarily from the Atlantic margin south of Flemish Cap, contributes to an excellent data set from which to deduce the rift, drift, and early subsidence history of a passive margin. Deep seismic reflection and refraction profiles provide structural control across the continental margin from the Scotian Shelf to Baffin Bay. The critical elements to be recognized from these data are the variations in thickness and physical properties of the crust across the margin. Seismic refraction data indicate that the transition zone between "typical" oceanic and continental velocity sections on the Scotian margin, perhaps the type case for Atlantic rift margins, is about 100 km wide, compared with the 50 km width of the transition zone for the transform margin along the southern Grand Bank. Some authors interpret seismic reflection data as showing that (continental) salt overlies what has been interpreted by others to be oceanic crust, and that seismic reflectors can be traced offshore from the continental shelf to underlie the level at which basaltic velocities are found. The solution to this conflict is being sought at present by large aperture seismic reflection experiments being carried out on the continental margin.

Magnetic lineations are the most universally used but most interpretationally suspect source of control over the early spreading history. The time scale for lineations recognizable in the deep sea has been well established by correlation with the paleontological control from IPOD and DSDP holes reaching basaltic basement in the oceans. However, recognition of these lineations by their character and extrapolation of the time scale to the oldest recognizable lineations at the ocean margins is fraught with uncertainties. During the early phases of opening, even after the plates had split apart, interaction between their edges still occurred. This was particularly true at such transform offsets as the southern margin of the Grand Banks, south of which lies a quiet magnetic zone, and north of which (east of the Grand Banks) the magnetic lineations appear to be oblique and ill defined. However, models have been developed that substantiate the formation of sea-floor spreading anomalies as soon as rifting is complete, and that account for the observations of *apparently* chaotic anomalies as a consequence of oblique spreading.

From these data sets, it can be summarized that rifting was initiated along the Scotian margin in late Triassic time, with sea-floor spreading starting about 30 Ma later. The oldest sea floor adjacent to the rifted margin is therefore early Middle Jurassic in age. The sea floor is progressively younger eastward along the transform margin and is mid-Cretaceous in age at its eastern end. The eastern margin of the Grand Banks was formed by the separation of Iberia from North America at that time. Sea-floor spreading between the northeast Newfoundland and British margins started at about 90 Ma. Although timing of the movement along each margin is somewhat complicated, it has been suggested that initial opening took place along the axis of Rockall Trough. The spreading axis jumped to the west, 75 Ma ago, coincident with the onset of sea-floor spreading in the southern Labrador Sea. Spreading progressed to the northern Labrador Sea about 70 Ma ago and into Baffin Bay about 60 Ma ago. Such spreading ceased 40 Ma ago, although it has continued to the present in the North Atlantic.

Mesozoic and Tertiary Development

Stratigraphic control on which to base paleoenvironmental models for the development of the margin is available from the analysis of well cuttings on the Scotian Shelf, Grand Banks, and Labrador margins. The interpretation for Baffin Bay and the deeper water areas along the southern part of the margin where there are no wells is much more conjectural.

Before the opening of the Atlantic Ocean, red-bed and salt sequences of Late Triassic and Early Jurassic age were deposited in the Scotian Basin. Salt deposition ceased abruptly in the Early Jurassic to be succeeded by a generally dolomitic sequence interpreted to have been deposited in a widening epicontinental sea under generally hot and arid conditions. The open marine environment after rifting resulted in deposition along the Late Jurassic margin of sediments ranging from shallow marine and locally deltaic clastics to shelf carbonates and marine basinal shales. The regional pattern of sedimentation changed during the Early Cretaceous when a crustal block trending southeast from Newfoundland was uplifted, possibly coeval with initiation of rifting between the Grand Banks and Iberia. South of the uplift, a massive influx of sandstones and shales terminated further development of the carbonate banks and resulted in the buildup of deltas across the Scotian Shelf. A regional transgression during middle to Late Cretaceous time gradually drowned the deltas. Sand, silt, and mud were deposited to build out the continental shelf as it exists today, and over 12 km of subsidence has occurred in the seaward portion of the Scotian Basin.

In the vicinity of the uplift, following the initiation of sea-floor spreading north of the transform margin, a complex structural and stratigraphic history is implied by the combination of block faulting, salt movement, and erosion. On the northern margin of the Grand Banks the Upper Jurassic–Lower Cretaceous section is relatively complete and is the current focus of the search for oil. There, a major mid-Cretaceous unconformity is overlain by up to 8 km of Late Cretaceous and Tertiary sedimentary rocks. The Charlie Fracture Zone marks the boundary between the northern Grand Banks tectonic zone and that of the Labrador Sea. A rift, or narrow continental seaway, as an epicontinental extension of the newly developed North Atlantic, was established in latest Jurassic or earliest Cretaceous time in the area now covered by the Labrador Sea. Volcanism occurred locally in this basin, which also received predominantly continental clastic sediments. The oldest marine sediments in the basin are probably Albian in age. By early Tertiary time, the margins of the Labrador sea had subsided and the basins of the Labrador Shelf were transgressed from the east as the result of a marine connection with the North Atlantic. Deep-water conditions were established in early Eocene time, but by middle Oligocene the rapid accumulation of sediment had resulted in shallow-water sand deposition over the western half of the shelf. Rapid subsidence took place during Miocene time and resulted in diapirism of the Eocene mudstones.

The absence of deep exploration wells in Baffin Bay means that limited stratigraphic information is available to constrain the diverse models proposed on the basis of geophysical data. Late Cretaceous shallow marine sediments are found on the west side of Baffin Bay and underlie Tertiary basalts in west Greenland. The sediments have faunal affinities with western Canada, but micropaleontological data also support the establishment of a seaway through Davis Strait at a time when oceanic lineations indicate that sea-floor spreading commenced in the southern Labrador Sea. These lineations indicate that the formation of Davis Strait, between 60 and 40 Ma ago, was associated with a drastic change in the direction of motion between the North American and Greenland plates. This resulted in highly oblique spreading in this region. However, the motion along Nares Strait, required as a consequence of opening Baffin Bay either by sea-floor spreading or by stretching or intruding continental crust, is not compatible with the geological knowledge of the Strait. Much attention will be focused on this problem in the next few years. The relative elevation of various occurrences of marine sedimentary rocks on Baffin Island and the adjacent shelf suggests that some 600 m of uplift of Baffin Island has occurred since Eocene time.

Quaternary History

About 97% of Canada has been glaciated, although only 1% remains under permanent ice. Many of the ice sheets terminated on the continental shelves, leaving detailed sedimentary records of fluctuations in the ice-sheet margins. In deeper water of the continental margin, there is a more continuous glacial sedimentation record in which microfossils provide clues to the nature of events that initiated or terminated glacial ages. The increase in the knowledge of the offshore surficial sedimentary record has led to an interpretation of Quaternary history that is not wholly consistent with previous land-based interpretations. These differences mainly concern the timing and extent of ice-sheet growth during the latest (Wisconsinan) glacial period. Reconciliation of the terrestrial and marine Quaternary interpretations will be a prime objective of this volume.

The surficial geology of the Scotian Shelf has been more thoroughly investigated and mapped than any other part of the Atlantic margin. Sub-bottom deposits of glacial material and reworked glacial debris on the sea-floor indicate that ice extended seaward at one time to the outer margin of the shelf. A discontinuous end-moraine complex extending along the entire Scotian Shelf possibly marks the lift-off point of a partly floating, retreating late Wisconsinan ice sheet. The occurrence of a submarine terrace at a depth of 115 to 120 m represents an old shoreline developed during a late Pleistocene lowering of sea level. The age of this terrace is uncertain, but it is likely that 15,000 to 19,000 years ago large areas of the shelf were subaerially exposed. This resulted in a poor exchange of water between the present coastline and shelf edge, which led to a cold brackish water environment over much of the shelf. As sea level rose during the Holocene transgression, the glacial sediments above the Wisconsinan shoreline were subject to reworking and redistribution, as a result of which lag deposits remained in the shallow areas and the finer sediments were redeposited in deeper water.

Surficial sediment units equivalent to those on the Scotian Shelf have been found elsewhere along the Atlantic margin of Canada, although their history of deposition and reworking differs. The conflicts between terrestrial and marine glacial interpretations become even more pronounced off the coast of Labrador where the arguments between maximum and minimum Wisconsinan ice-sheet limits have raged for a hundred years. It has been argued that the whole of the Labrador Shelf was glaciated by the Laurentide ice sheet during the late Wisconsinan. Ice on

Saglek Bank off northern Labrador extended to a grounding limit on the shelf and upper slope during glacial maximum. Basal melting of this ice sheet may have begun about 10,000 years B.P. This chronology is complicated, however, by uncertainties regarding the validity of radiocarbon dates obtained from the Saglek Bank sediments. Other data from the southern Labrador shelf show that grounded ice did not cover all parts of the shelf during the late Wisconsinan and that deglaciation of some areas started before 15,000 years B.P. Extensive studies of glacial deposits and raised marine sediments on Baffin Island show that late Wisconsinan ice did not cover the east coast except where local glaciers extended seaward through fjords.

Work is continuing on the Labrador and Baffin Island shelves in order to integrate in the volume the interpretation for these areas with that for the Scotian Shelf and Grand Banks. Concurrent with this work is the geodynamic modelling of the response of the earth to glacial loading and unloading, the effects of meltwater in the oceans, and consequent changes in sea level. Work is also in progress to confirm a preliminary chronostratigraphy based on microfossil, oxygen isotope, and volcanic-ash records in deep-water sediments from Baffin Bay, the Labrador Sea, and areas around the Grand Banks. These records indicate the frequency and intensity of alternating glacial and warm climatic intervals during the past half million years. Understanding of the relationship between the climatic events recorded in the Arctic and North Atlantic sediments will help to predict the causes of glaciation and to model the changes in oceanic and atmospheric circulation that are likely to accompany the next ice age. The consequences of glaciation, relative changes in sea level, and oceanic circulation will dominate the discussion of processes affecting the offshore Quaternary section.

Coastal and Modern Sedimentary Processes

The need for a description and understanding of Canada's coasts and of the processes by which they were formed and are now being modified is greater than ever before. The current interest is spurred on by oil and gas developments in the frontier regions offshore Eastern and Arctic Canada. Remedial measures must be available in the case of oil spills whose likelihood increases proportional to the amount of offshore drilling and tanker traffic. The landfall of subsea pipelines must be made in an environment that is not destructive to the pipeline and that is compatible with the development of new towns and harbours. Whereas classification schemes for geological mapping have existed for a long time, those for coastlines had to be developed so as to be applicable to all foreseeable usage. A system has been developed in which the coast is graphically subdivided into shore types and an alphabetically coded descriptor provides a summary of the physical character of each shore unit. Description of the entire east coast of Canada will not be completed by the planned publication date of the volume, but an up-to-date summary of coastal environments will be included.

In understanding the coastal processes of eastern Canada, five problems are of prime importance: (1) the rates of relative change in sea level and their geographic variation; (2) the rates and processes of coastal cliff erosion; (3) sedimentology and processes affecting the stability of barrier systems; (4) dynamics of sediment transport and facies characteristics of deposits in inner shelf, nearshore, and littoral environments; and (5) physical effects of ice and of freeze-thaw processes in nearshore and littoral environments. Some of these problems have been addressed in each of several areas such as the Miramichi Estuary of New Brunswick, the beaches of Prince Edward Island, the Canso Strait area in Nova Scotia, the potential sites for tidal power generation in the Bay of Fundy, and selected parts of the Atlantic coast of Nova Scotia.

Future work will undoubtedly be directed towards Arctic fjords where fresh-water input is cold and highly seasonal, sea-ice cover exists for much of the year, and in many cases glaciers enter the fjords.

Investigation of modern processes is not limited to the coastal zone. On the shelves, we know little about how much of the surface morphology is relict rather than the result of modern processes. Iceberg scours are prevalent, but we have little idea of their age, frequency of occurrence, and mode of formation. We know little about the bottom-current regime and consequent sediment dynamics. We also have little idea of the problems created by gravity-induced shelf-sediment instabilities. Industrial activity will soon move into deeper water where the instabilities are presumably even greater: When will there be a sequel to the great Grand Banks earthquake and turbidity current of 1929!?

Such questions about processes obviously require more than the reconnaissance level description of the surficial sediments that exists in most areas. Some questions will undoubtedly be tackled soon, but most will probably be unanswered by the end of the century.

Comparative Development of the Margins

The east coast of Canada has three distinct passive margins—the Nova Scotian, Labrador, and Baffin margins—each of which has had a similar history but different timetable of critical events. The effect of these critical events is recorded by unconformities, changes in lithology, or thermal maturity within the sedimentary section built up during the evolution of the margin. This section of the volume, rather than dealing with the specifics of any one margin, will be directed towards comparing their development in order to establish a set of fundamental principles that govern the evolution of passive margins.

During the early rifting stage in the evolution of a rifted passive margin, the continental crust is uplifted due to heating and thermal expansion within the mantle. This is followed by a spreading phase during which the continental margins, formed by the split between the continents, subside as the lithosphere cools and sediments are deposited upon them. Deep exploratory wells drilled into those sedimentary rocks have yielded biostratigraphic information from which the subsidence history can be deduced. The components of subsidence due to sediment loading and, where possible, changes in eustatic sea level have been removed to determine the amount of "tectonic subsidence," which is that subsidence caused by lithospheric cooling and other deep-seated processes. Tectonic subsidence has a linear relationship with the square root of the time since subsidence began, and this agrees with a model in which the basins are formed in response to sediment and water loading on a thermally contracting lithosphere. Other observations are compatible with such a model. For example, the lithosphere beneath marginal basins was, for a large part of the basin's history, young, hot, and thin and consequently had a small flexural rigidity and wavelength. The small, relatively short wavelength gravity anomalies and greater than normal geothermal gradient during the early history of such basins are compatible with this predominantly thermal model. Such quantitative modelling of Mesozoic activity can be extended in several directions in order to deduce changes in lithospheric properties along the margin. For example, the area between Orphan Knoll and the Grand Banks has continental crust that was thinned by the rifting process. Also, the sediments in basins adjacent to a transform segment of the margin, such as the southern margin of the Grand Banks, have a different history from those adjacent to a rifted segment of the margin. During the closure of an ocean, sedimentary basins at its margin are deformed, while others develop within the craton (e.g., the Carboniferous basin in the Gulf of St. Lawrence). Both basins may therefore provide evidence for the properties of the lithosphere and the mechanisms by which closure is effected.

Changes in relative sea level are partially the result of the response of the earth to changes in ice loading. The diversity of changes in sea level throughout Atlantic Canada therefore enables the variation in distribution of ice cover in Atlantic Canada during the late Wisconsinan to be deduced. Such models are important in describing the processes by which the offshore surficial sedimentary sequences have developed.

Resources

Hydrocarbons

The bulk of Canada's hydrocarbon potential lies in three frontier regions (the MacKenzie Delta–Beaufort Sea, the Arctic Islands, and the east coast offshore), mostly in offshore areas with very hostile environments and attendant high costs, logistical problems, and environmental risks. Reserves (including remaining reserves and undiscovered potential) have been calculated on the basis of the probabilities of size, thickness, porosity, and content of a prospect, which naturally change as exploration progresses. In 1975 up to 30% of Canada's oil and gas potential appeared to be in offshore eastern Canada. Since that time, discoveries in the Hibernia area and drilling of the Texaco Blue H-28 well north of the Grand Banks in a world-record water depth of 1,500 m have had a significant effect on the calculations of resource potential for Atlantic Canada. For example, the 1981 estimate at a 50% probability level of the oil reserve in the Hibernia area is nearly 2 billion barrels, representing approximately 5% of the total Canadian reserves. An up-to-date estimate of the resource potential for all of the eastern offshore Canada area will be included in the published volume.

Coal

The Sydney coalfield, by far the most important of the coalfields in eastern Canada, produced half a billion tons of coal during nearly 200 years of mining. A major drilling program to determine the offshore extent of the Sydney coalfield has recently increased the reserve estimates to 150 million tons of proven, recoverable coal, and 1.7 billion tons of measured, indicated and inferred reserves. The proven life of the coalfield could therefore extend beyond the 30 years predicted on the basis of an anticipated annual demand of 5 million tons.

Minerals

Excluding hydrocarbons, the most important mineral resource of offshore eastern Canada is that of surficial deposits as a potential source for construction aggregate. Previously, their use was not economically viable. However, increased environmental objection to onshore sand and gravel quarrying and additional demand for construction materials through growing urbanization and construction of immense coastal and offshore installations in remote areas (e.g., container piers, tanker terminals, tidal power barrages, offshore oil storage tanks, artificial islands) will probably change this perception. A reasonable minimum estimate of the volume and grade of existing aggregate reserves in the Bay of Fundy, Scotian Shelf, and Gulf of St. Lawrence could now be made. This should soon also be possible for the rest of offshore eastern Canada.

Apart from their usefulness as a source of aggregates, surficial deposits in the Gulf of St. Lawrence may also be mined locally for optical quality silica sand. Placer deposits such as gold or tin might be found in the alluvial sediments of drowned river deposits, but it has been considered unlikely that such deposits would be economic. Gypsum and salt are the most extensively mined nonmetalliferous minerals in Atlantic Canada, and their deposits are known to extend offshore. Although it is possible to predict the most likely location of in-situ metalliferous deposits offshore, the patchy nature of onshore deposits would be an extremely high-risk exploratory venture.

Summary and Outstanding Problems

Despite the information explosion of the past 15 years from which the evolution of the continental margin of eastern Canada can be surmised, many questions still remain. Quantitative modelling of the margins is still in its infancy; our appreciation of coastal types and processes is still only reconnaissance in nature; hypotheses for the development of Labrador Sea and Baffin Bay meet constraints in all directions; and a unified glacial history for onshore and offshore eastern Canada has not yet been produced. Each of these topics encompasses a multitude of specific questions that will be either answered or enumerated in the volume in order to provide a focus for future work in the area. Our capacity to answer them will undoubtedly be heavily influenced by the level of industrial activity: Offshore eastern Canada in the late 1970s had many similarities with the North Sea in the mid-1960s. And look what has happened there!

Acknowledgments

This summary was prepared from material provided in published or unpublished form by D. Buckley, R. Fillon, F. Gradstein, A. Grant, P. Hacquebard, L. Jansa, C. Keen, L. King, B. Loncarevic, D. Loring, B. MacLean, P. Mudie, E. Owens, D. Piper, D. Sherwin, S. Srivastava, D. Umpleby, J. Wade, and G. Williams. Reviews of the entire summary by Charlotte Keen, David Piper, and Brian MacLean are particularly appreciated.

Selected Bibliography

Fader, G. B., King, L. H., and MacLean, B., 1977, Surficial geology of the eastern Gulf of Maine and Bay of Fundy: Geological Survey of Canada Paper 76–17, 23 p.

Fillon, R. H., and Harmes, R. A., Northern Labrador Shelf glaciation: Chronology and limits: Canadian Journal of Earth Sciences, v. 19, p. 162–192.

Gradstein, F. M., and Srivastava, S. P., 1980, Aspects of Cenozoic stratigraphy and paleoceanography of the Labrador Sea and Baffin Bay: Palaeogeography, Palaeoclimatology, Palaeoecology, v. 30, p. 261–295.

Grant, A. C., 1980, Problems with plate tectonics: The Labrador Sea: Bulletin of Canadian Petroleum Geology, v. 28, p. 252–278.

Hacquebard, P. A., 1979, A geological appraisal of the coal resources of Nova Scotia: Canadian Institute of Mining Bulletin, February 1979, p. 76–87.

Haworth, R. T., and Keen, C. E., 1979, The Canadian Atlantic margin: A passive continental margin encompassing an active past: Tectonophysics, v. 59, p. 83–126.

Jansa, L. F., and Wade, J. A., 1975, Paleogeography and sedimentation in the Mesozoic and Cenozoic, southeastern Canada; *in* C. J. Yorath, E.R. Parker, and D. F. Glass, eds. Canada's continental margins and offshore petroleum exploration: Canadian Society of Petroleum Geologists, Memoir 4, p. 79–102.

Keen, C. E., and Peirce, J. W., The geophysical implications of minimal Tertiary motion along Nares Strait: Greenland Geoscience (in press).

Loring, D. H., and Nota, D.J.G., 1973, Morphology and sediments of the Gulf of St. Lawrence: Fisheries Research Board of Canada Bulletin 182, 147 p.

MacLean, B., Srivastava, S. P., and Haworth, R. T., Bedrock structures off Cumberland Sound, Baffin Island Shelf: Core sample and geophysical data, *in* Arctic geology and geophysics: Canadian Society of Petroleum Geologists, Memoir 8 (in press).

Owens, E. H., 1977, Coastal environments of Canada: The impact and cleanup of oil spills: Canadian Department of Fisheries and Environment, Environmental Protection Service, Economic and Technical Review Report EPS-3-EC-77-13, 403 p.

Piper, D.J.W., and Keen, M. J., 1976, Geological studies in St. Margaret's Bay, Nova Scotia: Geological Survey of Canada Paper 76-18, 18 p.

Sherwin, D. F., 1981, Geological setting: Eastcoast offshore oil and gas: Eastcoast Petroleum Operators' Association, Eastern Offshore News, v. 3, p. 8–10.

Srivastava, S. P., 1978, Evolution of the Labrador Sea and its bearing on the early evolution of the North Atlantic: Geophysical Journal of the Royal Astronomical Society, v. 52, p. 313–357.

Umpleby, D. C., 1979, Geology of the Labrador Shelf: Geological Survey of Canada Paper 79–13, 34 p.

Wade, J. A., 1981, Geology of the Canadian Atlantic margin from Georges Bank to the Grand Banks *in* J. Wm. Kerr and A. J. Fergusson, eds. Geology of the North Atlantic Borderlands: Canadian Society of Petroleum Geologists, Memoir 7, p. 447–460.

Atlantic continental margin of the United States

John A. Grow
U.S. Geological Survey
Woods Hole, Massachusetts 02543

Robert E. Sheridan
Department of Geology
University of Delaware
Newark, Delaware 19711

Introduction

The objective of this Decade of North American Geology (D-NAG) volume will be to focus on the Mesozoic and Cenozoic evolution of the U.S. Atlantic continental margin, including the onshore coastal plain, related onshore Triassic-Jurassic rift grabens, and the offshore basins and platforms. Following multiple compressional tectonic episodes between Africa and North America during the Paleozoic Era that formed the Appalachian Mountains, the Mesozoic and Cenozoic Eras were dominated by tensional tectonic processes that separated Africa and North America. Extensional rifting during Triassic and Early Jurassic times resulted in numerous tensional grabens both onshore and offshore, which filled with nonmarine continental red beds, lacustrine deposits, and volcanic flows and debris. The final stage of this breakup between Africa and North America occurred beneath the present outer continental shelf and continental slope during Early or Middle Jurassic time when sea-floor spreading began to form new oceanic crust and lithosophere between the two continents as they drifted apart. Postrift subsidence of the marginal basins continued in response to cooling of the lithosphere and sedimentary loading.

Geophysical surveys and oil-exploration drilling along the U.S. Atlantic continental margin during the past 5 years are beginning to answer many questions concerning its deep structure and stratigraphy and how it evolved during the rifting and early sea-floor-spreading stages of the separation of this region from Africa. Earlier geophysical studies of the U.S. continental margin used marine refraction and submarine gravity measurements. Single-channel seismic-reflection, marine magnetic, aeromagnetic, and continuous gravity measurements became available during the 1960s. Interpretations of the margin based on proprietary oil company multichannel data began to appear in the early 1970s. These interpretations suggested that basins containing as much as 12 km of sedimentary rocks underlie the continental shelves. In 1973 the U.S. Geological Survey (USGS) contracted the first publicly available multichannel seismic profiles. The first oil exploration leases for the Outer Continental Shelf off New Jersey were granted in 1976, and drilling of the first exploration wells began in 1978.

Between 1973 and 1978, the USGS collected approximately 20,000 km of multichannel seismic reflection data (Fig. 1), 185,000 km of aeromagnetic data (Klitgord and Behrendt, 1979), 39,000 km of marine gravity profiles (Grow and others, 1979), and three deep-refraction profiles (Sheridan and others, 1979). Drill-hole information from five Continental Offshore Stratigraphic Test (COST) wells has been released since 1978 (Fig. 1), and more than 25 commercial exploration wells have been drilled. Although many interpretations of individual basins and segments of the margin have been published, preliminary syntheses of the entire margin's structure and evolution are just beginning to be possible.

A preliminary synthesis of these data has delineated four major troughs or basins (Klitgord and Behrendt, 1979). From north to south, these basins are the Georges Bank Basin, Baltimore Canyon Trough, Carolina Trough, and the Blake Plateau Basin (Fig. 2). This paper is a brief summary of the geology and geophysical setting on each of these basins (modified from Grow and Sheridan, 1981), incorporating some of the most recently available seismic and drill-hole data. We have selected the most representative profiles through the center of each of the four major basins in order to demonstrate how the deep crustal and sedimentary structures across the ocean-continent transition zone evolved.

Based on this preliminary synthesis, some of the remaining unanswered geologic questions are identified in a section on future research needs. In this section, we make suggestions for viable research approaches to solve these remaining geologic questions utilizing the newest geophysical and geologic tools available.

A final section will outline the format tentatively planned for the D-NAG volume on the U.S. Atlantic continental margin and coastal plain.

The Carolina Trough

The Carolina Trough is approximately 80 km wide and 400 km long and contains as much as 11 km of Jurassic to Holocene sedimentary strata. It lies north of the Blake Spur Fracture Zone between the East Coast Magnetic Anomaly (ECMA) on the southeast and the Brunswick Magnetic Anomaly (BMA) on the northwest (Klitgord and Behrendt, 1979; Dillon and others, 1982; Hutchinson and others, 1982). We start our discussion with the Carolina Trough because (1) it has the narrowest and simplest transition zone from continental to oceanic crust; (2) the separation of North America and Africa here appears to have been a simple pulling apart of the continents without oblique or translational complications; (3) the Mesozoic lineations here are parallel to the pre-existing Paleozoic tectonic lineations; and (4) there is no evidence of secondary volcanism or tectonism that has affected the trough itself.

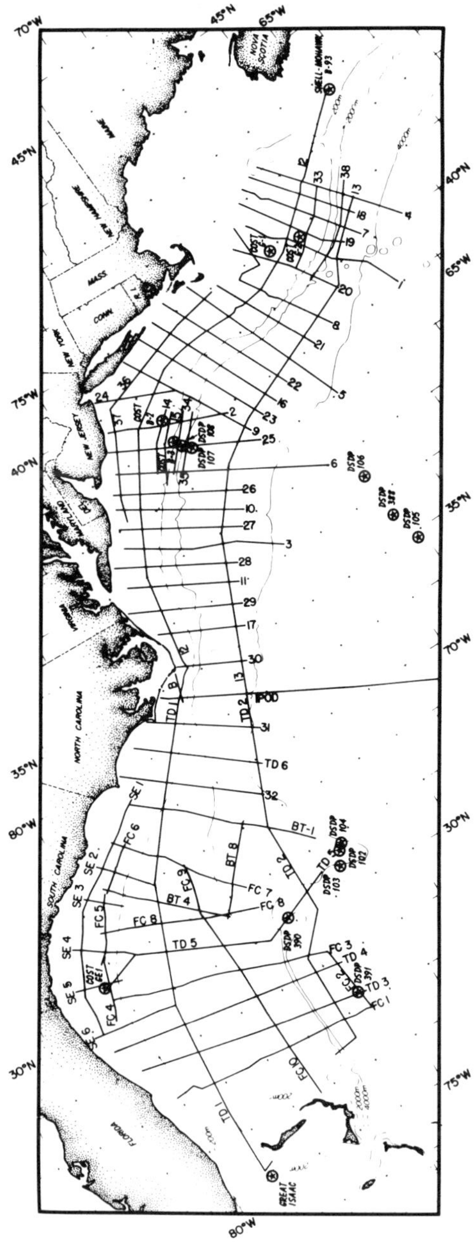

Figure 1. Locations of multichannel seismic reflection profiles collected for the U.S. Geological Survey between 1973 and 1978 (modified from Folger and others, 1979). Drill sites from DSDP and COST wells are also shown.

The first publicly available multichannel profile across the trough was the International Phase of Ocean Drilling (IPOD) line which crossed the northern part of the trough. The IPOD line showed three diapirs near the ECMA; approximately 20 more diapirs have since been found along the ECMA in the Carolina Trough. The diapirs are inferred to be salt on the basis of a small salinity anomaly detected over one diapir and the presence of salt along the margin from the Grand Banks to Nova Scotia and beneath Georges Bank. Three additional diapiric structures have been found along the ECMA in the Balitmore Canyon Trough, and salt has been drilled in one exploration well near these structures.

The best publicly available multichannel profile across the Carolina Trough obtained to date is line 32 southeast of Cape Fear (Fig. 3; see Fig. 1 for location). Line 32 displays a Triassic graben clearly cut by a "breakup unconformity" or "postrift unconformity". Prerift or synrift sedimentary units (sediments deposited before or during active rifting, respectively) appear to be present beneath the unconformity in a Triassic (and Early Jurassic?) graben. The postrift Jurassic and younger sedimentary units lie above the unconformity. A diapir is present at the ECMA on line 32, and the Jurassic shelf edge is about 15 km landward of the axis of the ECMA (Fig. 3).

A gravity model along USGS line 32 (Fig. 4) shows the position of the Carolina Trough in relation to the ECMA, the diapirs, and the inferred ocean-continent crustal structure (Hutchinson and others, 1982). Note that the Carolina Trough is underlain by rift-stage crust (8 to 15 km thick). The ECMA and a diapir occur at the boundary between rift-stage crust and the oceanic crust. The western flank of the Carolina Trough is marked by a hinge zone just east of the BMA and is underlain by an abrupt change in depth to Moho from continental crust on the west to rift-stage crust on the east.

A magnetic model along USGS line 32 (Fig. 5) demonstrates that the ECMA can be modeled by a simple edge effect between negatively magnetized (thermo-remnant) oceanic crust southeast of the ECMA and weakly positive

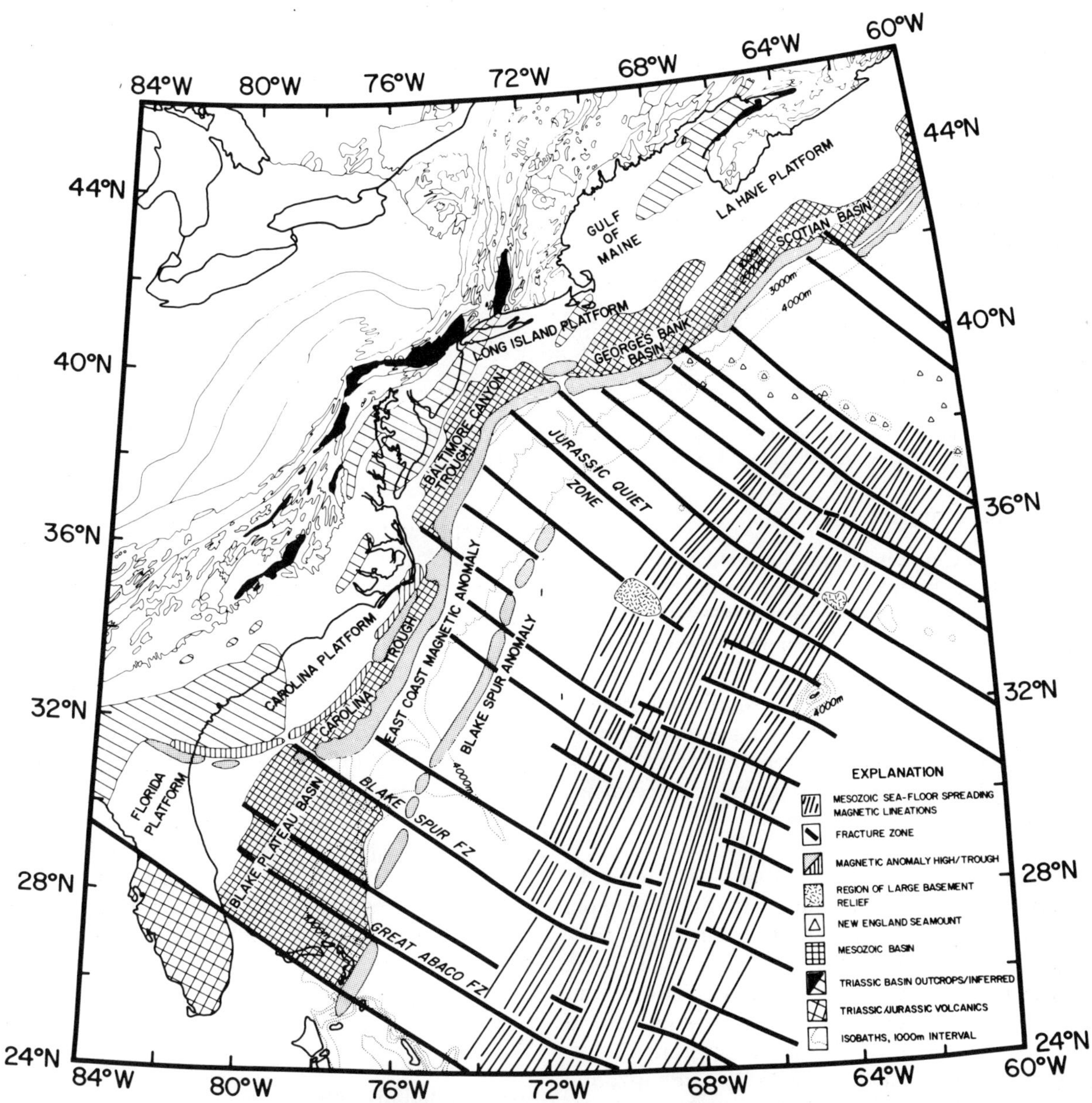

Figure 2. Location of Georges Bank Basin, Baltimore Canyon Trough, Carolina Trough, and the Blake Plateau Basin along the continental margin off the Eastern United States. Information based on multichannel seismic and aeromagnetic results (from Klitgord and Behrendt, 1979). Note correlation of basin boundaries with fracture zones. Major fracture zones also show abrupt changes in the free-air gravity anomalies along the length of the margin.

(induced) magnetization in the rift-stage crust (Hutchinson and others, 1982). The BMA can be modeled by depth-to-basement near the hinge zone and the Triassic graben. The short-wavelength anomalies landward of the BMA can be modeled by small fluctuations within the positively induced magnetization of the continental basement.

A composite geologic cross section across the Carolina Trough along line 32 summarizes the ocean-continent transition zone off the Eastern United States (Fig. 6). Although the wedge of postrift sediments appears to thin landward to less than 1 km at the coastline, it actually continues approximately 200 km farther landward as a very thin wedge of onlapping coastal plain sediments. The major sedimentary basin is underlain by rift-stage crust, approximately 8 to 15 km thick, bounded by abrupt transitions in depth to Moho—seaward into oceanic crust at the ECMA and landward into thick continental crust near the hinge zone. Note that the position of the present and past shelf edges has varied greatly during time with respect to the ECMA and hinge zone.

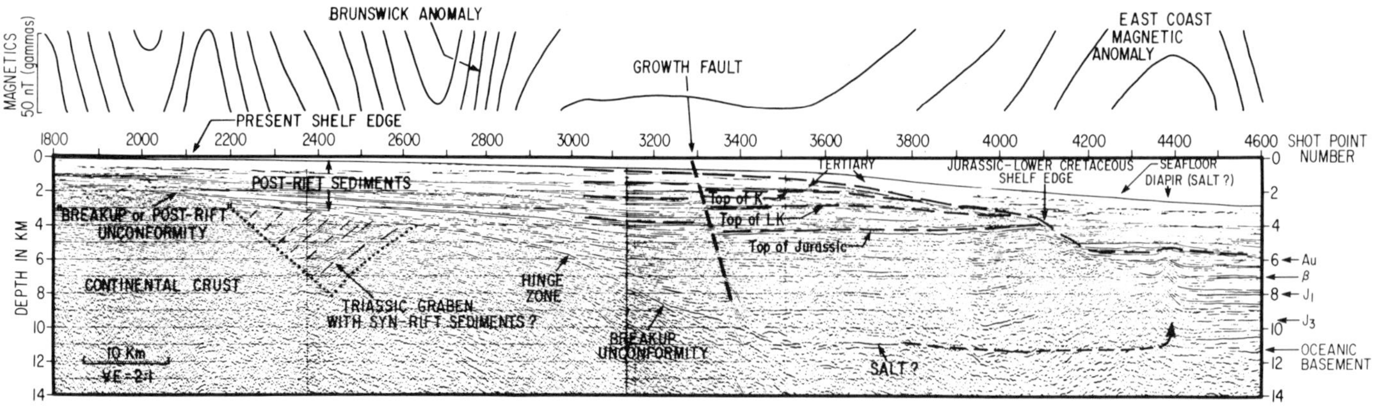

Figure 3. Multichannel seismic depth section across Carolina Trough along part of line 32 (from Grow and Sheridan, 1981).

Baltimore Canyon Trough

The Baltimore Canyon Trough lies beneath the Outer Continental Shelf between Virginia and New Jersey. It varies in width from 50 km off Virginia to 150 km off New Jersey. It is 10 km deep off Virginia and more than 16 km deep off New Jersey. An abrupt increase in width and depth takes place north of Delaware Bay. The southern Baltimore Canyon Trough is similar to the Carolina Trough in that the separation from Africa was a direct pulling apart, but the similarities decrease northward where the sediments become much thicker and the zone of rift-stage crust is wider.

Multichannel seismic profile 25 (Figs. 7, 8) crosses the Baltimore Canyon Trough at its widest and simplest region. In this area, a 40-km progradation of the shelf edge took place during Jurassic time and was followed by a 20-km retreat of the shelf edge during Tertiary time. The 4,823-m-deep COST B-3 well was drilled 10 km north of line 25 and penetrated Upper Jurassic sedimentary rocks near the bottom (Fig. 9, from Mattick and Bayer, 1980). Subhorizontal reflectors, presumed to represent undeformed postrift deposits, are present to 13 km depth (Fig. 8), and synrift (Triassic and Lower Jurassic?) deformed sedimentary rocks are interpreted to be present between 13 and 18 km depth.

A schematic cross section (Fig. 10) across Baltimore Canyon Trough summarizes the line 25 and COST B-3 data. Some salt structures and a mafic intrusion slightly northeast of line 25 have been projected on to the line in order to illustrate the types of structures in the trough. The hinge zone seen clearly in the Carolina Trough (Figs. 3, 4, 6) is just landward of the northwest end of line 25 (Fig. 7), near the coastline. A thin, onlapping wedge of coastal plain sediments extends landward of the hinge zone forming the Salisbury Embayment in southern New Jersey, Delaware, eastern Maryland, and eastern Virginia. Refraction profiles have established that a 7.1 to 7.2 km/s refractor lies at 13 to 15 km depth beneath the upper rise and slope; this refractor is interpreted to indicate the top of oceanic crust (layer 3). Note that the 7.1 to 7.2 km/s refractor continues northwestward to the ECMA (Figs. 7, 8), which marks the landward edge of oceanic crust.

The Baltimore Canyon Trough off New Jersey is the only area along the U.S. Atlantic margin where the Upper Jurassic and Lower Cretaceous shelf edge, created by a buried carbonate-bank margin, appears to have prograded seaward of the ECMA onto the Lower Jurassic oceanic crust. The carbonate bank environment ceased by the end of Early Cretaceous time, probably at a time of major relative sea-level change. Continued subsidence during Late Cretaceous and Cenozoic times was accompanied by more clastic sequences associated with more numerous transgressions and regressions. Eustatic sea-level changes have been interpreted in the COST wells (Fig. 11), but the data do not correlate in a simple one-to-one manner with the eustatic curve of Vail and others (1977).

Georges Bank Basin

The Georges Bank Basin formed as a result of an oblique separation from Africa; the ocean-continent boundary marked by the ECMA is at a 30- to 40-degree angle with respect to Paleozoic and Triassic tectonic lineations (Fig. 2). The basin is elongate along its southern border, near the ECMA, but its width ranges from 60 to 140 km because of several Triassic synrift grabens that splay off the main trough at various angles. Although the postrift sedimentary strata reach a maximum thickness of about 7 km, synrift grabens containing another 5 km of prerift and/or synrift sediments underlie parts of the basin. The COST G-1 well penetrated to 4,900 m and passed through Middle and probably Lower Jurassic dolomitic sandstones and entered lower Paleozoic metamorphic basement rocks. The COST G-2 well penetrated to 6,669 m, passed through Upper and Middle Jurassic carbonate deposits, and termi-

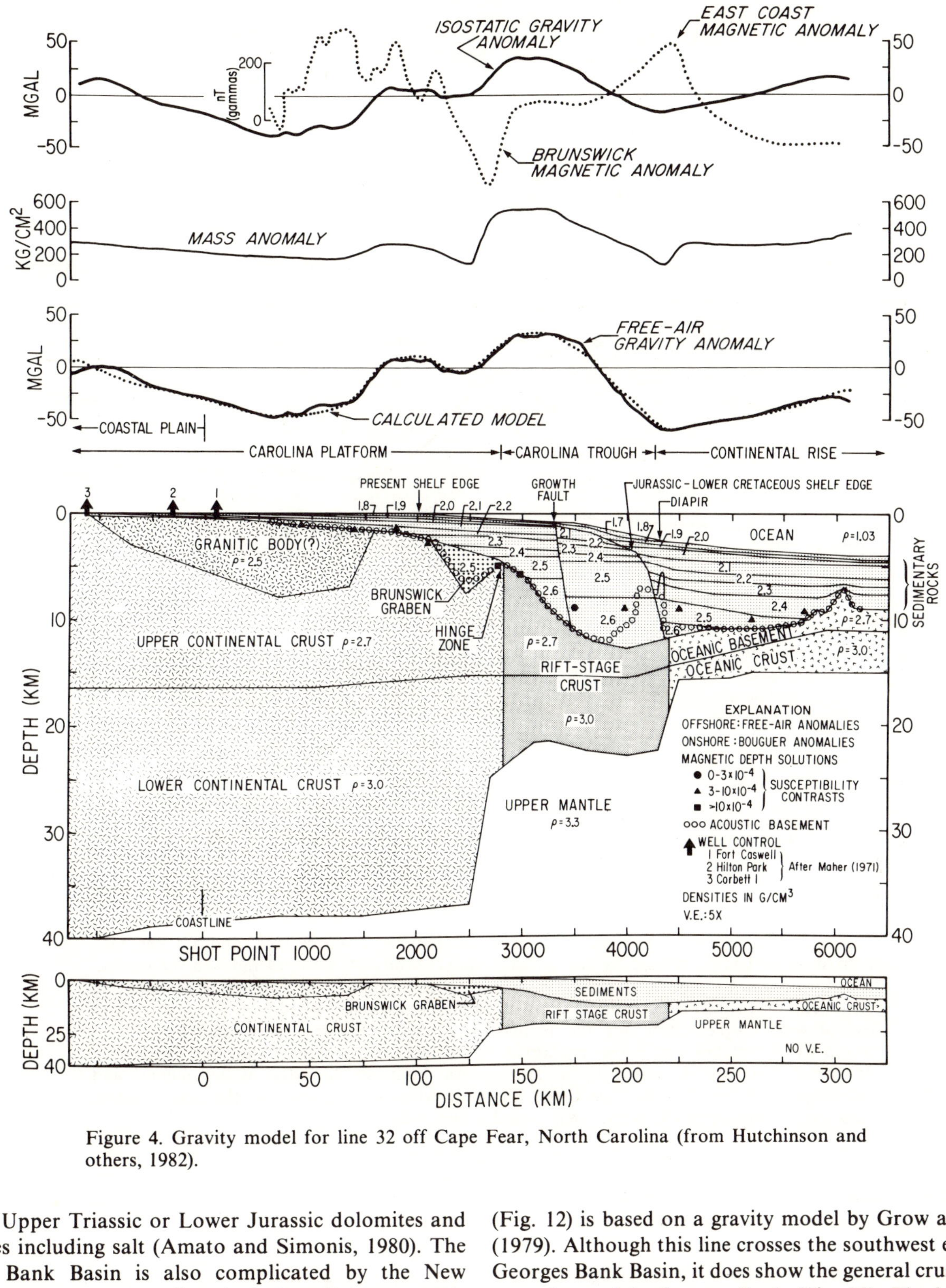

Figure 4. Gravity model for line 32 off Cape Fear, North Carolina (from Hutchinson and others, 1982).

nated in Upper Triassic or Lower Jurassic dolomites and evaporites including salt (Amato and Simonis, 1980). The Georges Bank Basin is also complicated by the New England Seamounts that disturbed the adjacent oceanic crust and nearby parts of the continental margin because of local secondary volcanism and secondary tectonic movements in Jurassic through Late Cretaceous time. A schematic cross section along multichannel line 5 off Cape Cod (Fig. 12) is based on a gravity model by Grow and others (1979). Although this line crosses the southwest end of the Georges Bank Basin, it does show the general crustal structure observed in Georges Bank (Fig. 12). The hinge zone along the northwestern margin of the Georges Bank Basin is similar to the hinge zone beneath the Nova Scotian Shelf. Here, as in the Carolina Trough, our Georges Bank profile (Fig. 12) shows a noticeable hinge zone at the 100-km mark

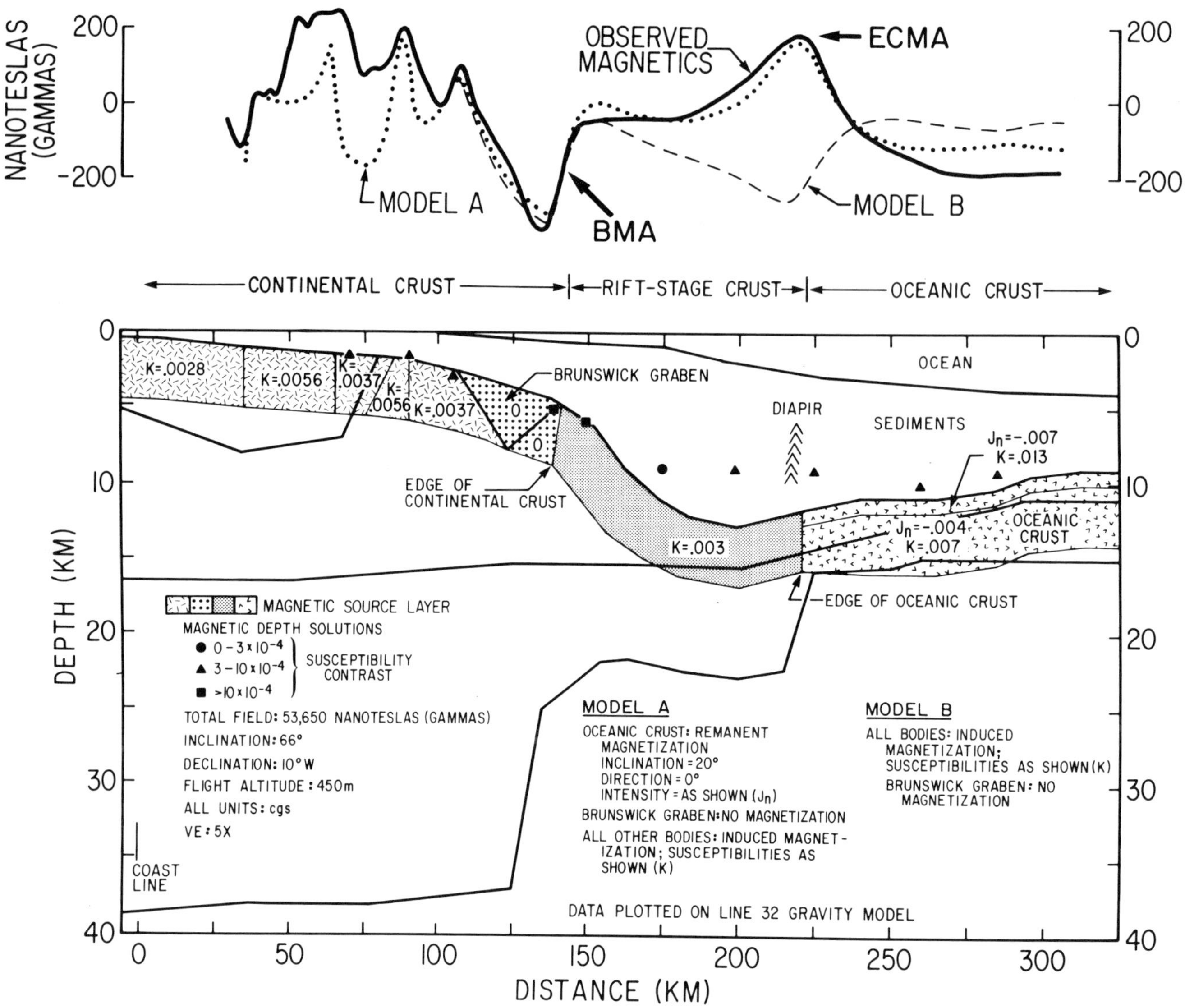

Figure 5. Magnetic model of line 32. The model assumes a source layer about 4 km thick consisting of bodies defined by the gravity model. Model A assumes induced magnetization for continental and rift-stage crust, zero magnetization for the Brunswick graben, and remanent magnetization for oceanic crust. Model B assumes all induced magnetizations except zero magnetization for the Brunswick graben. The Brunswick and East Coast Magnetic Anomalies are easily explained by Model A, suggesting that the Brunswick Magnetic Anomaly is caused by a nonmagnetic body at the edge of continental and rift-stage crust, and that the East Coast Magnetic Anomaly originates as the edge effect of magnetic oceanic crust (from Hutchinson and others, 1982).

on the distance scale. The transitional thickness or "rift-stage crust" is present between 100 and 180 km on the same scale.

The Early Cretaceous and Jurassic shelf edge on line 5 (Fig. 12) is directly over the boundary between the thin continental crust and the oceanic crust and about 20 km seaward of the present shelf edge. This position suggests that the shelf edge did not migrate significantly either landward or seaward of the initial ocean-continent boundary during Jurassic and Cretaceous times but that it did retreat during Tertiary times.

Blake Plateau Basin

The Blake Plateau, a 350-km-wide marginal plateau between water depths of 500 and 1,500 m, is underlain by postrift sedimentary rocks that are as thick as 12 km. The plateau is south of the Blake Spur Fracture Zone and the

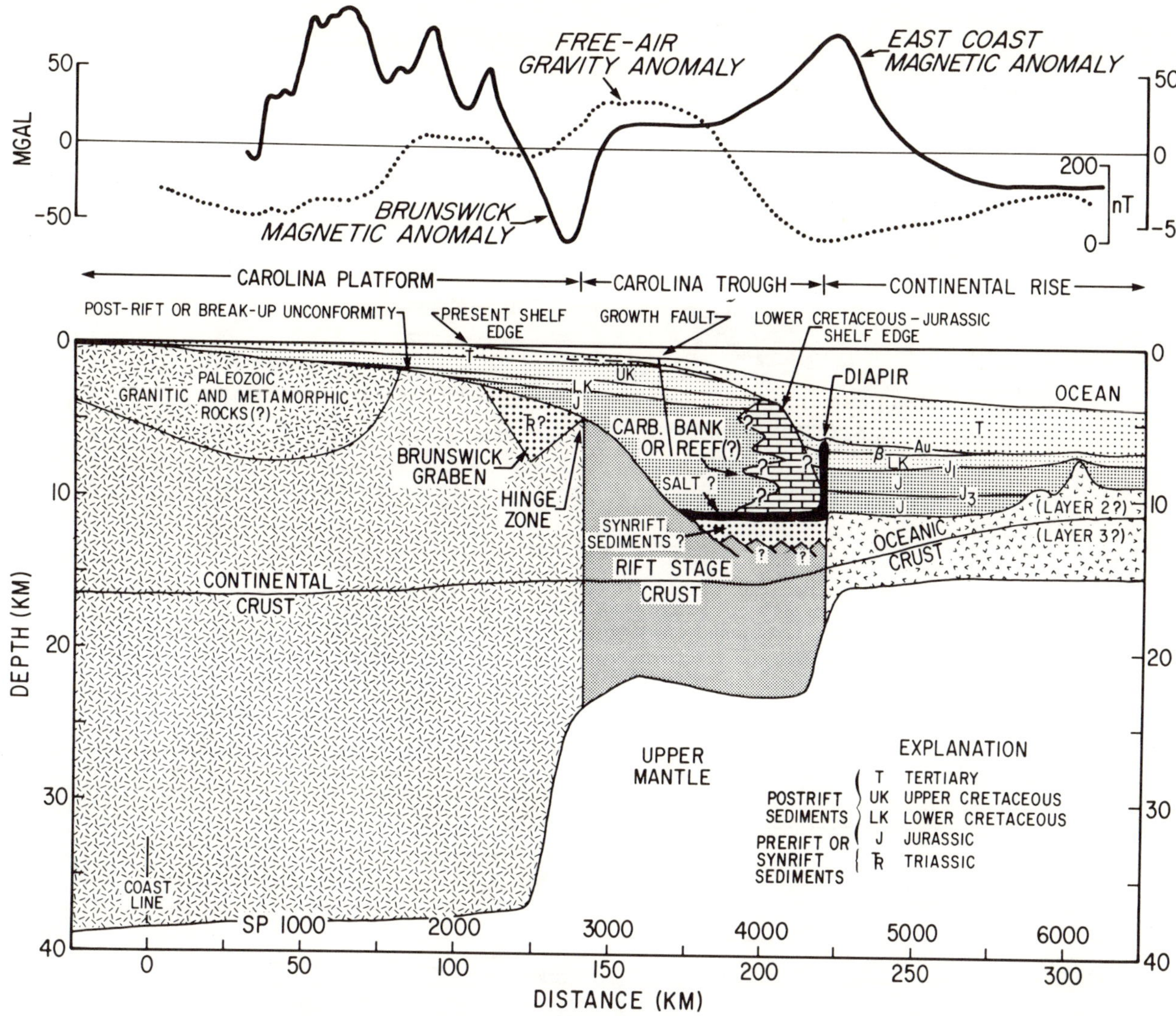

Figure 6. Composite geologic section across Carolina Trough along multichannel line 32 (Figs. 3, 4) (from Hutchinson and others, 1982).

southern termination of the East Coast Magnetic Anomaly (Fig. 2). The thickest part of the basin is beneath the western part of the plateau. The sedimentary strata beneath the middle and outer parts of the plateau dip gently toward the west (Fig. 13). Landward of and flanking this deep depocenter, the basement beneath the coastal plain is gently downwarped into the Southeast Georgia Embayment, which contains 2 to 3 km of sedimentary rocks.

Gravity modeling across the plateau indicates that a transitional crystalline crust 15 to 20 km thick underlies the western Blake Plateau Basin and that a transitional crystalline crust nearly 25 km thick underlies the eastern Blake Plateau Basin (Fig. 13). A very abrupt transition takes place beneath the Blake Escarpment, and a thin oceanic crust underlies the Blake Basin.

The configuration of the sedimentary and crustal units, combined with the absence of the ECMA south of the Blake Spur Fracture Zone, indicates a broad zone of extensional rifting and volcanism between Africa and North America in this area. Although the processes of Triassic rifting and crustal thinning beneath the western plateau probably were similar to those beneath the Carolina Trough and other basins to the north, the Early Jurassic history of the Blake Plateau Basin was different. Normal sea-floor spreading between the ECMA and the Blake Spur Magnetic Anomaly began possibly in Early Jurassic time north of the Blake Spur Fracture Zone, but contemporaneous volcanic intrusions appear to have penetrated the thinned continental crust over a broad zone beneath the middle and western Blake Plateau for the first 10 to 15 m.y. After an eastward jump of the extension and volcanic axis from the western plateau to the Blake Escarpment, normal

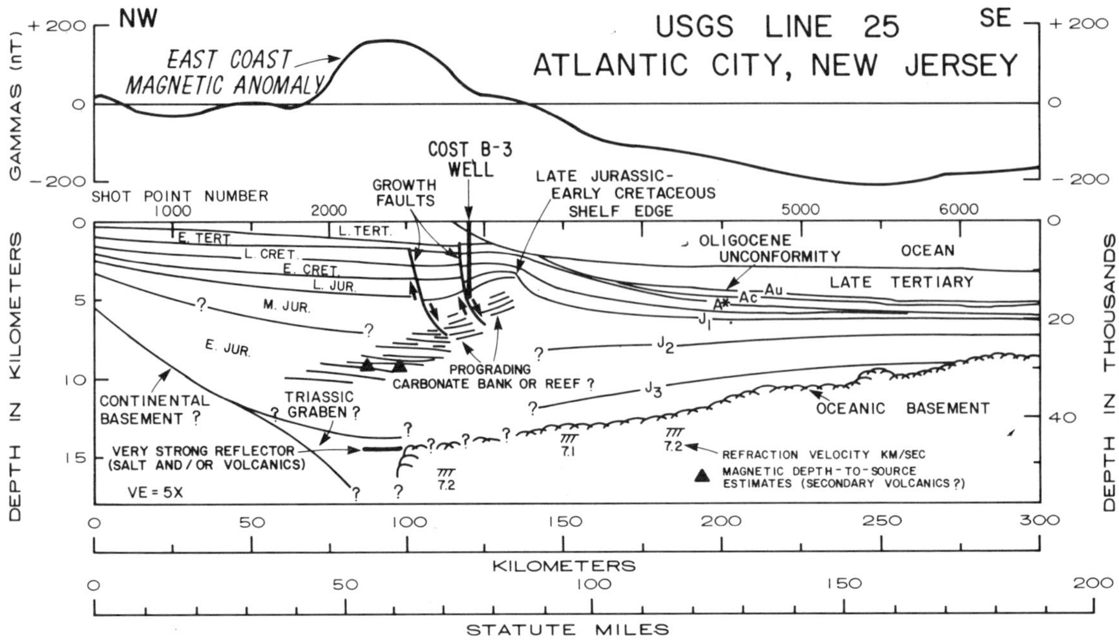

Figure 7. Structural cross section along USGS multichannel seismic line 25 off Atlantic City, New Jersey. The actual seismic record between shot points 1900 and 3600 is shown in Figure 8 (from Grow, 1980).

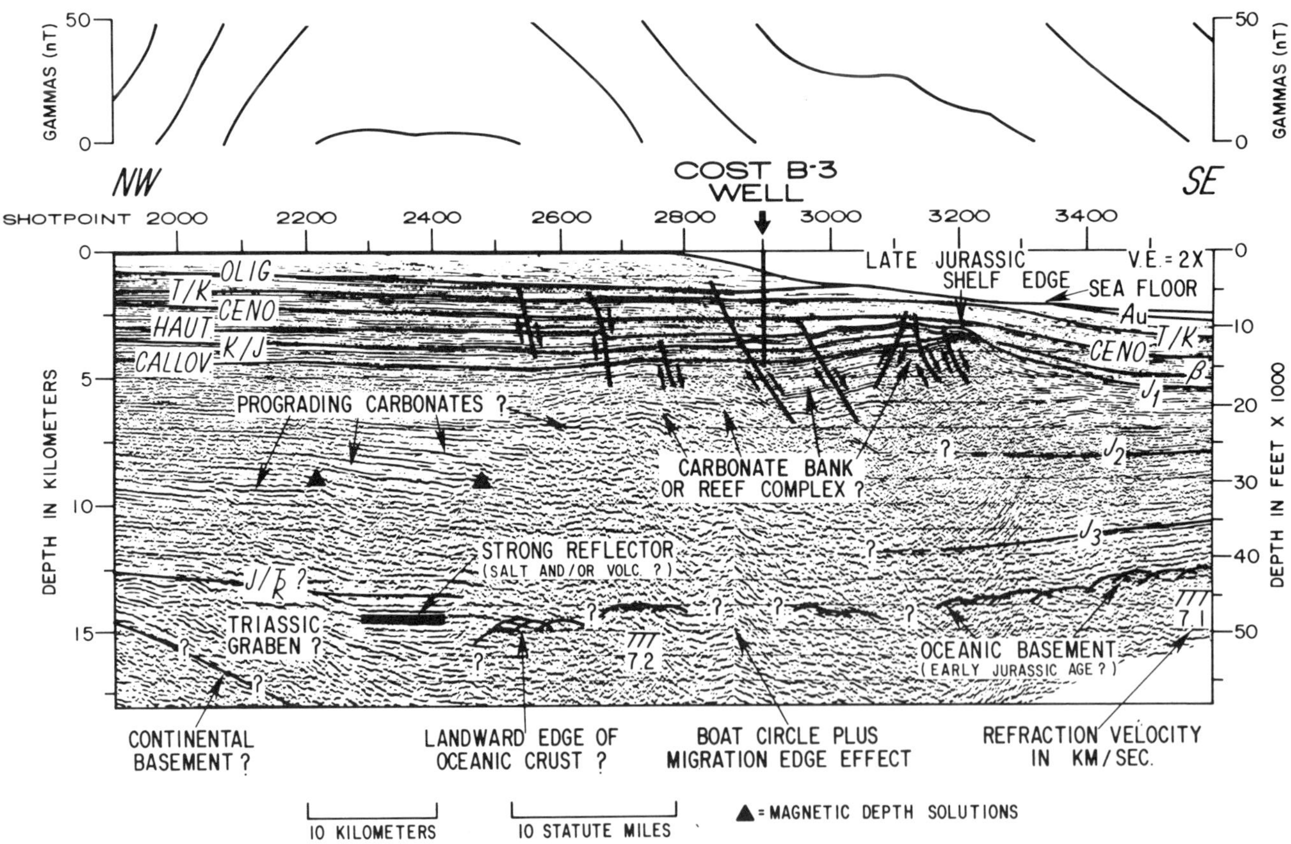

Figure 8. Annotated multichannel depth section (migrated) for USGS seismic line 25 with magnetic anomalies plotted at the top. Line 25 is the first seismic profile to achieve penetration deeper than about 6 km in the vicinity of the ECMA and the first profile to record reflected energy from deeper than 9 to 10 km in the region immediately seaward of the Late Jurassic–Early Cretaceous shelf edge (from Grow, 1980).

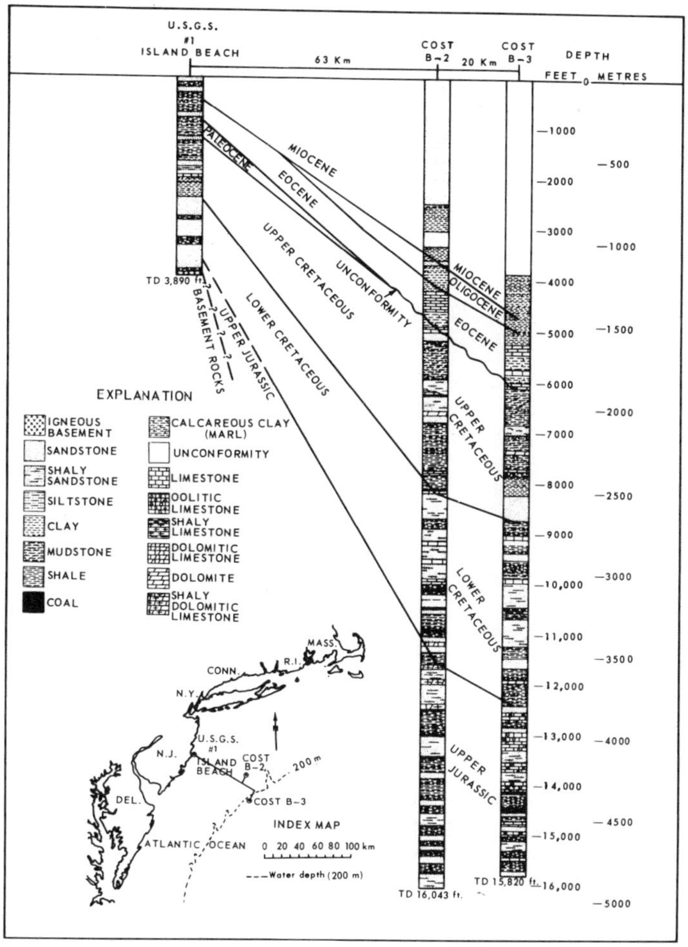

Figure 9. Stratigraphic correlations from the Island Beach No. 1 well to the COST B-2 and B-3 wells across Baltimore Canyon Trough (from Mattick and Bayer, 1980).

sea-floor spreading began in the Blake Basin with the formation of typical thin oceanic crust (Fig. 13). The above-mentioned conclusions are consistent with earlier interpretations for sea-floor spreading or extension-center jumps beneath the Blake Plateau.

Summary

1. Four major sedimentary basins formed along the U.S. Atlantic continental margin in response to Triassic–Early Jurassic continental rifting and Early to Middle Jurassic sea-floor spreading between Africa and North America: Georges Bank Basin, Baltimore Canyon Trough, Carolina Trough, and Blake Plateau Basin.

2. Postrift sedimentary strata in the four basins are as thick as 7, 13, 11, and 12 km from north to south, respectively. The postrift units are typically separated from basement or synrift sedimentary rocks by a breakup or postrift unconformity underlain by synrift grabens, which may contain as much as 5 km of sedimentary rocks.

3. Upper Triassic and/or Lower Jurassic salt has been drilled at 6.7 km in the COST G-2 well within Georges Bank Basin in what appears to be late synrift or earliest postrift sedimentary units (Pl. 4, Amato and Simonis, 1980.) Three salt diapirs along the ECMA in the Baltimore Canyon and 23 diapirs in the Carolina Trough also appear to rise from near the base of the postrift sedimentary rocks. Therefore, conditions favorable for evaporite deposits appear to have existed during late rift stage and/or into the earliest postrift stage (i.e., at the time sea-floor spreading began between North America and Africa).

4. The Jurassic sedimentary rocks that have been analyzed to date in the Baltimore Canyon Trough (COST B-2 and B-3 wells) and Georges Bank Basin (COSTG-1 and G-2 wells) were of shallow-marine and nonmarine origin. This fact supports an inference that these basins subsided during the postrift stage in response to sediment loading and lithospheric cooling.

5. Gravity models across the three northern basins indicate that the deepest parts of these basins are underlain by transitional crust 8 to 15 km thick, which appears to have formed primarily by extension and thinning of the preexisting continental crust.

6. Beneath the Blake Plateau Basin, the 15- to 25-km-thick transitional crust probably formed via a complicated mixing of continental fragments and volcanic intrusions over a zone 350 km wide during the first 10 to 15 m.y. of continental drift. Formation of this crust was followed by an abrupt eastward jump of the spreading center to beneath the Blake Escarpment where normal sea-floor spreading began to form thin oceanic crust beneath the Blake Basin.

Future Research Directions

1. Although analysis of conventional multichannel seismic reflection, gravity, and magnetic data has given us a preliminary model of the rift-stage crust underlying the major sedimentary basins, more sophisticated seismic experiments are needed to delineate this structure. Multiple-ship Large Aperature Seismic Experiments (LASE) using multichannel seismic streamer offsets of 10 to 20 km are among the most promising new methods to probe the deep crust and upper mantle. Ocean-bottom hydrophone and seismograph refraction experiments over the rift-stage crust, ECMA, and inner edge of the oceanic crust will also yield extremely valuable geophysical constraints for improving our geologic models.

2. Theoretical models for crustal thinning during rifting, initial subsidence due to rifting, subsidence due to sediment loading, and subsidence due to thermal evolution of continental margins have offered valuable new insights into the mechanisms governing margin development. An important limiting factor in most subsidence models has been the absence of drill holes that penetrate through both the early

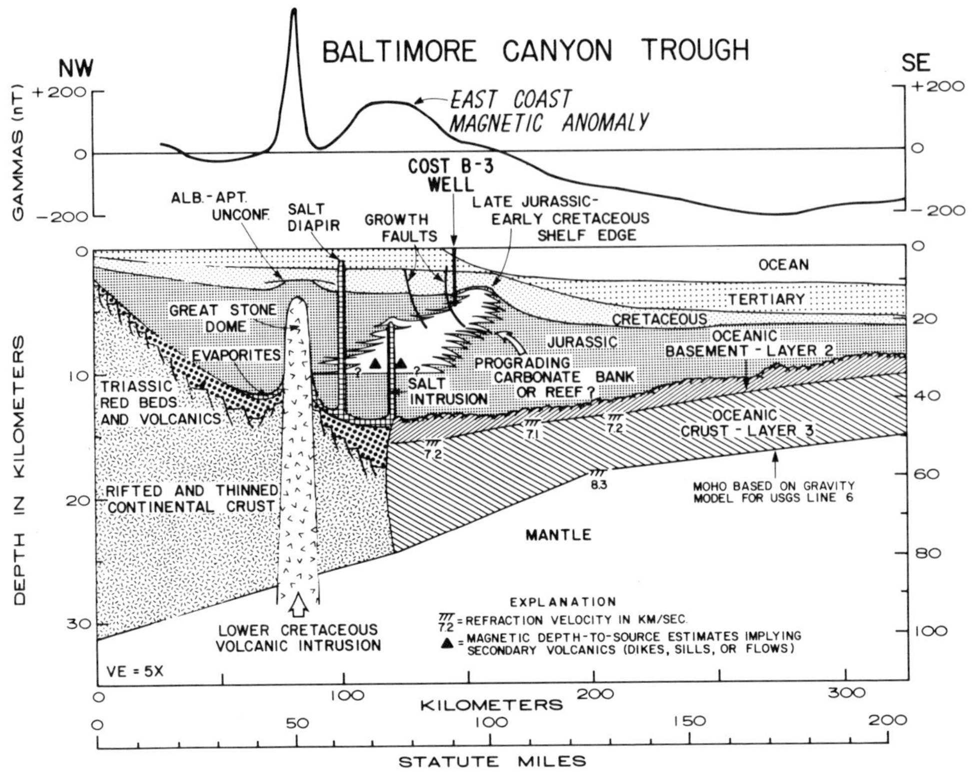

Figure 10. Schematic cross section through Baltimore Canyon Trough approximately along seismic line 25 (Figs. 7, 8) with nearby COST B-3 well, geologic features, and geophysical parameters projected into the profile. Refraction data are from Sheridan and others (1979). The Moho configuration is projected from a gravity model for line 6 (from Grow, 1980). See Schlee and others (1976) and Schlee (1981) for additional information on Baltimore Canyon Trough.

postrift and synrift sedimentary sequences. It is critical that more exploration drilling data be obtained so that biostratigraphic and paleobathymetric interpretations can be provided for improving the subsidence models.

3. Renewed interest in the coastal plain embayments and intervening platforms is developing as land and marine reflection profiling techniques begin to supplement outcrop, gravity, magnetic, and drill-hole data. Buried Triassic basins and faults within the postrift sediments have been found which are probably reactivated Paleozoic and Triassic-Jurassic fault systems. More reflection profiling and drill holes are needed both on land and within the major bays, estuaries, and rivers.

4. New geophysical techniques are also being developed to reveal modern geologic processes on the continental slope and rise. Submarine landslides as well as less catastrophic erosional and depositional processes on the slope and rise are being studied with a variety of new tools including multibeam sonars for detailed bathymetry, medium-range sidescan-sonars that provide imaging similar to aerial photographs used by land geologists, near-bottom reflection profiling devices, and deep-diving research submarines. These technologies are evolving rapidly, and studies based on them are just beginning to appear.

5. Eustatic sea-level changes during the Mesozoic and Cenozoic and their relation to transgressions, regressions, and margin subsidence are other areas of important current and future research. Sea-level curves proposed by Vail and others (1977) on the basis of oil-company seismic stratigraphic analyses have triggered many new studies aimed at calibrating sea-level curves. These sea-level fluctuations appear to be dependent in part on climatic factors (including polar ice), worldwide sea-floor spreading rates, and regional tectonic effects. Calibration of regional sea-level curves and separation of these components are essential for future improvements in our understanding of the evolution of continental margins.

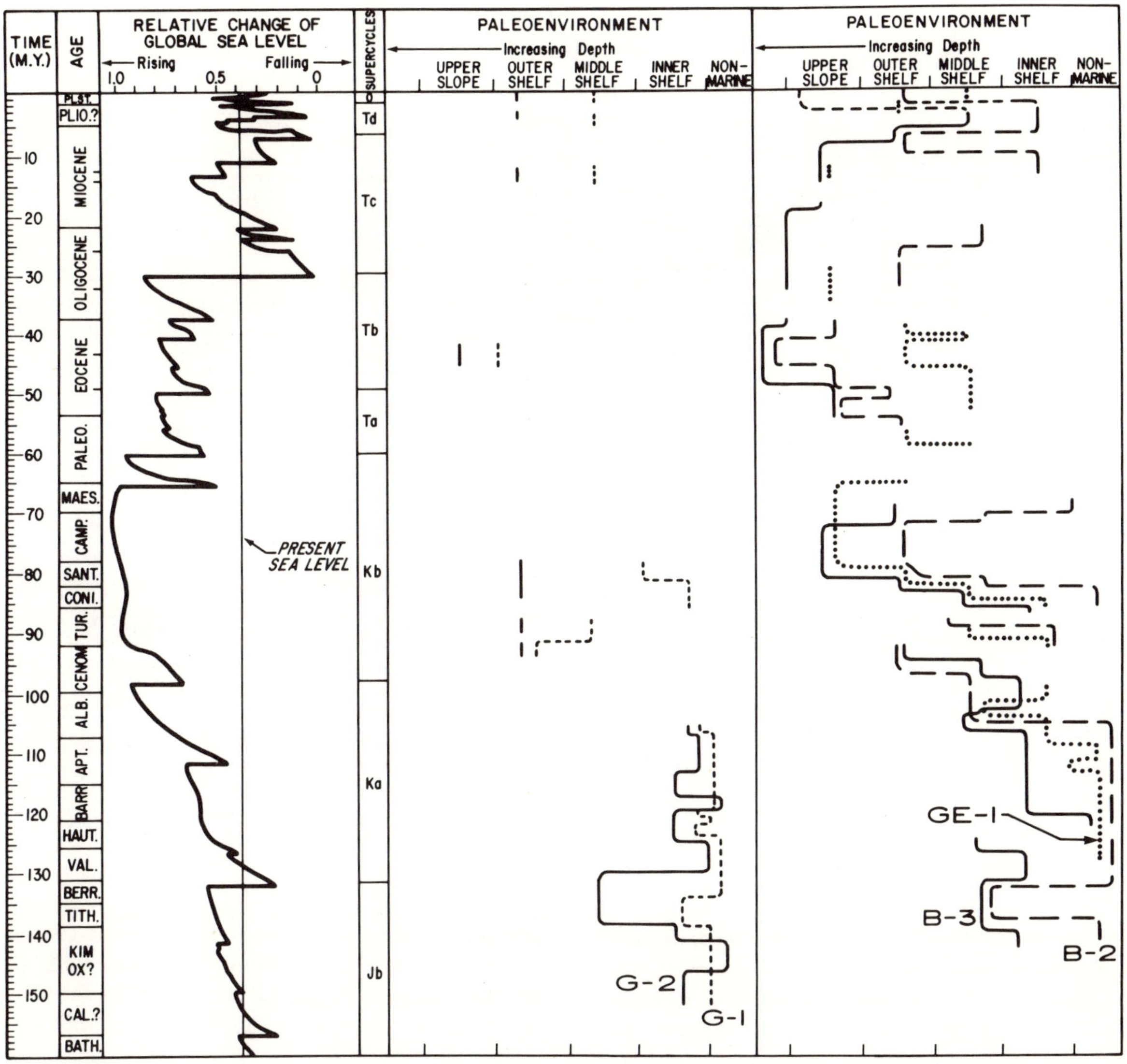

Figure 11. Comparison of global sea-level changes (Vail and others, 1977) with paleobathymetry and sediment accumulation rates at the COST G-1, G-2, GE-1, B-2, and B-3 well sites (from Poag, 1982).

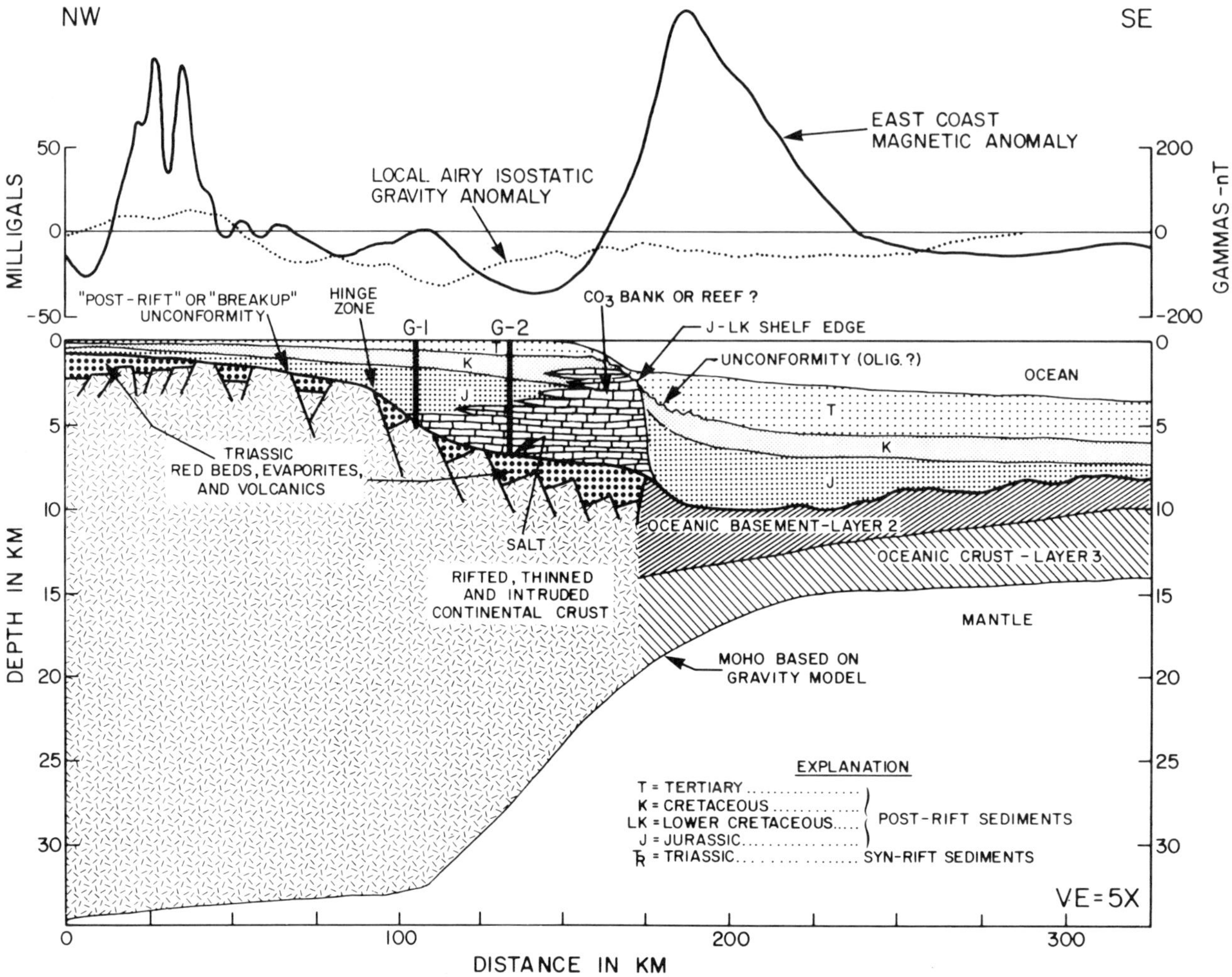

Figure 12. Composite geologic section across southwest end of Georges Bank Basin along multichannel seismic line 5 with recently available drill-hole data projected from the COST G-1 and G-2 wells (Amato and Bebout, 1980; Amato and Simonis, 1980). Moho structure based on gravity model along line 5 (Grow and others, 1979). Note hinge zone and underlying change in depth to the Moho; the structure is similar to the gravity model across the Carolina Trough (Fig. 4) (from Grow and Sheridan, 1981).

Tentative Table of Contents for D-NAG Synthesis Volume

Atlantic Continental Margin of the United States

I. Introduction

II. Physiographic provinces

III. History of geological and geophysical exploration

IV. Synthesis of relevant data bases
- A. Outcrop geology
- B. Onshore well data
- C. Offshore well data
- D. Offshore core, dredge, and submarine data
- E. Magnetics data
- F. Gravity data
- G. Heat-flow data
- H. Seismicity data
- I. Refraction data
- J. Reflection data

V. Basin syntheses, structure maps, and cross sections
- A. Onshore prerift and synrift grabens
- B. Traissic-Jurassic volcanism
- C. Georges Bank Basin
- D. Baltimore Canyon Trough
- E. Carolina Trough
- F. Blake Plateau Basin
- G. Bahama Platform
- H. Paleoshelf edges

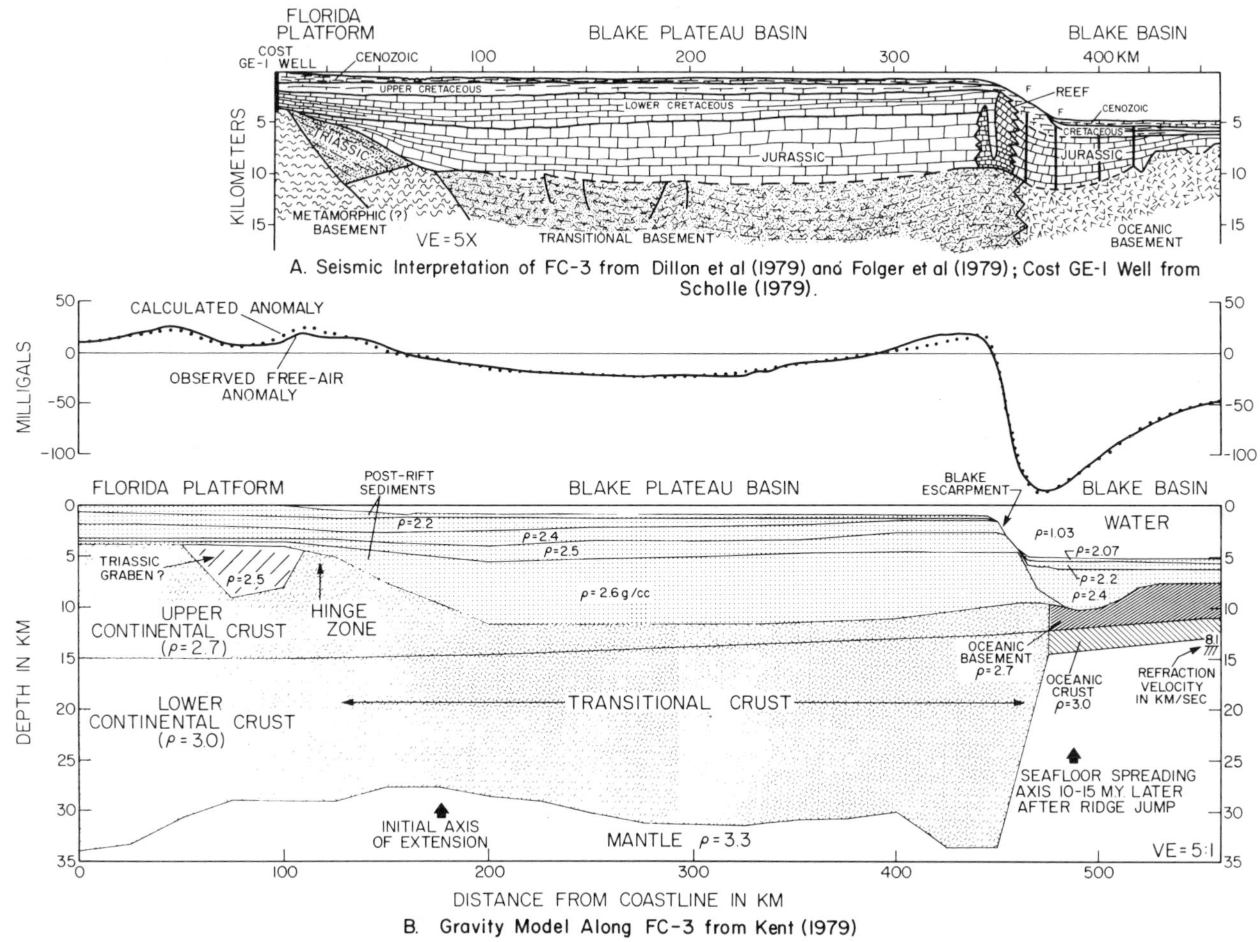

Figure 13. Composite geologic section across the Blake Plateau Basin along multichannel seismic line FC-3 (modified from Dillon and others, 1979, and Folger and others, 1979); Moho configuration is based on gravity model of Kent (1979). The COST GE-1 well data are from Scholle (1979). Thicker sediments and thinner transitional crust are present beneath the western Blake Plateau where initial rifting probably began. A seaward jump in the extension axis from the western plateau to the Blake Escarpment probably took place 10 to 15 m.y. after sea-floor spreading began in the region north of the Blake Spur Fracture Zones (from Grow and Sheridan, 1981).

VI. Deep crustal structure of the margin
- A. Gravity models
- B. Magnetic models
- C. Large aperature seismic experiments—preliminary results

VII. Subsidence and thermal evolution of the Atlantic margin
- A. Mechanism of rifting and crustal thinning
- B. Models of subsidence
- C. Thermal evolution
- D. Sea-level variations, transgressions, and regressions

VIII. Mineral and energy resources
- A. Petroleum
- B. Groundwater
- C. Sand and gravel
- D. Hard minerals
- E. Geothermal

IX. Environmental problems and land use
- A. Petroleum exploration
- B. Earthquake hazards
- C. Waste disposal

X. Definition of major unanswered geologic questions
- A. Geophysical
- B. Tectonic
- C. Stratigraphic
- D. Economic

XI. Summary and conclusions

ACKNOWLEDGMENTS

Although the references have been kept to a minimum in this D-NAG outline, the reader may find a more complete listing in Grow and Sheridan (1981). We thank K. D. Klitgord, W. Poag, and E. Winget for critically reviewing an earlier version of the manuscript and making valuable suggestions.

References Cited

Amato, R. V., and Bebout, J. W., 1980, Geological and operational summary, COST No. G-1 well, Georges Bank area, North Atlantic Outer Continental Shelf: U.S. Geological Survey Open-File Report 80–268, 111 p.

Amato, R. V., and Simonis, E. K., 1979, Geological and operational summary, COST No. B-3 well, Baltimore Canyon Trough area, mid-Atlantic Outer Continental Shelf: U. S. Geological Survey Open-File Report 79–1159, 118 p.

—— 1980, Geological and operational summary, COST No. G-2 well, Georges Bank area, North Atlantic Outer Continental Shelf: U.S. Geological Survey Open-File Report, 80–269, 116 p.

Dillon, W. P., Paull, C. K., Buffler, R. T., and Fail, J. P., 1979, Structure and development of the Southeast Georgia Embayment and Northern Blake Plateau: Preliminary analysis, *in* Watkins, J. S., Montadert, L., and Dickerson, P. W., eds., Geological and geophysical investigations of continental margins: American Association of Petroleum Geologists Memoir 29, p. 27–41.

Dillon, W. P., Popenoe, P., Grow, J. A., Klitgord, K. A., Swift, B. A., Paull, C. K., and Cashman, K. V., 1982, Growth faulting and salt diapirism—their relationship and control in the Carolina Trough, eastern North America, *in* Watkins, J. S., and Drake, C. L., eds., Hedberg Symposium Volume: American Association of Petroleum Geologists Memoir (in press).

Folger, D. W., Dillon, W. P., Grow, J. A., Klitgord, K. D., and Schlee, J. S., 1979, Evolution of the Atlantic continental margin of the United States, *in* Talwani and others, eds., Deep drilling results in the Atlantic Ocean—Continental margins and paleoenvironments: American Geophysical Union, M. Ewing Series 3, p. 87–108.

Grow, J. A., 1980, Deep structure and evolution of the Baltimore Canyon Trough in the vicinity of the COST B-3 well, *in* Scholle, P. A., ed., Geologic studies of the COST B-3 well, U.S. mid-Atlantic continental slope area: U.S. Geological Survey Circular 833, p. 117–125.

Grow, J. A., and Sheridan, R. E., 1981, Deep structure and evolution of the continental margin off the Eastern United States, *in* Geology of Continental Margins Symposium: Oceanologica Acta, Proceedings, International Geological Congress, 26th, Paris, p. 11–19.

Grow, J. A., Bowin, C. O., and Hutchinson, D. R., 1979, the gravity field of the U.S. Atlantic continental margin: Tectonophysics, v. 59, p. 27–52.

Hutchinson, D. R., Grow, J. A., Klitgord, K. D., and Swift, B. A., 1982, Deep structure and evolution of the Carolina Trough, in Watkins, J. S., and Drake, C. L., eds., Hedberg Symposium Volume: American Association Petroleum Geologists Memoir (in press).

Kent, K. M., 1979, Two-dimensional gravity model of the Southeast Georgia Embayment–Blake Plateau [Master's thesis]: Newark, University Delaware, 89 p.

Klitgord, K. D., and Behrendt, J. C., 1979, Basin structure of the U.S. Atlantic margin, *in* Watkins and others, eds., Geological and geophysical investigations of continental margins: American Association of Petroleum Geologists Memoir 29, p. 85–112.

Mattick, R. E., and Bayer, K. C., 1980, Geologic setting and hydrocarbon exploration activity, *in* Scholle, P. A., ed., Geological studies of the COST No. B-3 well, U.S. mid-Atlantic continental slope area: U.S. Geological Survey Circular 833, p. 4–12.

Mattick, R. E., Foote, R. Q., Weaver, N. L., and Grim, M. S., 1974, Structural framework of United States Atlantic Outer Continental Shelf north of Cape Hatteras: American Association of Petroleum Geologists Bulletin, v. 58, p. 1179–1190.

Poag, C. W., 1982, Foraminiferal and seismic stratigraphy, paleoenvironments and depositional cycles in the Georges Bank Basin, *in* Scholle, P. A., and Wenkam, C. R., eds., Geological studies of the COST Nos. G-1 and G-2 wells, United States North Atlantic Outer Continental Shelf area: U.S. Geological Survey Circular 861 (in press).

Schlee, J. S., 1981, Seismic stratigraphy of Baltimore Canyon Trough: American Association of Petroleum Geologists Bulletin, v. 65, p. 26–53.

Schlee, J., Behrendt, J. C., Grow, J. A., Robb, J. M., Mattick, R. E., Taylor, P. T., and Lawson, B. J., 1976, Regional geologic framework off northeastern United States: American Association of Petroleum Geologists Bulletin, v. 60, p. 926–951.

Scholle, P. A., ed., 1979, Geological studies of the COST GE-1 well, United States South Atlantic Outer Continental Shelf area: U.S. Geological Survey Circular 800, 114 p.

Sheridan, R. E., Grow, J. A., Behrendt, J. C., and Bayer, K. C., 1979, Seismic refraction study of the continental edge off the Eastern United States: Tectonophysics, v. 59, p. 1–26.

Vail, P. R., Mitchell, R. M., Jr., and Thompson, S. III, 1977, Seismic stratigraphy and global changes of sea level, Part 4: Global cycles of relative changes of sea level, *in* Payton, E. C., ed., Seismic stratigraphy—applications to hydrocarbon exploration: American Association of Petroleum Geologists Memoir 26, p. 83–97.

The Gulf of Mexico Basin

Amos Salvador
Department of Geological Sciences
University of Texas
Austin, Texas 78712

R. T. Buffler
Institute for Geophysics
University of Texas
700 The Strand
Galveston, Texas 77550

Introduction

The accumulation of abundant geologic and geophysical information about the Gulf of Mexico Basin during the last few decades has contributed significantly to the understanding of the geologic history of this important geologic province—its birth during Late Triassic–Jurassic time, and the subsequent geologic events that have given the basin its present configuration.

Much of this information has been the product of extensive surface and sub-surface geologic studies, geophysical surveys, and drilling around the periphery of the basin, mainly in connection with the exploration for oil and gas. More recently, invaluable data have been obtained in the previously little-known central part of the basin by regional geophysical surveys—particularly seismic reflection and refraction—and the results of the Deep Sea Drilling Project. The geophysical surveys were conducted by academic institutions, the U.S. Geological Survey, and the petroleum industry.

The results of these geophysical surveys have shed much light on the seismic nature of the crust underlying the Gulf of Mexico Basin—"oceanic" in the deep central part and "transitional" around the periphery of the deep basin. (The term "transitional" is applied to crust intermediate in thickness between oceanic and continental crust, possibly the result of stretching during episodes of rifting or pull-apart.) The surveys have also for the first time contributed information concerning the distribution, thickness, and structural configuration of the extensive salt deposits, as well as of the thick sedimentary column that fills the central part of the basin.

This large volume of geologic and geophysical information is now scattered throughout the extensive literature on the Gulf of Mexico Basin. Nowhere is there available in a single publication an integrated summary and an interpretation of the basin—its origin and subsequent development, its sedimentary record, its energy and mineral resources. It will be the primary objective of the GSA's Decade of North American Geology (D-NAG) volume on the Gulf of Mexico Basin, therefore, to provide such a synthesis.

Limits of the study area

The Gulf of Mexico Basin is a roughly circular structural and sedimentary basin approximately 1,500 km in diameter. The limits of the basin, for the purpose of the D-NAG volume, are shown in Figure 1.

The Florida and Yucatan carbonate platforms border the basin to the east and south and are included within the limits of the area to be described. The western limit of the basin has been placed roughly at the foot of the Sierra de Chiapas and the Sierra Madre Oriental of Mexico. To the north, the structural limits of the basin correspond from west to east with the basinward flanks of the Marathon Mountains, the Ouachita Structural Belt, the Ouachita Mountains, the Central Mississippi Uplift, and the Appalachian Mountains. Along the northwestern and northern borders of the basin, however, the Gulf of Mexico Basin volume will include the description of the Cretaceous and the Tertiary rocks to their pinchout beyond the structural limits of the basin. The D-NAG volumes concerning the adjoining areas will treat the pre-Cretaceous section below the Cretaceous/Tertiary overlap.

Geologic setting

The Gulf of Mexico Basin is a relatively simple struc-

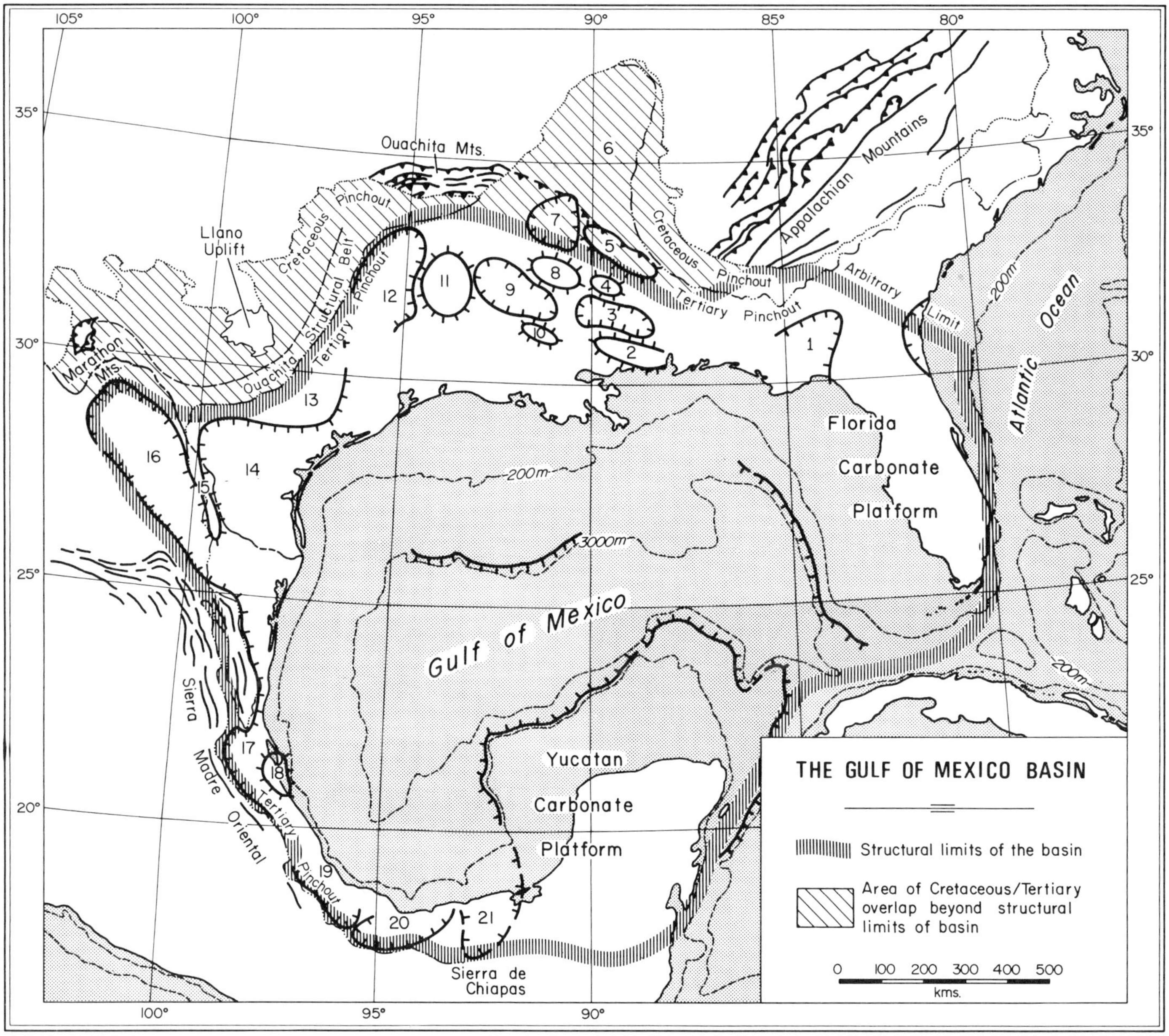

Figure 1. The Gulf of Mexico Basin showing boundaries of area covered by the D-NAG volume. Second-order structural elements: 1. Apalachicola (Southwest Georgia) Embayment. 2. Wiggins Uplift. 3. Mississippi Salt Basin. 4. Jackson Dome. 5. Central Mississippi Uplift. 6. Mississippi Embayment. 7. Desha Basin. 8. Monroe Uplift. 9. North Louisiana Salt Basin. 10. LaSalle Arch. 11. Sabine Uplift. 12. East Texas Basin. 13. San Marcos Arch. 14. Rio Grande (Maverick, Burgos) Embayment. 15. Burro-Picachos-Tamaulipas Arch (Peninsula). 16. Sabinas Basin. 17. Tampico-Misantla Basin. 18. Tuxpan Platform. 19. Veracruz Basin. 20. Comalcalco (Saline) Basin. 21. Macuspana Basin.

tural basin filled with 10 to 15 km of sedimentary rocks ranging in age from Late Triassic to Holocene. The central part of the basin is occupied by the Gulf of Mexico, which attains water depths of 3,750 m and where the basement has been estimated to reach a depth of about 12 to 15 km below sea level. Geophysical surveys indicate that the deep central part of the basin is underlain by an oceanic-type crust, which is surrounded in places by transitional crust inferred to represent rifted or stretched continental crust. Normal continental crust has been interpreted to be present under the outer, peripheral parts of the basin. The continental crust and the transitional crust consist of a variety of igneous, metamorphic and older sedimentary rocks. Two cross-sections, one east-west and one north-south, showing the

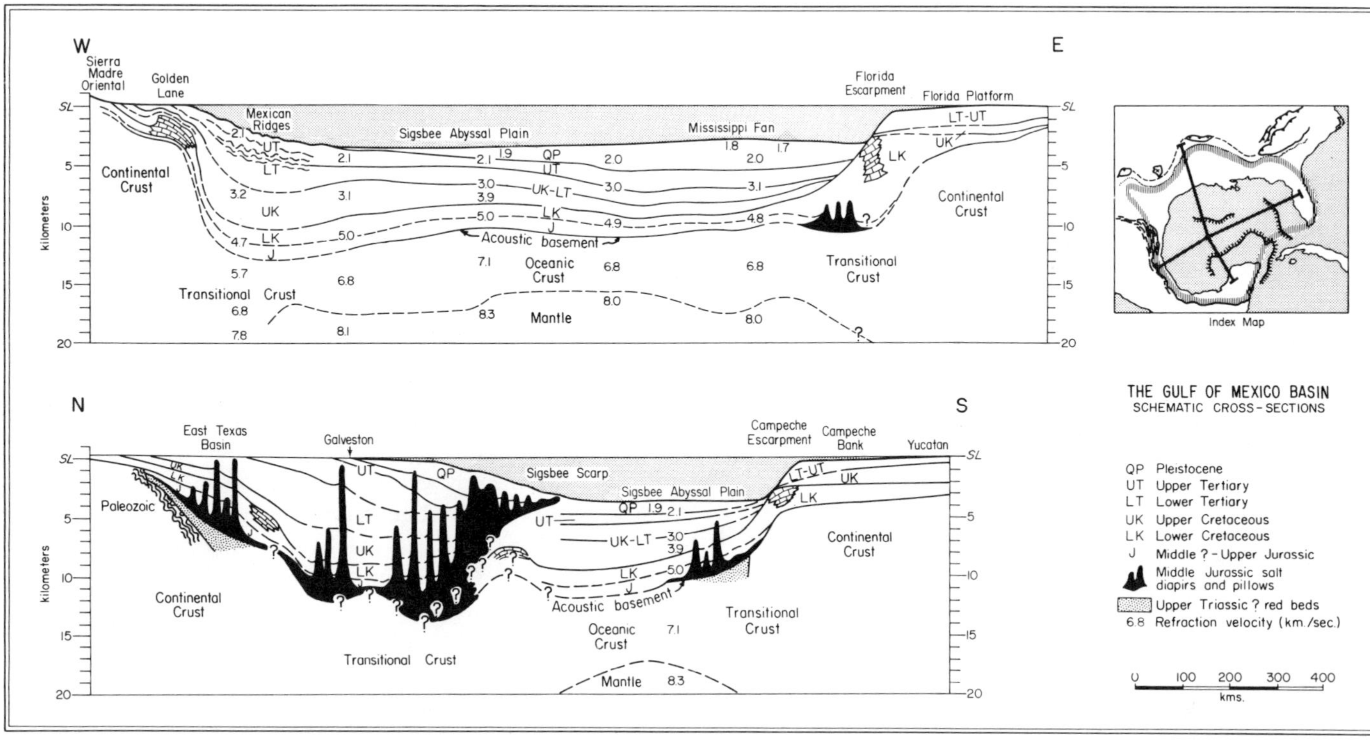

Figure 2. Schematic cross-sections, Gulf of Mexico Basin.

generalized geologic setting of the Gulf of Mexico Basin, are depicted in Figure 2.

Paleozoic rocks within the Gulf of Mexico Basin are known only from sparse outcrops and from only modest penetration by a limited number of wells around its rim. The geologic history of the basin during Paleozoic time needs to be inferred, therefore, from the study of neighboring areas. Similar statements apply to the Triassic, the Lower Jurassic, and most of the Middle Jurassic.

The earliest stratigraphic interval of which more is known is the uppermost Middle Jurassic, represented over most of the Gulf of Mexico Basin by extensive salt deposits. The plastic nature of this salt interval under the load of the overlying sediments has resulted in the formation of many diapiric salt domes, pillows, and ridges throughout the basin, some of which are still active.

Overlying the salt, or the oceanic "basement" in the central part of the basin, is a very thick sequence of sediments, deposited for the most part in marine environments—predominantly clastics to the north and mostly carbonates to the east and south. To the west the Jurassic and Cretaceous are predominantly composed of carbonates and the Tertiary and Quaternary are mainly clastic. These sediments filled the basin as it subsided persistently from Late Jurassic time to the present day. The thickest sedimentary fill is found along the north flank of the basin where up to 15 km of sediments, mainly Tertiary clastics, have prograded into the basin (Fig. 2).

The broad outline of a simple structural basin has been locally modified by salt and shale tectonics and by growth-faulting around the margin of the Tertiary depocenters. A number of large, second-order structural elements have been recognized around the flanks of the basin. Some of the most prominent among them are shown in Figure 1.

Origin and Development of the Gulf of Mexico Basin

Little is known about the geologic history of the Gulf of Mexico area before Late Triassic time. Some authors have postulated the presence of a basin in the area during most of Paleozoic time, but present evidence seems to indicate that the area was, at the end of Paleozoic time, part of the large supercontinent of Pangea. In any case, the present Gulf of Mexico Basin is believed to have originated in Late Triassic time as the result of rifting within the North American Plate at the time it began to crack and drift away from the African and South American plates. Rifting probably continued through Lower and Middle Jurassic time with the formation of "stretched" or "transitional" continental crust throughout the central part of the basin. Intermittent advance of the sea into the basin from the west during late Middle Jurassic time resulted in the formation of the extensive salt deposits known today in the Gulf of Mexico Basin. It has been suggested that the main rifting episode, which resulted in the southward separation of the Yucatan block from the North American Plate and the formation of true oceanic crust in the central part of the basin, took place during late Middle or early Late Jurassic time, after the formation of the salt deposits.

Since Late Jurassic time, the basin has been a stable geologic province characterized by the persistent subsidence of its central part, probably due to thermal cooling and sediment loading, as the basin filled with thick prograding clastic wedges along its western and northern margins. To the east and south, the stable Florida and Yucatan platforms were not covered by the sea until the beginning of Cretaceous time; since then, the formation of platform carbonates has characterized the geologic history of these two stable areas. The main influx of clastics was from the north and northwest, particularly during Tertiary and Quaternary times.

Approach and Organization of the Gulf of Mexico Basin Volume

The D-NAG volume on the Gulf of Mexico Basin will be divided into 19 chapters organized around four main themes, as follows (Table I):

1. The first part of the volume (7 chapters) will provide the background necessary for the description of the geology and geophysical composition of the Gulf of Mexico Basin: an introduction, a description of the physiographic and tectonic features of the basin, a discussion of the crust under the basin, and reviews of the sedimentary and structural processes that have persisted during most of the history of the basin and are still active today.

2. Following is a description of the rock sequences forming the Gulf of Mexico Basin, from the "basement" to the recent sediments, and the interpretation of their geologic history (5 chapters)

3. The third part (3 chapters) is interpretative: a description of the stratigraphy of the central part of the basin based on seismic information, a speculative review of the geologic processes controlling sedimentary cycles, and a summary chapter discussing the origin and evolution of the Gulf of Mexico Basin.

4. The last part will cover fossil fuels and mineral resources (3 chapters).

The volume will end with a chapter in which the most important unanswered questions concerning the geology of the Gulf of Mexico Basin will be discussed.

To amplify: The volume will lead off with an introduction in which the limits of the area covered by the synthesis and the scope of the project are discussed. The next two chapters will cover a description of the physiography of the land area surrounding the Gulf of Mexico as well as a description of the bathymetry under the Gulf, and a review of the Quaternary sedimentary processes that have given the basin its present physiographic and bathymetric configuration: the formation of deltas, carbonate platforms, and so forth. These same or similar sedimentary processes, it is believed, have been active during most of the geologic history of the basin, and their description and analysis early in the volume will provide the basis for understanding better the sedimentology, stratigraphy, and paleogeography of the pre-Quaternary sedimentary sequences.

Additional background will be provided by the next chapters. Chapter 4 will describe the present-day structural architecture of the Gulf of Mexico Basin, its overall framework as well as the second-order structural features that locally modify the regional configuration. Chapter 5 will discuss the origin, age, composition, and distribution of the crustal types (oceanic, transitional, and continental) that underlie the Gulf of Mexico Basin. The discussion will cover the geophysical evidence (magnetics, gravity, refraction, and reflection seismic) and the interpretation of this evidence used to establish the distribution in the basin of the various crustal types. Chapter 6 will be devoted to a review of the structural evolution of the Gulf of Mexico Basin, emphasizing the mechanics of the development of some of its most typical structural styles—growth faults, salt and shale tectonics, as well as the causes controlling the persistent subsidence that the basin has suffered since its birth. Chapter 7 will summarize the igneous activity in the basin. As in the case of the sedimentary processes, the review of the structural development in the basin in the first part of the volume is expected to enhance the understand-

TABLE 1. OUTLINE FOR D-NAG VOLUME ON GULF OF MEXICO

Chapter	Subject
1	Introduction
2	Physiographic provinces and bathymetry
3	Quaternary sedimentary processes
4	Tectonic features
5	Crust and mantle, geophysics (magnetics, gravity, refraction, and deep reflection)
6	Evolution of structural features
7	Igneous activity
8	Pre-Upper Triassic—nature and age of "basement"—Paleozoic
9	Upper Triassic–Jurassic
10	Lower Cretaceous
11	Upper Cretaceous
12	Cenozoic
13	Seismic stratigraphy of the Gulf of Mexico
14	Sedimentary cycles
15	Origin and evolution of the Gulf of Mexico Basin
16	Oil and gas resources
17	Lignite, uranium, phosphates, sulfur and other minerals; geopressured-geothermal resources
18	Water resources
19	Summary—outstanding problems
	Bibliography

ing of the stratigraphy of the sedimentary section, as some of the structural features—growth faults, salt and shale tectonics, regional subsidence—influenced directly the sedimentary history of the basin.

The second part of the volume on the Gulf of Mexico Basin—the geologic-geophysical description and historical synthesis of the rock sequences filling the basin—will be approached in a chronological manner rather than region by region. This was deemed as the most logical and appropriate approach since the area covered by the volume, the present-day structural Gulf of Mexico Basin, appears to have been a single depositional basin virtually since its birth, a basin characterized by a persistently subsiding central part surrounded by stable shelves over which the sea periodically advanced and retreated.

The first chapter (8) of this second part will discuss the pre–Upper Triassic rocks known from the Gulf of Mexico Basin. It will cover the composition and age of the "basement" rocks and will attempt to summarize what little is known about the Paleozoic, based on the few and scattered occurrences of rocks of this age within the basin. An attempt will be made in this chapter to summarize some of the previous ideas concerning the geologic history of the area during pre–Late Triassic time.

The next chapter (9) will cover the Late Triassic through Jurassic, the time of the early and most active geologic development of the Gulf of Mexico Basin. It will contain a discussion of the stratigraphy of the rocks of this age, the igneous and tectonic activity of the time, and an attempt to reconstruct the paleogeographic development of the area. Considerable emphasis will be placed on the development of graben systems—an indication of an episode of rifting in the area that has been interpreted as the first evidence of the birth of the Gulf of Mexico Basin. The chapter will also include a detailed discussion of the origin, age, and distribution of the extensive Jurassic salt deposits in the basin. These salt deposits probably provide the best testimony concerning the origin and early development of the Gulf of Mexico Basin.

The structural and depositional framework of the Gulf of Mexico Basin at the end of Jurassic time was not appreciably different from that of today; the basin was well established by then, and during the rest of its geologic history the basin was characterized by essentially continuous deposition. This history of undisturbed, stable deposition will be covered in three chapters on the Lower Cretaceous, the Upper Cretaceous, and the Cenozoic (10, 11, 12). Each chapter will discuss the stratigraphy, the contemporaneous igneous and tectonic activity, and will attempt to reconstruct the evolution of the paleogeography of the basin.

The selection of time intervals to be covered by each of the chapters is not entirely arbitrary. The Lower Cretaceous, for instance, has definite geologic characteristics (lithologic, paleogeographic) that separate it from the underlying Jurassic and the overlying Upper Cretaceous. The Early Cretaceous was a time during which widespread reefs grew all around the margins of the broad, stable shelves that rimmed the Gulf of Mexico Basin; the Upper Cretaceous, on the other hand, was characterized by relatively thin sections of carbonates and fine-grained clastics during a period of major transgression of the sea.

How to treat the Cenozoic was more of a problem. It was thought originally that this part of the stratigraphic section might be discussed in four separate chapters: Paleocene-Eocene, Oligocene-Miocene, Pliocene-Pleistocene, and Holocene. There was some logic to this subdivision. During Paleocene and Eocene times thick coarse clastics were deposited in extensive delta systems, which attained great thickness in south Texas and northeast Mexico. The clastic depocenters shifted gradually during Oligocene and Miocene times from south Texas to south Louisiana, then to offshore Louisiana and southeast Texas during Pliocene and Pleistocene times.

But there were problems, too. For instance, it could be argued, on the one hand, that the Pliocene and the Pleistocene should not be treated in the same chapter, because the Pliocene does not reflect, as the Pleistocene does, the influence of the glacial events taking place to the north. On the other hand, it seemed logical to discuss the Pliocene and Pleistocene in the same chapter because both attain their maximum development offshore along the northern part of the Gulf of Mexico and depend mainly for their study on subsurface information, and because there is at present a complete lack of agreement concerning the placement of the Pliocene-Pleistocene boundary, locally as well as internationally.

Many other similar problems arise in trying to treat the Cenozoic in several separate chapters: The entire section is composed of a laterally variable, monotonous sequence of interbedded sandstones and shales with no easily recognizable regional breaks or lithologically distinct units; the boundaries between units are transitional and in most cases vague, subjective, and the topic of durable controversies, all of this the result of the alternation of closely related environments of deposition that persisted in the Gulf of Mexico Basin during the Cenozoic; and the nature of the fauna is generally undiagnostic for accurate age determination. It was finally decided, therefore, to discuss the entire Cenozoic in a single chapter, in this manner avoiding having to designate arbitrary boundaries and having to repeat in each of several separate chapters the same descriptions of the environmental and tectonic controls of deposition, of the lithologic properties, and other stratigraphic characteristics common to the entire Cenozoic sequence.

The discussion of the regional stratigraphy will be followed by three chapters (13, 14, 15) devoted to the interpretation of the data reviewed previously. One chapter is on the seismic stratigraphy of the undrilled central part of the Gulf of Mexico Basin, an area where seismic data provide the only source of information for interpreting the structure and stratigraphy of the basin. This will be followed by a chapter in which the geologic processes controlling sedimentary cycles in the basin will be reviewed. Included in this review will be a discussion of the evidence that may support or contradict the speculation that some of the stratigraphic and depositional features recognized in the sedimentary section (unconformities, onlaps, offlaps) may be the result of eustatic sea-level changes that took place during the evolution of the basin.

The origin and subsequent evolution of the Gulf of Mexico Basin will be traced in the next chapter. It will serve as summary and synthesis of the various aspects of the geology of the basin discussed in previous chapters.

A geologic synthesis of the Gulf of Mexico Basin would not be complete without a discussion of its energy and mineral resources. Three chapters (16, 17, 18) will be devoted to this subject: one will cover oil and gas; one will discuss lignite, uranium, phosphates, sulfur and other minerals, as well as geopressured-geothermal resources; and a third one will review water, a resource that may be critical in determining the future of the area.

The volume will conclude with a summary chapter that will discuss the most significant among the yet unanswered questions concerning the geologic and geophysical composition of the basin, and with a bibliography of the most significant publications concerning the geology of the Gulf of Mexico Basin.

Conclusion

This synthesis and interpretation of the geologic and geophysical information now available on the Gulf of Mexico Basin should serve to clarify some problems and to answer many questions, to support some assumptions, hypotheses, and ideas, and to contradict or weaken others. But it should not be expected to answer all questions or clarify all problems. Additional geophysical surveys (seismic as well as magnetic and gravity) will make important contributions to answering questions and solving problems, but only deep drilling in the central undrilled part of the basin can provide unequivocal answers to the fundamental questions concerning the geologic history of the Gulf of Mexico Basin.

Probably most important among these questions are those concerning the nature and origin of the older rocks—the "basement" and the Upper Triassic and Jurassic sections. Most fundamental are the following:

1. What are the origin, age, composition, tectonic history, and distribution of the crustal types (oceanic, transitional, continental) that floor the Gulf of Mexico Basin?

2. What are the origin and age of the extensive salt deposits in the Gulf of Mexico Basin?

3. What are the geologic characteristics of sedimentary and igneous rocks that underlie these salt deposits?

4. What are the age, nature, and depositional history of the post-"basement" or post-salt deposits in the central, deep part of the Gulf of Mexico Basin?

Not until reasonable answers to these questions can be given will it be possible to postulate a coherent explanation of the geologic history of the Gulf of Mexico Basin.

The integration and interpretation of the great volume of geologic and geophysical information on the Gulf of Mexico Basin now scattered in the literature will undoubtedly represent not only a much-needed summary of what we now know about the basin but also a statement of what we do not know. Further, it will undoubtedly suggest what programs of geologic and geophysical research may be needed to answer some of the most important questions concerning the origin and development of the Gulf of Mexico Basin and, one hopes, to stimulate the initiation of such programs.

In addition, a better description and understanding of the origin and development of the Gulf of Mexico Basin will serve as an example that may be used to better interpret the history of other basins and other areas developed under similar sedimentary and tectonic environments.

The Caribbean Region

J. E. Case
U.S. Geological Survey
345 Middlefield Road
Menlo Park, California 94025

Gabriel Dengo
Apartado 468, Guatemala City
Guatemala

Introduction

The Caribbean region has been an attractive area of the world since Columbus rediscovered it in 1492. In colonial times the Caribbean was an area of settlement, economic exploitation, and intense competition between the powerful nations of Spain, England, France, and Holland. Geological interest in the region, in a sense, started with the search for gold, silver, and precious stones, and the quest for the sources of the El Dorado legend. Many accounts of exploration during the conquest not only mention fabulous mines but also describe volcanoes, thermal springs, and the effects of earthquakes. It was only natural that in the initial days of modern geology the region should be intriguing to earth scientists. Thus, geologic accounts appear in the writings of such naturalists as von Humboldt, Karsten, De La Beche, Montessus de Ballore, and others. Within a global context, the first attempt to portray the geology of the region was by Suess in 1885.

In the early part of the 20th century, scientific interest, and the construction of the Panama Canal and exploration for oil greatly advanced geologic knowledge of the region. Many European scientists worked in the Caribbean during their *Wanderjahren*. Much of this knowledge was masterfully summarized by Schuchert (1935), and for the Central American part by Sapper (1937). Until that time, scientific contributions were mostly by geologists residing in countries outside the region; only a few indigenous geologists conducted Caribbean research, but some of their contributions were outstanding.

Long-range scientific programs were initiated shortly before World War II by Utrecht and Princeton Universities, but the work of both institutions was interrupted during the war years. After the war, a true international community began cooperating toward a common goal: understanding the geology of the Caribbean region and its relations to the surrounding areas. With the rapid improvement in and use of geophysical methods at sea, intensive offshore studies of the Caribbean were pioneered by the Lamont-Doherty Geological Observatory of Columbia University, and followed by numerous institutions in the United States, Holland, Great Britain, France, and the Soviet Union. On land, in cooperation with various Caribbean nations, several European and American universities carried out long-range programs of geologic investigations. A new generation of native Caribbean geologists joined the efforts of those from abroad, and the geological surveys of several Caribbean countries were strengthened and became active in research.

The formulation of new geologic concepts and ideas, some of worldwide importance, was significantly aided by investigations in the Caribbean. Confirming data for the plate-tectonic hypothesis were provided by the regional patterns of Caribbean seismicity, offshore structure, and active volcanoes, especially in Middle America and the Lesser Antilles.

Modern sources of data, which also fully credit the work of contemporary investigators, are the tectonic map of northern South America by Martín (1978), the geologic-tectonic map of the Caribbean by Case and Holcombe (1980), a recent volume on the geology of Central America by Weyl (1980), another volume on the structural geology of the Caribbean by Butterlin (1977), a geologic review of the Andes by Zeil (1979), a volume on the Gulf of Mexico and the Caribbean edited by Nairn and Stehli (1975), and a volume on the geology, geophysics, and resources of the Caribbean edited by the late John D. Weaver (1977). Volumes of the Caribbean Geological Conferences, and Central American Geological Conferences, and publications of the national geological surveys, especially those of Venezuela and Colombia, provide much of the foundation for our present level of geological knowledge of the Caribbean region. Part of the material in this chapter is a modification and update of a geologic summary prepared by Case for the Weaver (1977) volume.

General Distribution of Lithologic-Deformational Superterranes

Theories of the origin of the Caribbean Sea and its marginal lands are highly varied: Some investigators have proposed schemes involving tensional rifting and extreme rotation of various Caribbean blocks as part of a continental drift process; others regard the Caribbean region, especially the southern part, as part of a Pacific plate that has moved eastward or remained relatively static while North and South America drifted westward at a faster rate; still others have regarded the Caribbean as a foundered continental mass, possibly formed by a process of oceanization or basification; and others have invoked north-south extension combined with left-lateral shear as a formative process. Many investigators since 1960 have assumed that the volcanic Lesser Antilles are a product of westward underthrusting of Atlantic oceanic crust beneath the island arc, and that the Middle American volcanoes result from northeastward underthrusting of Pacific Ocean crust. Some geologists have postulated that the Caribbean basin has existed since Jurassic time and was bordered on its northern and southern sides by orthogeosynclines, and is thus related to global contraction.

In broadest plate-tectonic terms, the Caribbean (Fig. 1) is bordered on its northern and southern sides by wide plate-boundary zones. The northern zone shows aggregate left-lateral movement and the southern zone shows right-lateral movement. The Cocos and Nazca plates are being subducted eastward beneath Middle America and northern South America, and the Atlantic Ocean sectors of the North American and South American plates are being subducted westward beneath the Antilles. Within this broad framework, many small terranes have been and are still moving independently of their neighbors, forming an "ice-floe" mosaic of small blocks along the borders of the Caribbean Sea.

The numerous lithologic-deformational terranes of the Caribbean region can be grouped into several superterranes of presumable tectonically related units, as shown schematically in Figure 1. The distribution of these various units is briefly described here in terms of their *present* geography. Our future task will be to trace their movements and deformation through time, by means of lithofacies maps, and paleomagnetic and paleobiogeographic data.

Precambrian and Paleozoic Cratonic Complexes

Two main exposed continental nuclei of Precambrian and metamorphosed Paleozoic rocks frame the Caribbean region: (1) the Guyana shield in the southeast, consisting mainly of Precambrian metamorphic and igneous rocks but including substantial areas of unmetamorphosed sedimentary and volcanic rocks, and (2) nuclear Central America in the northwest, composed of Precambrian(?) and Paleozoic metamorphic and igneous rocks. Structurally isolated blocks of Precambrian and metamorphosed Paleozoic rocks in northern Venezuela and northern Colombia (not shown on Fig. 1) are shield outliers. Isotopic ages for rocks of the Guyana shield indicate a time span of 3.4 to about 1.0 b.y. B.P. for major geologic events. In the outliers of the Venezuelan ranges and northern Colombia, isotopic ages on rocks range from 600 to about 1,400 m.y. Important deposits of iron, bauxite, diamonds, and gold have been found in Precambrian rocks of the northern Guyana shield.

In nuclear Central America, reported ages on rocks range from 600 to about 1,100 m.y. Metamorphic events evidently occurred into Permian time for some Paleozoic rocks of northern South America, and into pre-Carboniferous and possibly Permian time in nuclear Central America. However, Paleozoic metamorphism did not occur everywhere. In some areas, metamorphic rocks are unconformably overlain by unmetamorphosed pre-Mesozoic rocks and in other areas metamorphic facies can be traced laterally into unmetamorphosed sequences (Merida Andes), or against fault contacts with unmetamorphosed Paleozoic rocks. Correlations of sedimentological, magmatic, metamorphic, and deformational episodes in the ancient rocks with those of North America, Africa, and Europe will appear in the Caribbean volume of this series.

Paleozoic Unmetamorphosed Sequences and Plutons

Pennsylvanian and Permian sedimentary rocks (not shown in Fig. 1), mainly of shelf or miogeoclinal facies, overlie the older crystalline basement of nuclear Central America in southern Mexico, Guatemala, and Belize in the northwest. In Venezuela and Colombia, all Paleozoic systems are represented at various localities, and most of the unmetamorphosed sedimentary sequences are in shelf or miogeoclinal facies, although some deposits are of deep-water facies. Small areas of late Paleozoic volcanic rocks of diverse composition have been reported in both northern and southern provinces.

Paleozoic(?) plutons occur in the Maya Mountains (Belize) and may be present in Mexico, just northwest of the Motagua-Polochic fault zone. Other Paleozoic plutons have been reported from nuclear Central America in this fault zone. In northern South America, Paleozoic plutons crop out in the Laramide deformed belt of the Cordillera Central, the Sierra Nevada de Santa Marta, and the Guajira Peninsula, Colombia, and in the Merida Andes and Baul uplift of Venezuela. Whether these Paleozoic igneous rocks are subduction related is unknown at present.

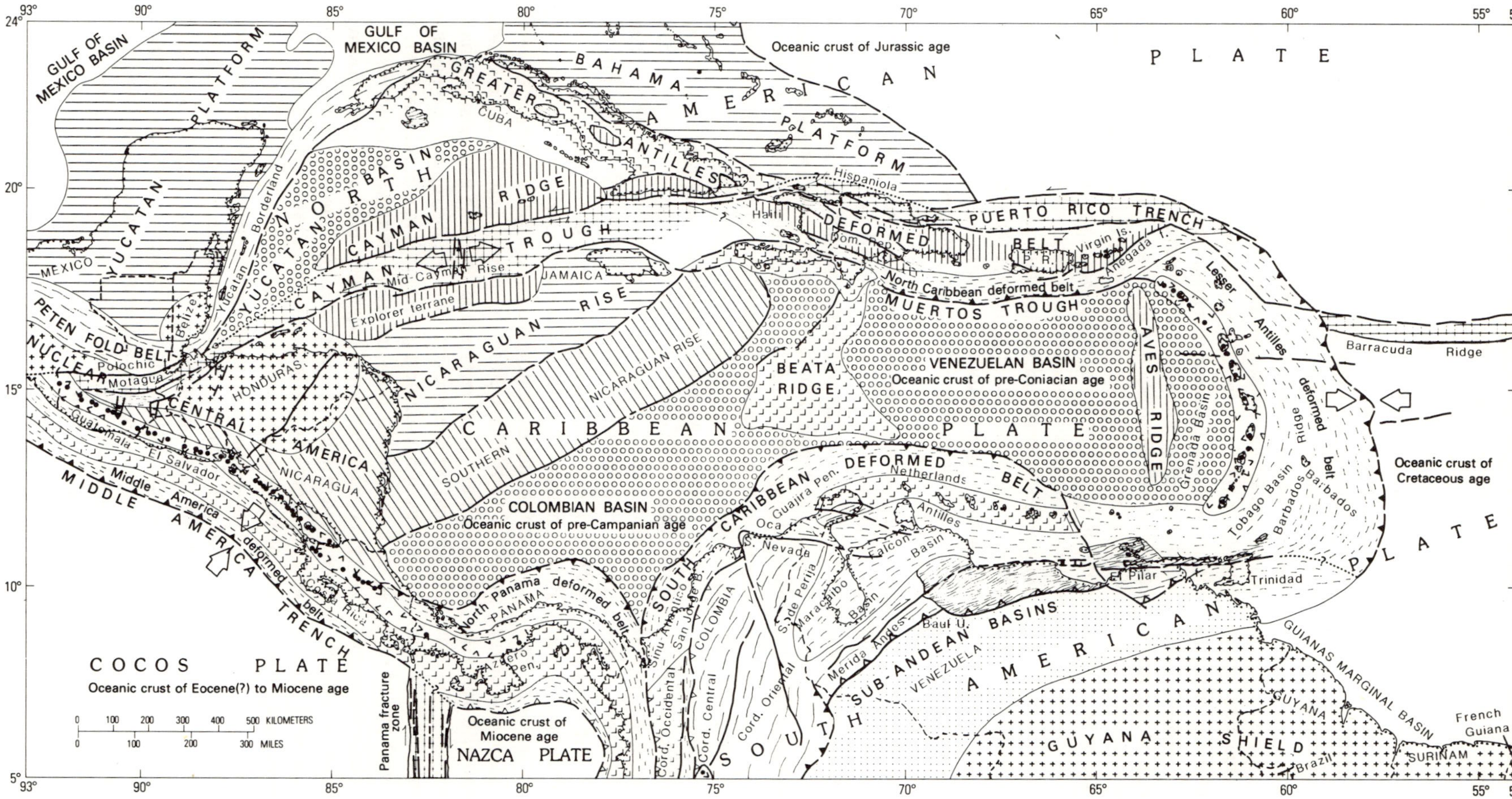

Figure 1. Schematic map of Caribbean region, showing exposed and inferred tectonostratigraphic superterranes, compiled by J. E. Case. Plus pattern, Precambrian and Paleozoic cratonic areas. Horizontal rule pattern, Mesozoic and Cenozoic moderately deformed carbonate platforms (includes local clastic and evaporite deposits). Long dashed pattern, "Laramide" deformed belts involving Cretaceous and Paleogene strata and various older sedimentary and crystalline rocks. Upright V pattern, areas of oceanic crust, upper mantle(?), and primitive island arcs, mainly of late Mesozoic and Paleogene age, deformed in "Laramide" but also in Neogene time. Lazy-S pattern, Mesozoic metamorphic belts. Vertical rule pattern, Late Cretaceous and Paleogene volcanic complexes, mainly calc-alkalic. Open dot pattern, moderately deformed sub-Andeans basins that contain thick Cretaceous and Tertiary marine and continental strata and varying thicknesses of pre-Cretaceous strata. Open circles, deep interior Caribbean basin, relatively undeformed, where pelagic, hemipelagic, and turbidite strata overlie oceanic crust. Short dashed pattern, Neogene deformed belts that may also involve Cretaceous and Paleogene rocks and deformational events. Diagonal pattern, complex multiterrane including carbonate banks, clastic strata, pelagic strata, and local volcanic and intrusive rocks resting on oceanic, transitional, and continental crust. Random V and large black dot pattern (Holocene volcanoes), Neogene volcanic belts. Cross-hatch pattern, complex transform systems; some are "leaky" and constructional, others are transtensional and form graben. Unpatterned areas within the Caribbean region are undefined terranes at present. Unpatterned areas outside the Caribbean region are deep ocean basins covered by varying thicknesses of pelagic, turbidite, and deep-sea fan deposits. Note that any individual terrane may include elements of adjacent terranes and that many terranes actually overlap in their times of deformation. Phanerozoic plutons (not shown) occur in many deformed belts.

No Paleozoic or older rocks have been positively identified in the Greater and Lesser Antilles or in southern Nicaragua, Costa Rica, Panama, and westernmost Colombia. A controversy currently surrounds the critical question of Mesozoic versus Paleozoic ages for certain metamorphic rocks in Cuba, and for those dredged from the Yucatan borderland. However, early Paleozoic metamorphic rocks were drilled on Deep Sea Drilling Project (DSDP) Leg 77 in the southeastern Gulf of Mexico.

The relation of Paleozoic rocks of the Caribbean region to those of Mexico, Florida, the Ouachita-Marathon fold belt, the Appalachian Mountains, and the Paleozoic systems of Africa and Europe constitute key problems in global reconstructions, especially on testing the validity of Pangea A versus Pangea B paleomagnetic reconstructions. These problems will be addressed in the Caribbean volume of this series.

Triassic and Jurassic Redbeds, Volcanic Rocks, and Granitoid Plutons

Triassic and Jurassic rocks of shelf or continental (red-bed) facies, as well as local silicic and some mafic volcanic rocks, are exposed in nuclear Central America and northern South America. The Jurassic rocks in both areas grade laterally into thick marine sequences. Jurassic evaporites occur beneath the Gulf of Mexico, in the subsurface of Mexico and northern Cuba, and possibly in Colombia. Marine Jurassic clastic and carbonate sequences are exposed in thrust slices in Cuba. Elsewhere, Jurassic deep-water cherts are exposed in Puerto Rico and Costa Rica, but Jurassic age assignments reported from Panama and westernmost Colombia are controversial at present. Intermediate to silicic plutons of Triassic and Jurassic age occur in the Cordillera Central, Cordillera Oriental, and Sierra Nevada de Santa Marta of Colombia, in the Sierra de Perija of Venezuela, and possibly in Belize. Some porphyry copper deposits of Colombia are associated with Jurassic plutons.

A critical regional question concerns the tectonic implications of the magmatism associated with the Mesozoic red-bed sequences: Was this igneous activity related to tensional opening of the Gulf of Mexico and the Caribbean, or was the activity (especially in northern South America) subduction-related? Partial answers may be discussed in the Caribbean volume of this series.

Cretaceous and Lower Tertiary Rocks

Gently deformed Cretaceous and lower Tertiary (mainly pre-middle Eocene) carbonate, local evaporite, and clastic deposits occur on the Yucatan and adjacent Florida and Bahama platforms, and will be treated in other D-NAG volumes. Thicknesses exceed 5 km in places. A thick sequence of miogeoclinal, shelf, and local continental deposits has been deformed in the Peten fold-belt of southern Mexico and Guatemala. Cretaceous and Paleogene rocks are major hydrocarbon reservoirs in the Yucatan platform and Peten fold-belt. This folded sequence may continue offshore in the horst and graben terrane of the Yucatan borderland, which strikes toward the late Mesozoic and early Cenozoic thrusts and fold-belts of Cuba. Although thick Cretaceous carbonate deposits occur along the north margin of Cuba, clastic, volcaniclastic, and volcanic rocks predominate in central, southern, and eastern Cuba. Discontinuous Cretaceous carbonates which extend eastward into Hispaniola and southward into Jamaica are less common in Puerto Rico. Carbonate and local clastic rocks of Cretaceous and Tertiary age cap the northern Nicaraguan Rise and Jamaica (bauxite deposits of Jamaica have developed mainly on Tertiary limestones). Volcanic and volcaniclastic rocks of Cretaceous and early Tertiary age constitute major units throughout the Greater Antilles and across the Virgin Islands.

In northern South America, great thicknesses (up to 10 km) of Cretaceous clastic and carbonate rocks occur east and south of the Mesozoic metamorphic belts. These miogeoclinal sequences were deposited in a crudely arcuate pattern around the northern Guyana shield. Cretaceous (and possibly older) mafic volcanic rocks, locally with associated deep-water deposits, are exposed in eastern Panama, the Cordillera Occidental of Colombia, and the Netherlands Antilles.

In Central America, south of the Motagua-Polochic fault zone, Cretaceous deposits include continental, shelf, and miogeoclinal sequences as far east and south as Nicaragua. Farther south, however, mafic volcanic sequences with local associated deep-water deposits occur along the Pacific margin of Costa Rica and Panama.

The belts of mafic volcanic rocks in the Greater Antilles, southern Central America, and northern South America have been interpreted by some geologists as having formed on oceanic crust or even as representing actual exposures of Mesozoic oceanic crust in certain localities. It should be pointed out, however, that the Cretaceous and lower Tertiary volcanic sequences tend to grade upward from tholeiitic basalts into calc-alkaline basalts and andesitic sequences on both the north and south margins of the Caribbean Sea.

Cretaceous and Tertiary plutons have been reported throughout Central America from Mexico to Panama, but as yet the data permit only limited geographic zoning by age and composition. Many metallic ore deposits, including Pb-Zn-Ag, Cu-Au, and Au, are associated with this group of plutons. Cretaceous plutons occur in the Cordillera Central, Cordillera Oriental, and Sierra Nevada de Santa Marta of Colombia, and lower Tertiary plutons are found in the Sierra Nevada de Santa Marta and on the Guajira

Peninsula. In Venezuela, Cretaceous to middle Tertiary plutons are exposed in the Caribbean coastal ranges and on the offshore islands. Cretaceous plutons occur in the Netherlands Antilles. In the Greater Antilles, most isotopic ages of intermediate to silicic plutons are Cretaceous and early Tertiary. Cretaceous and early Tertiary plutonic rocks have been dredged from Cayman Ridge, the walls of Cayman Trough, and Aves Ridge.

In the Venezuelan and Colombian Basins, a Mesozoic (probably Cretaceous) oceanic crust is overlain by Upper Cretaceous "flood" basalts, and by Upper Cretaceous and Tertiary pelagic or hemipelagic strata. The Yucatan Basin has an oceanic crust that is Eocene or older, which is overlain by a sequence of apparent pelagic or hemipelagic strata. Little significant intraplate deformation has occurred in these deep basins, although much deformation has taken place around their margins.

Mesozoic Metamorphic Belts

Assemblages of metamorphic rocks of probably Mesozoic age occur mainly in the Greater Antilles and northern South America, although scattered complexes of possible Mesozoic age have been identified in Central America. Most of these metamorphic belts are of greenschist to amphibolite facies. Some marble, in metamorphic sequences south of the Motagua fault in Guatemala, is of Cretaceous age. Cretaceous metamorphic rocks have been reported from northeastern Honduras, but whether these rocks are regionally or contact metamorphosed is not clearly defined in the literature. Metamorphic rocks of varying grade occur in western Panama in the vicinity of the Azuero Peninsula. Poorly known metamorphic rocks of probable Mesozoic age occur in the Cordillera Occidental of western Colombia and may be present along the west flank of the Cordillera Central. Metamorphosed Cretaceous rocks—the Ruma metamorphic zone—occur in the Sierra Nevada de Santa Marta and in the Guajira Peninsula of northern Colombia and extend eastward to the Netherlands Antilles. Metamorphosed Jurassic and Cretaceous rocks that have been faunally and isotopically dated make up the Caribbean coastal ranges of Venezuela and Trinidad. Some of these rocks are metamorphosed shelf and slope sedimentary assemblages, and others are metavolcanic rocks of amphibolite and blueschist facies; most of the sequences are thought to be allochthonous. Rocks of blueschist facies likewise occur in central and eastern Cuba, Jamaica, Hispaniola, and Guatemala. Nearly all of the metamorphic belts are spatially associated with mafic and ultramafic bodies—gabbros, tholeiitic basalts, serpentinites, and peridotites, most of which are allochthonous but some of which may be intrusive. Major deposits of nickel and chromite and minor deposits of manganese are found in the mafic-ultramafic belts.

Again, key questions regarding the metamorphic belts include depth of metamorphism and polarity of subduction and whether the metamorphism is genetically related to subduction or loading phenomena. Field relations of the metamorphic belts to accompanying mafic and ultramafic masses are poorly understood. Some of these questions will be addressed in the Caribbean volume of this series..

Mid-Cretaceous to Middle Eocene Deformed Belts

Two main groups of structures and rocks are involved in mid-Cretaceous to middle Eocene episodes of deformation. One group includes the folds in the thick shelf and slope depositional sequences (miogeoclinal) that generally occur along the margins of relatively stable cratons, for example, the Peten fold-belt of Mexico and northern Guatemala, and folds of the Andean and sub-Andean basins of northern South America. The second group of structures comprises the adjacent deformed belts that include mafic to andesitic volcanic complexes, local deep-water sedimentary deposits, ultramafic masses, and local metamorphic rocks, including some blueschist facies. The volcanic belts generally occur on the Caribbean side of the miogeoclinal belts.

The northerly deformed belt probably extends from the Peten area along the Yucatan Borderland and across the Greater Antilles. The southerly belt extends along the Andean complexes of Colombia and Venezuela and forms a broadly arcuate pattern around the west, northwest, and north sides of the Guyana shield. These deformed belts are superimposed on a wide variety of older deformed terranes.

Other areas of late Mesozoic and early Cenozoic deformation that do not include elongate fold-belts are found in the complex basins and uplifts of nuclear Central America in southern Guatemala, Honduras, and Nicaragua. Here the structural style is one of horsts and grabens, with local folds of varying trend. This structural style may extend northeastward to Jamaica along much of the Nicaraguan Rise.

The extent of late Mesozoic and early Cenozoic deformation in the Lesser Antilles is unknown because only a few scattered outcrops of Eocene or older rocks have been found, but the older rocks generally are more highly deformed than the Neogene rocks.

Neogene Deposits

Neogene carbonate deposits characterize the Yucatan and Bahama platforms and much of the Nicaraguan Rise. Marine and paralic carbonate and clastic deposits occur in a wide variety of basins that are superimposed on Laramide deformational belts throughout the margins of the region. Thick continental deposits occur in the sub-Andean basins of South America and in marginal basins around the entire periphery of the Caribbean Sea. Major oil and gas resources occur in some of these beds, especially in northern South America, Trinidad, Barbados, and locally in the Greater Antilles. Coal deposits occur in northern South America and locally elsewhere.

Neogene Deformed Belts

Several major Neogene fold-belts and melange belts can now be defined, many of which can be traced from the land to the sea floor of the Caribbean region. Cenozoic and possibly older rocks are present in these deformational belts. In many of the belts, deformation has continued, probably intermittently, from late Eocene to Holocene time. The offshore fold terranes have four common characteristics: (1) most are accompanied by conspicuous negative free-air and isostatic anomalies (caused by thick sedimentary accumulations); (2) they involve great thicknesses of low-velocity material; (3) reflection data indicate apparent thrusts, and incoherent seismic returns in some belts suggest the presence of melange units; and (4) most exhibit a distinctive hummocky topography offshore, interspersed with smooth elongate ridges and valleys. In addition, the Middle America and Lesser Antilles deformed belts are characterized by a high level of seismicity and by Benioff zones. Some of the convergent belts, however, are accompanied by a low level of seismicity and by a near-absence of magmatic activity, possibly reflecting extremely low rates of convergence.

One of the largest folded terranes, the South Caribbean deformed belt, extends eastward and offshore from the Sinu-Atlantico basin of northern Colombia, where as much as 10 km of Tertiary rocks are exposed, nearly to Aves Ridge. This belt coincides with a prominent negative gravity anomaly along much of its extent. The northern border of this zone, near the base of the continental slope, is a zone of apparent underthrusting. A second major belt, characterized by incoherent seismic reflectors, that lies east of the Lesser Antilles has been interpreted as a thrust-thickened sedimentary wedge related to westward underthrusting of Atlantic oceanic crust beneath the Lesser Antillean arc, a concept that has been partly confirmed by recent DSDP drilling. Included within this second deformed belt is the structural high of Barbados Ridge—extending from Tobago to Barbados and beyond—and by the broad, relatively undeformed Tobago basin, which contains a thick sequence of Tertiary sediments between the volcanic Lesser Antilles and Barbados Ridge. The principal part of the second deformed belt extends northward and westward around the Lesser Antilles to about the extension of the Anegada fault system near latitude 18°N. Several geologists have proposed that dominant underthrusting of the Antillean arc passes into dominant left-lateral strike-slip displacement north of about latitude 19°N.

A highly folded sedimentary sequence has not yet been identified on the south wall of the Puerto Rico Trench, where, instead, block faulting seems to predominate. Deformed deposits typical of convergent boundaries, however, have been identified south of Puerto Rico and north of the Muertos Trough (North Caribbean deformed belt). Relative northward underthrusting of the Greater Antilles by the Venezuelan Basin is suggested by some seismic profiles. Although a direct continuity has not been established with certainty, the North Caribbean deformed belt apparently continues onshore in southern Hispaniola as a terrane of deformed rocks in and near the Enriquillo-Cul-de-Sac graben system. Some geologists have emphasized that this fold-belt in Hispaniola is one of the few examples of intense post-Eocene orogeny in the Greater Antilles.

Another apparent belt of deformed rocks and associated negative gravity anomalies curves concentrically around northern Panama. This fold-belt trends perpendicularly into the South Caribbean deformed belt of northern Colombia, a circumstance that poses a major tectonic question: Do these two fold-belts represent a single but oroclinally bent deformed zone, or was each fold-belt formed independently? The fold-belt off Panama continues westward to emerge on land as the Limon basin near the border between Panama and Cost Rica. Offshore multichannel data suggest westward thrusting of Caribbean crust beneath Panama, and many of the thrusts mapped on land in the Limon basin in fact dip westerly or southwesterly. Thus, Panama appears to be an independently moving block between the Nazca plate and the Colombian Basin.

The deformed belt, forearc ridge, forearc basin, and Neogene volcanic belt of Middle America are genetically associated with convergence at the Middle America Trench. A few structural and stratigraphic details of the offshore areas have become known through DSDP drilling and seismic-reflection and -refraction data.

Along the Pacific margin, relations at the intersection of the Middle America Trench, Panama Fracture Zone, the Cocos and Nazca plates, and the buried Panama-Colombia trench system are still poorly known and merit comprehensive multidisciplinary studies.

Summary of Some Younger Igneous and Sedimentary Terranes

Analysis of the contents of the various tectonostratigraphic terranes of the Caribbean region leads to recognition of four main unmetamorphosed late Mesozoic and younger magmatic and sedimentological assemblages in the region:

1. Mesozoic to Paleogene terranes of oceanic crust and possible primitive island arcs, and associated pelagic, turbiditic, and seamount-associated reefal deposits, occur along the Middle America forearc ridge, in eastern Panama, western Colombia, and the Netherlands and

Venezuelan Antilles; on the Beata Ridge, possibly the southern Nicaraguan Rise, possibly Aves Ridge, the south Haiti Peninsula, and in central and eastern Cuba.

2. Large Paleogene flysch basins are known to occur in the San Jorge terrane of Colombia, in the Falcon Basin and in the northern margin of the sub-Andean basins of Venezuela, on the Barbados Ridge, and in southern Hispaniola and Cuba. Paleogene flysch basins probably are present in the North Panama deformed belt, the South Caribbean deformed belt, the Lesser Antilles deformed belt, and the North Caribbean deformed belt, although they are largely concealed at depth. Some of these basins are geographically and temporally associated with calc-alkaline magmatic belts. Key questions are: which of the Paleogene flysch basins are forearc, backarc, or trench deposits?

3. Belts of late Mesozoic to Paleogene calc-alkaline volcanic rocks, commonly associated with primitive island-arc volcanic rocks, and accompanying volcanogenic and forearc- or backarc-basin deposits are widespread. These occur on Aves Ridge, in Puerto Rico and the Virgin Islands, central Hispaniola, probably Jamaica and in eastern and central Cuba, probably along the Cayman Ridge and the Explorer terrane, and in nuclear Central America south of the Motagua-Polochic terrane, as represented by scattered plutons and minor (volumetrically) extrusive rocks (not shown in Fig. 1). Some geologists have postulated that the northern Nicaraguan Rise is underlain by Cretaceous magmatic arcs, although platform carbonates and clastic deposits characterize most of the terrane. Calc-alkaline rocks are significant elements of Panama and western Colombia and major deposits of copper and manganese are associated with some of these rocks.

4. Subduction-related Neogene magmatic belts have formed in Middle America, the Lesser Antilles, and in the northern Andes of Colombia and Ecuador. Some of these belts are characterized by major metallic deposits, for example, the porphyry copper deposits of western Panama.

Key questions are: how many distinct magmatic arcs can be identified, and if they are subduction related, what was the polarity of the subduction during specific intervals of geologic time? For example, a continuing controversy exists for the Greater Antilles—whether subduction was principally from the south or from the north, or both.

Great Neogene strike-slip or transform faults along the northern and southern plate-boundary zones constitute major tectonic features of the Caribbean plate. The general Motagua–Cayman Trough–Puerto Rico Trench system on the north is characterized by left-lateral slip, and the general Oca-El Pilar system on the south is characterized by right-lateral slip. A pre-Miocene boundary of the Caribbean plate extends across northern Cuba and probably along the Yucatan borderland into Guatemala. The present northern boundary zone, the Motagua-Cayman Trough system, contains within it a mini-spreading center, the mid-Cayman Rise, which has probably been active since Eocene time. Many of the sedimentary basins and local uplifts associated with these complex transform zones are regarded as pull-apart transtensional or transpressional features. Some of these basins are sites of major accumulations of hydrocarbons.

One of the outstanding tectonic problems of the region, not yet properly addressed in the literature, is that of the enormous vertical displacements and depressions that occurred during the Neogene. Many uplifted areas (Hispaniola, and the Sierra Nevada de Santa Marta and Cordillera Occidental of Colombia) are not isostatically compensated and thus are being forcefully held up at present, whereas other uplifts (nuclear Central America, the Central and Eastern Andes of Colombia and the Merida Andes) appear to be isostatically compensated. Tectonic boundaries between oceanic, continental, and transitional crust are not yet well defined because of a scarcity of seismic-refraction lines. Some boundaries appear to bear little relation to the present boundaries between land and water: for example, nuclear Central America has a continental crust, whereas southern Central America, in part at least, has a foundation of oceanic crust. Again, some of these problems will be addressed in the Caribbean volume of this series.

Caribbean Region Volume Topics

To summarize in one volume the enormous amount of information gathered by numerous scientists from a wide international community poses a major problem in maintaining a common approach to the analysis of the regional geologic aspects. Nonetheless, we recognize that the thematic constraints of such a volume will require a common effort on the part of the various authors not only to synthesize all pertinent geologic information but also to define contemporary problems clearly enough to encourage and guide future research. The Caribbean volume of this series will include discussions of the geology from Cuba and Guatemala and southward into the northern Andes and northern Guyana shield.

The following topics have been selected for the Caribbean volume in the D-NAG series.

1. Introduction: History of geologic exploration
2. Tectonic evolution of the Greater Antilles
3. Precambrian of northern South America
4. Venezuelan Andes
5. Caribbean Mountain System (Venezuela, Trinidad, and Tobago)
6. Netherlands and Venezuelan Antilles
7. Colombia (stressing relations to the Caribbean)
8. Lesser Antilles
9. Basement of northern Central America
10. Paleozoic to Cenozoic tectonic evolution of northern Central America

11. Neotectonics and volcanism in Middle America
12. Southern Central America
13. Magmatic evolution of the Caribbean region
14. Seismicity of the Caribbean region
15. Metallogenesis of the Caribbean region
16. Energy resources of the Caribbean region
17. Caribbean sea floor
18. Physical properties of the Caribbean crust and upper mantle
19. Tectonic models: Fixistic views
20. Tectonic models: Mobilistic views
21. General summary and state of the art

Acknowledgments

Helpful reviews of the manuscript were provided by David Scholl and Andrew Griscom.

References Cited

Butterlin, Jacques, 1977, Géologie structurale de la région des Caraïbes (Mexique-Amérique Central-Antilles-Cordillere Caraïbe): Paris, Masson, 259 p.

Case, J. E., and Holcombe, T. L., 1980, Geologic-tectonic map of the Caribbean region: U.S. Geological Survey Miscellaneous Investigations Series Map I-1100, scale 1:2,500,000.

Martín F., Cecilia, 1978, Mapa tectónico, Norte de América del Sur: Venezuela Ministerio de Energía y Minas, scale 1:2,500,000.

Nairn, A.E.M., and Stehli, F. G., eds., 1975, The Ocean basins and margins, v. 3, The Gulf of Mexico and the Caribbean: New York, Plenum Press, 706 p.

Sapper, K., 1937, Mittelamerika—Handbuch der Regionalen Geologie: Heidelberg, Steinmann and Wikkens, 160 p.

Schuchert, Charles, 1935, Historical geology of the Antillean-Caribbean region: New York, John Wiley & Sons, Inc., 811 p

Weaver, J. D., ed., 1977, Geology, geophysics, and resources of the Caribbean; Report of the IDOE Workshop on the Geology and Marine Geophysics of the Caribbean Region and Its Resources, Kingston, Jamaica, Feb. 17–22, 1975: Mayagüez, University of Puerto Rico, 150 p.

Weyl, Richard, 1980, Geology of Central America, 2nd edition: Berlin, Gebrüder Borntraeger, 371 p.

Zeil, Werner, 1979, The Andes, a geological review: Berlin, Gebrüder Borntraeger, 260 p.

The Eastern Pacific and Hawaii

D. M. Hussong
Hawaii Institute of Geophysics
2525 Correa Road
Honolulu, Hawaii 96822

E. L. Winterer
Scripps Institute of Oceanography
University of California, San Diego
La Jolla, California 92093

R. W. Decker
U.S. Geological Survey
Hawaii Volcano Observatory
Hawaii National Park
Hawaii 96718

Introduction

This volume will synthesize and describe the geology of the eastern northern Pacific Ocean Basin, a particularly large, variable, and dynamic region that is far less studied than most continental areas. Interpretation of the geology and geophysics of oceanic regions has been remarkably useful for elucidating the geometry and history of large-scale tectonics, especially lithospheric plate tectonics. Because it is accomplished by remote sensing techniques that generalize sea-floor structures and features, the structural complexities that are observed onshore at road-cut scales are smoothed out in marine studies. Thus, in many ways the success of marine studies is that they observe the forest without the confusion of looking at each tree, while on land the geologist must try to interpret the forest by the more difficult task of generalizing the study of features with the relative size of the trees.

In the past few years the scales of onshore and offshore geologic studies have been converging. Continental geologists are using more geophysics and remote sensing, while marine geologists have tried to conduct higher-resolution studies.

This volume will strive to provide a comprehensive description of our present state-of-knowledge of marine geology. At the same time, we will emphasize two general objectives in our field, as follows:

1. We will take particular note of efforts to design and conduct increasingly higher-resolution studies that can resolve structure to scales comparable to "road-cut" geology. This philosophy will make it easier to effectively compare geology above and below sea level. It will also help bridge the gap between conventional marine geology that deals with isolated hand specimens from cores and rock dredges and conventional marine geophysics that rarely can resolve features with dimensions less than kilometers.

2. We will also strive to highlight work that is underway to define many of the processes involved in marine geology, such as tectonic processes at plate boundaries, geothermal and density modeling for plate motion studies, and crustal generation and alteration. This shift in emphasis from descriptive marine geology to necessarily more quantitative and kinematic studies is, of course, also driving the field toward higher-resolution studies.

The volume will be organized to first describe and synthesize available data from the entire eastern Pacific. In most cases this will not require publishing new maps, as recent synthesis maps will have already been published. The volume will then describe the history, structure, and tectonics of sea-floor spreading centers, lithospheric plate interiors, convergent and transform margins, seamounts, and sediments. We will also devote a significant part of the volume to the geology of the Hawaiian Islands and associated volcanic and geothermal processes. The only part of the volume that may be organized in time slices will be that devoted to a plate-tectonics history of the eastern Pacific. The balance of the volume will consider contemporary problems of resources development and environmental protection. The volume, as presently conceived, will end with a summary chapter on major remaining problems and a discussion of future perspectives.

Region Covered

The geographical extent of the region to be covered by the volume is still not rigidly defined, and probably will not be until each chapter is outlined by the prospective authors. Our intention is to try to keep the region as small as possible while still including as many components of Pacific Ocean Basin geology as possible.

A first cut at defining our region is shown in Figure 1. Our boundary, starting in Southern California, is inland far enough to include the marine borderlands and the Franciscan ophiolites. Farther north it remains inland so as to include the coast ranges and the subduction-related andesitic volcanoes of Oregon, Washington, and southwestern Canada. We realize this will probably overlap the Cordilleran and other volumes. However, we intend to deal only with those aspects of the land geology that relate directly to offshore geology, in this case to obduction and subduction processes. To the north the boundary of the region we will cover is offshore, along the shelf break, to the eastern Aleutians where we intend to encompass the Arc Islands. The boundary then runs to the south just west of the Emperor Seamount chain, then west of the Line Islands and into the

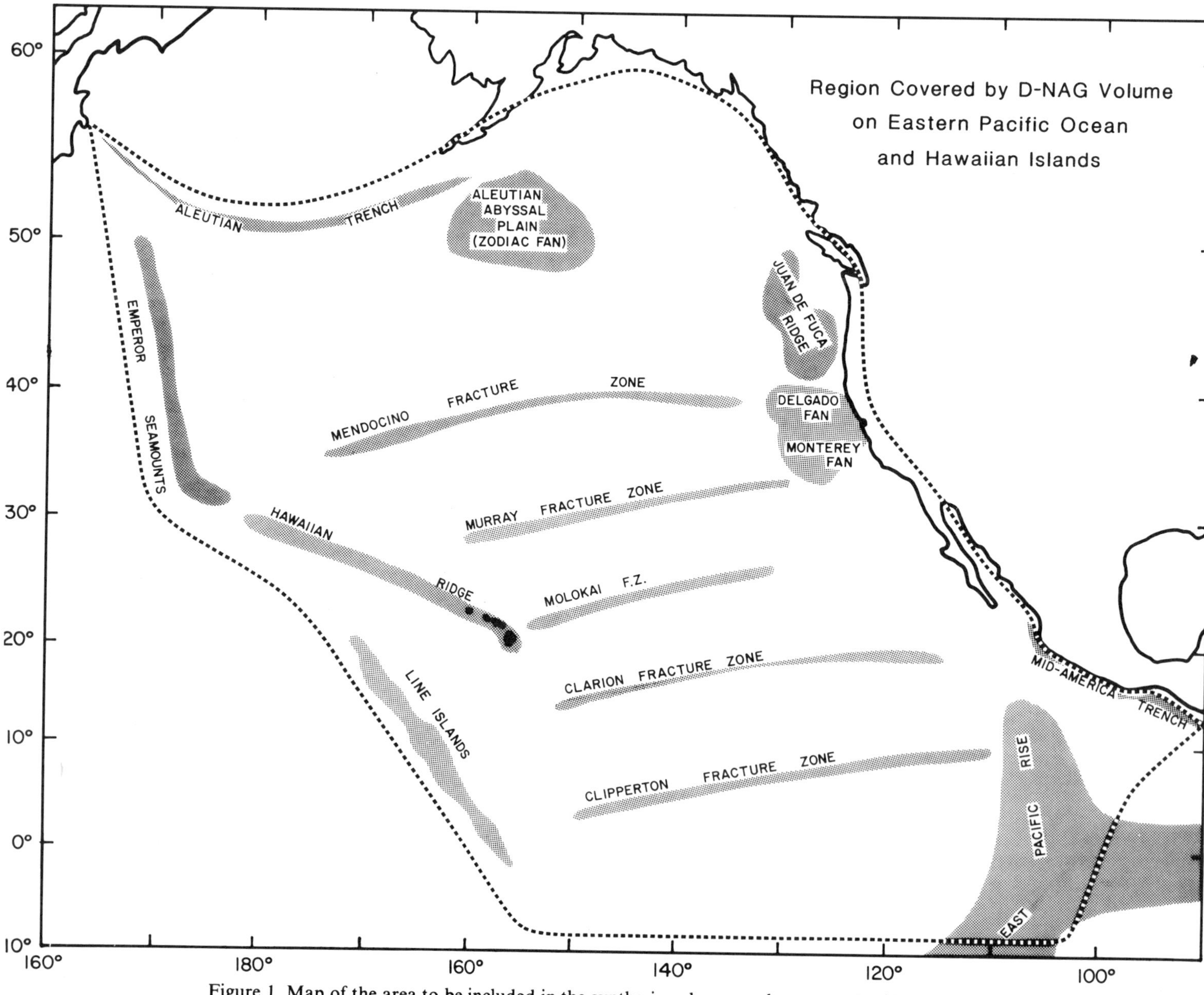

Figure 1. Map of the area to be included in the synthesis volume on the eastern Pacific Ocean.

Penrhyn Basin to just south of the equator. The boundary stays just south of the equator to the East Pacific Rise (EPR), where it trends north along the rise crest (or perhaps just east of the rise crest to include the symmetry of this fast spreading center), past the Galapagos/EPR Triple Junction to the Middle America Trench. This region includes the northwestern part of the trench, the Gulf of California spreading center, and trends back inland in Southern California.

The region we are considering is huge and may be too complex for this short volume. For this reason we will not attempt to include descriptions of everything in the area. Instead, we will select representative, well-studied, "type" examples of the major features such as spreading centers, mid-plate volcanoes, and subduction zones.

Tentative Description of Volume

Introduction

After a statement on the organization and philosophy of this particular volume, we will briefly describe our rationale for subdividing the region and choosing some geologic regions to emphasize at the expense of others.

A chapter on the history and present state of scientific exploration will also not attempt to be comprehensive. This section will try to first describe the shifting philosophies and hypotheses that governed the way in which marine geologic research was and still is conducted. It will also provide a feel for the amount and distribution of available data and try to relate this data density to the degree of coverage on land in order to demonstrate the paucity of marine data and the generalities that result.

The last part of the introductory chapter will describe some of the new exploration tools available to marine geologists and geophysicists, including multichannel seismic reflection acquisition and processing techniques, high-resolution multibeam bathymetry, ocean-bottom seismometers, and deep-towed sensors. Again, the emphasis of this discussion of new techniques will be geared to the increased resolution that will allow marine studies to approach the scale of onshore studies.

Synthesis of Available Data

By 1984 several new maps that synthesize bathymetry, magnetics, and sediment type and thickness will be available. We do not expect to include preparation of any new full-size maps for this volume, except possibly heat flow, crustal structure, and geochemistry. This chapter of the volume will describe the major features of the available data map compilations and perhaps produce a few page-size maps of particularly salient regions. This chapter and the first introductory chapter will be kept short.

Evolution of Oceanic Lithosphere

This critical chapter will comprise nearly 30% of the volume and will try to trace the formation and alteration of oceanic crust and sea-floor morphology. The first section of the chapter will describe the formation of crust at representative spreading centers, such as the East Pacific Rise adjacent to the Nazca Plate (fast spreading; ~8 cm/yr), the East Pacific Rise adjacent to the Cocos Plate (medium spreading rate; ~4 cm/yr), and the Gorda–Juan de Fuca region (slower spreading; ~3.5 cm/yr).

A summary of our knowledge of the tectonics (including recent low-magnitude local seismicity studies using hydrophones and ocean-bottom seismometers) and petrology/geochemistry of the various types of spreading centers will introduce this section. The implications on the thermal and geochemical budget of the upper crust and the oceans will be considered in light of the recently discovered hot springs on some of these rise crests. The section will end with a synthesis of the available data to derive spatial and temporal models of crustal generation at the various spreading centers.

Following the analysis of crustal generation, we anticipate including a description of the alteration of the crust and lithosphere with age. The velocity structure and age/depth/heat-flow relationships of the crust and upper mantle have previously been described; these compilations will only be updated as appropriate. The principal objective of the section on alteration will be to describe what petrologic and physical (crack infilling, etc.) changes might reasonably be responsible for the geophysically observed alterations. Then, how do these changes affect sea-floor morphology? Can they be observed in uplifted and subaerially exposed oceanic rocks? This section on crustal aging should also include a description of sediment patterns in the region, including any regional hiatuses or periods of unusually rapid sedimentation (such as might be caused by proximity to the equator). What are the dynamics of deposition? How do sediment distribution patterns affect alteration of underlying crust?

The third major topic in Chapter III is devoted to a summary of the development of other sea-floor features. Where and why do various types of seamounts form? How many of the seamounts and seamount chains originate on spreading axes, and why should there be concentrations of magma erupted at isolated parts of spreading centers? Although only the eastern fringes of the extensive fields of Cretaceous volcanic seamounts occur in our region, the whole question of the apparent increase in ocean basin volcanism toward the end of Mesozoic time is of tremendous general importance to geologic history and should be addressed in this volume. Are, for instance, the Line Islands a hot-spot trace, as suggested by their alignment along an apparent path for plate motion during their formation, or are they nearly simultaneous Cretaceous eruptions along some zone of weakness such as an old plate boundary, as suggested by radiometric dating of dredged rocks from the island chain?

Although transform faults, fracture zones, and propagating (or jumping) rifts have been adequately described in the literature, some of the effects of these tectonic features on crustal structure and petrology are not well described. From a few isolated studies in the western Pacific and the Atlantic it seems that crustal structure and petrology are significantly different in transforms and fracture zones; such relationships should be tested in the exceptionally well-defined fracture zones of the eastern Pacific. We also wish to consider what happens where these features go ashore, such as where the East Pacific Rise spreading center/transform system of the Gulf of California extends into the North American continent at the Salton Sea. What similarities and/or differences occur between the same tectonic features in continental versus ocean lithosphere?

The last part of this chapter will be a review of recent sediment stratigraphy efforts with descriptions of progress and remaining major problems in paleogeography, paleooceanography, and paleoclimatology. This section will likely be drawn heavily from the results of the hydraulic piston coring program presently underway on the National Science Foundation Drilling Vessel Glomar Challenger.

Geology of the Hawaiian Islands and Emperor Seamount Chain

This topic will be approached from at least two directions: (1) A general investigation and report by a two-person team composed of a subaerial geologist and a submarine geologist. The key questions are: What are the products and volumes that comprise the chain? What are the major geologic processes now affecting the chain? What is the geologic history of the chain (including a thorough investigation and discussion of the origin and age of the Hawaiian-Emperor bend)? What is the origin of the volcanism that has generated and is still generating the chain? (2) The parts of the chain that form islands should be investigated and reported on in a more conventional manner by individual geologists who will summarize what has been done, make new investigations if possible, and prepare summary geologic maps and reports where appropriate. These studies will not ignore the submarine parts of the chain, but emphasis will be on the subaerial studies. The main thrust of the submarine investigations should be in part (1). Subtopics to be covered include the following.

Topography. What types and how much of the geomorphology is constructional? What types and how much is destructional? What are the principal changes and rates of changes in topography at present? Are there actually two closely spaced parallel chains of volcanoes?

Geophysics. What is the present seismicity, heat-flow, and magnetic pattern of the chain?

Composition. How much of the entire island-seamount chain is volcanic versus limestone or other sedimentary rocks? What types of volcanic rocks are present, and what was their petrogenesis? What are the weathering products? Are there any mineral deposits, petroleum, or geothermal resources?

Structure. What is the crustal structure of the island chain? What are the internal structures of the various seamounts and islands?

Age. What are the ages of the various seamounts and islands? What are the age ranges? What is the age of the bend? How much time did it take the entire bend to form?

Origin and Evolution. Is there a hot spot? If not, what else? How does a hot spot work? What data support conventional plate-tectonics theory? What data do not? Has the origin been steady, episodic, waxing, waning? How do progressive island-seamount chains start? How do they end? What is the relation of the Hawaiian–Emperor Chain to other volcanic chains?

Subdivisions. The theme here is to investigate and report in increasing detail through time. The past is destroyed by time, but the "present is the key to the past." Dredging on Loihi Seamount has brought up some very young alkalic basalts, strongly challenging the conventional wisdom that the young to mature volcanic rocks are all tholeiitic basalts. The detailed description of each subdivision of the island chain might include, for instance for Kure Island to Kihoa Island, what is the subaerial and submarine topography? What is the areal geology? What are the internal structures and compositions? What is the geologic history? What geologic processes are most important at present? What are the natural resources (including groundwater)? Similar description would follow for Kure to Nehoa, Kaula Niihau, Kauai, Oahu, Molokai, Lanai, Maui and Kahoolawe. The "Big Island" of Hawaii will likely have an increased amount and subdivision of investigations reflecting the fact that Hawaii has as much area as all the rest of the islands and that volcanic activity is still vigorous on this island. The subdivisions may include the following.

Kohala Volcano. What is the origin and age of the deep valley erosion on the windward flank? At what stage of petrogenesis are the most recent lavas? Is this volcano extinct?

Mauna Kea Volcano. What is the nature, extent, and age of glaciation on this mountain? At what stage of petrogenesis are the most recent lavas? What is the average repose time between eruptions in the last 50,000 years? Why are the more recent vents so scattered on the summit and flanks? What is the origin of occasional seismic swarms beneath Mauna Kea?

Hualalei Volcano. What is the average repose time between eruptions in the last 50,000 years? What is the origin of the trachyte lavas? At what stage of petrogenesis are the most recent lavas? How did the complex pit craters form? What is the origin of the many xenolith nodules in the lava flows? What are the eruptions hazards to populated areas on the Kona coast? What the the geothermal power prospects?

Mauna Loa Volcano. What are the nature and dynamics of eruption? What determines the geomorphology of this mountain? What are the eruption hazards to Hilo and other populated areas on the flanks? What are the characteristics and amounts of volcanic gases? What are the geothermal power prospects? What is the nature of seismicity of Mauna Loa?

Kilauea Volcano. What is the nature and frequency of eruptions in the last few thousand years? What are the structure and dynamics of the rift zones? What is the nature of seismicity and ground surface deformation on Kilauea? What are the geothermal power prospects? Can the eruptions be forecast?

It is planned that this chapter will end with a discussion of eruption mechanics, from magma generation in the upper mantle to the surface.

Geology of the Ocean Margin

Subduction processes have been a topic of increased interest recently, largely because of unexpected and variable interpretations of convergent margins sampled by the D/V Glomar Challenger. It has been thought that subduction of an oceanic plate resulted in accretion of oceanic material from the plate onto the frontal part of the overriding plate. By this model, ocean sediment and even seamounts and slivers of crystalline oceanic crust are constantly augmenting and producing uplift of the overriding island arc or continent. Recent results show this to be a vast oversimplification. Convergent margins seem to have complex histories including periods of tectonic erosion and subsidence as well as periods of accretion and uplift. Volcanism and back-arc extension appear related to the variable large-scale stresses that must be associated with the different modes of subduction. In this volume we hope to compare the subduction tectonics, structure, and associated seismicity and volcanism of the northern Middle America Trench, the Cascades Convergence Zone, and the Aleutian Trench and Arc in an attempt to categorize various subduction processes.

Related to this topic will be a discussion of the allochthonous terranes of western North America. Why were these apparent oceanic rocks scraped from the subducting Pacific Plate and incorporated into the North American Plate? Are these ophiolitic rocks ocean basin crust, back-arc basin crust, island arc crust? Does convergence of certain types of crust favor offscraping, or obduction? How significant is this process toward the building of the continents?

A brief survey of mineralization and the economic potential of subduction zone and arc-related minerals, including hydrocarbons, will be included in this chapter.

The chapter will then provide a description of the structure and tectonics of the transform plate margin of California, including economic potential and earthquake hazards. Included in this will be a description of the continental shelf, slope, and rise off California.

The final topic in this chapter will be an overview of the development and significance of deep-sea fans such as the Monterey and Delgada fans off Southern California and the Aleutian fan off the eastern Aleutians.

Plate-Tectonics History of the Region

This relatively short chapter will deal with the plate kinematics and paleogeography of the eastern Pacific. We hope that recently acquired data and improved analysis techniques will permit a meaningful attempt to relate plate interaction (particularly between the Pacific and North American Plates) and events in the geologic history (vertical tectonics, volcanism, apparent changes in regional stress) of the plate.

Contemporary Problems

This chapter will be concerned with resource potential and development of the eastern Pacific, beginning with a section on current interpretation of the origin and distribution of manganese nodules. Other obvious topics include the economic potential of possible sources of energy, particularly geothermal energy conversion systems.

Integrated with this brief discussion of economic marine geology will be an evaluation of associated environmental problems. The two topics of greatest immediate concern in this regard are the potential pollution and disturbance of delicate biologic environments by what amounts to shallow penetration strip mining of manganese nodules in deep water, and the safety of disposal or long-term storage of nuclear wastes in the deep ocean.

An important topic of less direct geologic relevance is ocean and sea-floor pollution caused by runoff of waste from coastal mineral exploration, oil spills, discharge of polluted rivers, and the like. Although a topic of great concern to any earth scientist, we are uncertain of its relevance in this volume and invite comment.

Perspectives and Major Unanswered Problems

This summary chapter will attempt to bring together

all the preceding chapters of the volume and to interface this volume with others in the D-NAG series. This chapter will be organized in a manner dependent on how the rest of the volume takes shape, but will surely consider (and perhaps be subdivided into sections on) major unanswered problems of geologic, geophysical, tectonic, stratigraphic, and economic nature.

Outline

I. Introduction
 1. General introduction to volume—organization, philosophy
 2. physiographic provinces
 3. history and present state of scientific explora- of the area
 4. new techniques now employed for marine geologic and geophysical exploration

II. Synthesis of available data
 1. bathymetry
 2. gravity, geoidal height
 3. magnetics, including age distribution
 4. heat flow
 5. crust and mantle seismic velocity structure
 6. sediment thickness, type, geochemistry

III. Evolution of oceanic lithosphere
 1. formation of ocean crust (Juan de Fuca Ridge, Gorda Ridge, East Pacific Rise) and lithosphere
  a. tectonics, seismicity
  b. petrologic evolution
  c. geothermal activity
  d. models, different spreading rates
 2. alteration of crust and lithosphere with age
  a. velocity structure
  b. age/depth/heat-flow relationships and exceptions
  c. sea-floor sedimentation/erosion
 3. development of sea-floor features
  a. seamounts and guyots
  b. linear volcanic chains (hot spots, Cretaceous volcanism)
  c. transform faults, fracture zones
  d. propagation rifts, spreading-center jumps
 4. biostratigraphy
  a. paleogeography
  b. paleo-oceanography
  c. paleoclimatology

IV. Geology of the Hawaiian Islands and Emperor Seamount Chain
 1. summary of structure and evolution, Loihi seamount to the Emperor chain
 2. synthesis of volcanic processes
  a. eruption mechanics, from asthenosphere to surface
  b. petrology
  c. seismicity
 3. geothermal processes on Hawaii

V. Geology of the ocean margin
 1. subduction processes
  a. structure, tectonics, volcanism in Aleutians
  b. structure, tectonics, volcanism in Cascades region
  c. origin of allochthonous terranes in western North American, continental accretion
  d. mineralization, including hydrocarbons, in subduction zones
 2. structure, tectonics of transform margin of California, including the continental shelf, slope, and rise
 3. development of large turbidite fans

VI. Plate-tectonics history of region
 1. plate kinematics
 2. absolute motions
 3. paleogeography

VII. Contemporary problems
 1. resource potential and development
  a. minerals (Mn nodules?)
  b. energy (geothermal, ocean thermal energy, hydrocarbons, etc.)
 2. environmental problems
  a. sea-bed exploitation
  b. pollution related to mineral exploitation, drilling, land development, river pollution (Columbia River)
  c. potential for nuclear waste disposal

VIII. Perspectives and major unanswered problems
 1. geologic
 2. geophysical
 3. tectonic
 4. stratigraphic
 5. economic

Typeset by WESType Publishing Services, Inc., Boulder, Colorado

Printed in U.S.A. by Malloy Lithographing, Inc., Ann Arbor, Michigan